Mathematical Introduction to Data Science

Sven A. Wegner

Mathematical Introduction to Data Science

Springer

Sven A. Wegner
Hamburg, Germany

ISBN 978-3-662-69425-1 ISBN 978-3-662-69426-8 (eBook)
https://doi.org/10.1007/978-3-662-69426-8

The original submitted manuscript has been translated into English. The translation was done using artificial intelligence. A subsequent revision was performed by the author(s) to further refine the work and to ensure that the translation is appropriate concerning content and scientific correctness. It may, however, read stylistically different from a conventional translation.

This Springer imprint is published by the registered company Springer-Verlag GmbH, DE, part of Springer Nature.
The registered company address is: Heidelberger Platz 3, 14197 Berlin, Germany

Preface

Knowledge in the areas of data science and machine learning is increasingly expected from mathematics graduates and consequently, students of mathematics ask for these subjects to be included in the standard curricula. The idea behind this textbook is to present canonical data science and machine learning topics in a form tailored to the target audience of mathematics students. In doing so, our number one priority is a rigorous treatment that fosters profound understanding of the methods discussed. This includes in particular to always work out why exactly a method succeeds and to outline its limitations.

This book is based on several courses that the author has taught during the past few years, for students of mathematics both in Germany and in the UK. It assumes that the reader has a good knowledge of real analysis, measure theory, linear algebra, and probability theory. Some chapters also require knowledge in optimization and functional analysis. Prerequisites beyond the aforementioned standard undergraduate topics are indicated at the beginning of each chapter and can also be seen in the diagram on the following page. Knowledge in computer science or numerical methods is of course helpful, but not necessarily required.

We largely follow a theorem-proof style common in the mathematical literature, supplemented by detailed explanations in prose. The latter is complemented by 121 classroom-tested exercises. These include both theoretical tasks and tasks that require coding.

What follows is a brief outline of the topics covered in this book. We start with an introductory Chapter 1, in which we explain what is to be understood by *data*, and what kind of methods we seek in order to gain what kind of insights from data. The first proper Chapter 2 then deals with classic regression methods; notice that this possibly mundane-looking topic already introduces many general ideas that will prove fundamental throughout the whole book. In Chapter 3 we attend to the very simple and illustrative k-NN algorithm, discuss several preprocessing methods, and apply both to examples from text mining and product reviews. In Chapters 4 and 5 we deal with clustering methods for datasets in metric spaces and then on graphs. This is followed by Chapters 6 and 7, in which best-fit subspaces, their connection with the singular value decomposition of matrices (SVD), and, as an application thereof, principal component analysis (PCA), dimensionality reduction, and collaborative filtering are discussed. In Chapters 8–12 we turn to high-dimensional datasets.

First, we illustrate peculiarities of the uniform and the Gaussian distributions in high-dimensional spaces, before discussing strategies with which data originating from several independent Gaussian distributions can be separated and the underlying parameters can be estimated. In Chapter 13 we deal with the perceptron algorithm, followed by support vector machines (SVMs) in Chapter 14 and the kernel method for SVMs in Chapter 15. Finally, we arrive at neural networks, where we first deal with their so-called expressivity in Chapter 16.1 and then treat the backpropagation algorithm in Chapter 16.2. In the final Chapter 17 we discuss the gradient descent method in the context of convex functions. The following diagram shows the dependencies between the chapters.

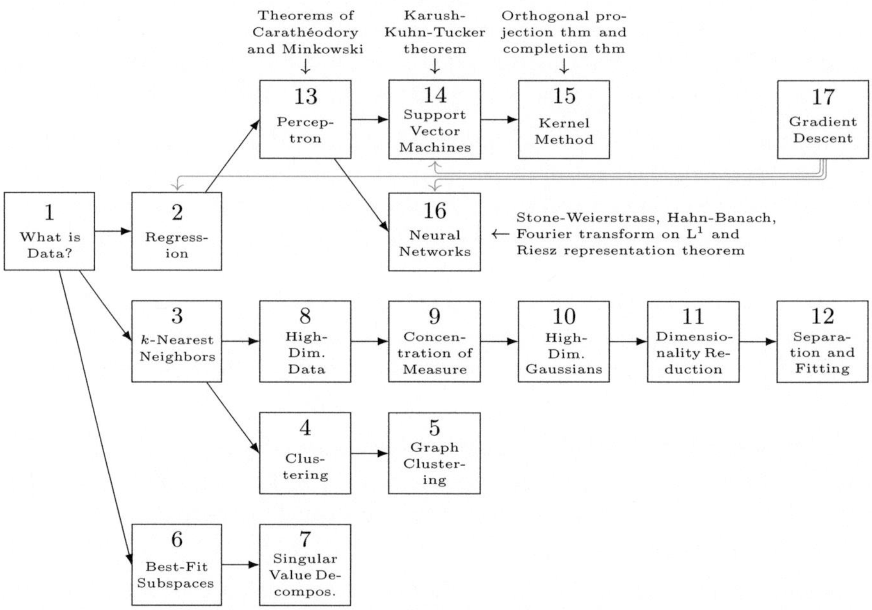

As indicated above, results from Chapter 17 are used in several of the earlier chapters. Some readers may therefore prefer to start with this last chapter. On the other hand, only two results from there, about the existence and uniqueness of minimizers, are used in Chapters 2 and 14. These can be easily looked up on the spot. In Chapter 16, we refer to a detailed discussion of the gradient descent method under favorable conditions in Chapter 17.

More important are the following prerequisites, also indicated in the diagram, for Chapters 13–16, which we will state without proof at the appropriate places: In Chapter 13, we will need Carathéodory's characterization of the convex hull and Minkowski's theorem on the separation of compact convex sets. Chapter 14 heavily uses the theory of Karush-Kuhn-Tucker for optimizing convex functions under inequality constraints. Finally, Chapter 15 requires two results from

Hilbert space theory and Chapter 16 several results from functional analysis. Notice however, that these are only needed in the second half of Subchapter 16.1 on expressivity.

The results used without proof in Chapters 13 and 15 are intuitively accessible. In contrast, it must be said — without wanting to deter any readers — that in particular the proof of Theorem 14.11 on how to calculate an SVM by solving a quadratic optimization problem, as well as the proofs of the expressivity results 16.20, 16.21, 16.22 and 16.25 for neural networks, should best be preceded by an optimization or functional analysis course if one wants to understand them fully.

I would like to express my sincere thanks to all participants of my courses and seminars on data science, as well as to my colleagues for numerous interesting and helpful discussions. I also would like to thank the entire data science and machine learning community, whose books, lecture notes, scientific articles, blogs, videos, and forum contributions inspired me a lot, and from whom I have learned everything that is covered in this text. A special thank you goes to the authors of the books [BHK20, LRU12, Ver18, SSBD14], on whose groundwork the current text largely builds. Each chapter contains a short section with more detailed references at the end. Finally, I would like to thank Andreas Rüdinger and Bianca Alton from Springer for their support throughout the publication process and for many concrete suggestions for improvement, as well as Marianne Lawson for her careful review of my manuscript and the correction of numerous errors.

Additional material, as well as a list of errata, if applicable, will be provided on the page `https://mathematicaldatascience.github.io`.

Hamburg, April 2024 Sven A. Wegner

Contents

1

What is Data (Science)?

We start by explaining what we will understand by the term *data* throughout this text. Simultaneously we issue a warning to that the exact definition may slightly change from chapter to chapter.

Definition 1.1. Let X and Y be sets.

(i) A finite subset $D \subseteq X$ is called an *unlabeled dataset* in X, its elements are called *data points*.

(ii) A finite set $D \subseteq X \times Y$ is called a *labeled dataset*. If $(x, y) \in D$, then x is the *feature part* of the data point (x, y) and y is its *label*.

(iii) A labeled dataset $D \subseteq X \times Y$ is called *categorically labeled*, if Y is finite and *continuously labeled*, if Y is a continuum.[1]

(iv) If $X = X_1 \times \cdots \times X_d$ and $x = (x_1, \ldots, x_d)$, we call the x_i's the *features* of x. If $Y = Y_1 \times \cdots \times Y_m$, we speak of *multidimensional labels*.

The set X in the above Definition 1.1 is often equal to the space $\mathbb{R}^d$ and then always equipped with the Euclidean norm and the standard scalar product. If $X \subseteq \mathbb{R}^d$ is not a vector space, then we furnish X with the Euclidean metric. In some chapters, however, (X, ρ) is an abstract metric space and in some cases ρ does not even need to be a metric, but only a so-called distance measure. In the categorical case, we may assume $Y = \{1, \ldots, m\}$ without loss of generality.

Remark 1.2. In some situations, it is necessary to allow duplicate data points. This can be achieved by replacing X with $X \times \mathbb{N}$. In a subset $D \subseteq X \times \mathbb{N}$, $x \in X$ can then appear twice, for example in the form of $(x, 1)$ and $(x, 2)$. To avoid unnecessary complications, we write x_i or $x^{(i)} \in X$ instead of $(x, i) \in X \times \mathbb{N}$ in the following. When we speak of a dataset $D = \{x^{(1)}, \ldots, x^{(n)}\} \subseteq X$, we tacitly assume that the same point, with a different index, can occur multiple times.

[1] Here, one should think of a closed, open, half-open, bounded or unbounded interval in $\mathbb{R}$ that has at least two points. Occasionally, we may however also consider more general subsets of $\mathbb{R}$.

S. A. Wegner, *Mathematical Introduction to Data Science*,
https://doi.org/10.1007/978-3-662-69426-8_1

In concrete applications, it is a question of modeling whether one or more coordinates of a data vector $(x_1, \ldots, x_d)$ should be designated as labels and if so, which ones. We consider the following examples.

Example 1.3. (i) Consider ten students taking an exam for the course "Mathematical Introduction to Data Science". For each student, we record the preparation time for the exam in hours, the time spent on social media, also in hours, and finally the exam result in percent. We can consider the following table as an unlabeled dataset $D \subseteq \mathbb{R}^3$, or for example as a labeled dataset $D \subseteq [0, 168]^2 \times [0, 100]$ with the preparation time and the time on social media as features and the exam result as a continuous label.

Student	Preparation time in h	Social media time in h	Exam result in %
1	0.0	20.0	0.0
2	1.5	8.5	2.0
3	2.0	6.0	7.0
4	2.0	6.0	10.5
5	8.0	10.0	29.5
6	8.5	3.0	49.0
7	9.5	0.0	59.5
8	12.0	2.0	63.5
9	18.0	4.0	85.0
10	19.0	0.5	98.0

(ii) We consider handwritten letters, which can be written as (7×5)-matrices of ones and zeros as illustrated in the following pictures. Assuming we know which letters the pictures suppose to represent, we obtain a labeled dataset $D \subseteq \mathbb{R}^{7\times5} \times \{\mathrm{a}, \mathrm{b}, \mathrm{c}, \ldots\}$.

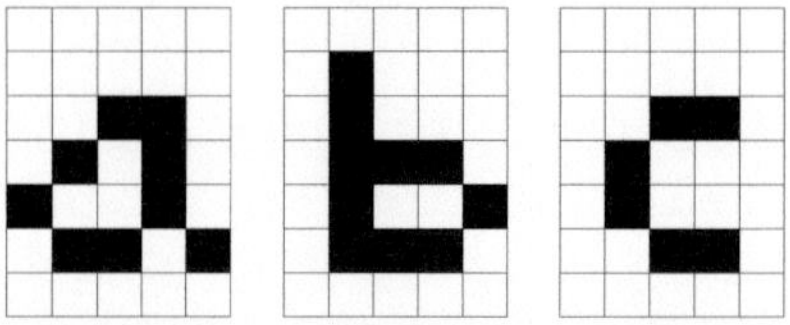

(iii) Let $X \subseteq \mathbb{R}^2$ be a cuboid and $Y = \{1, 2, 3\}$. The following picture represents a categorically labeled dataset, where each data point has two features. In the picture, label 1 corresponds to a white, label 2 to a gray, and label 3 to a black point.

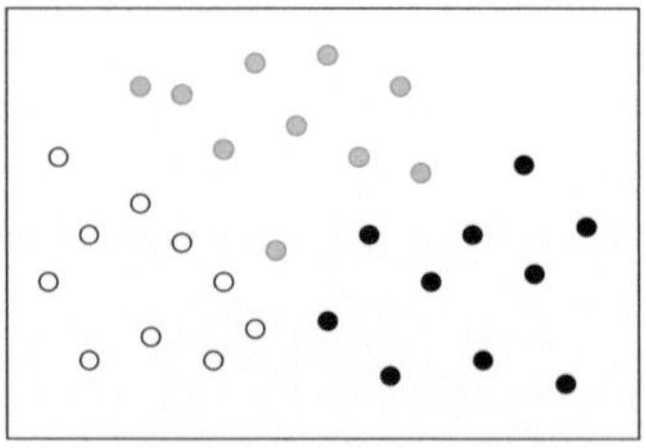

(iv) We again consider a course "Mathematical Introduction to Data Science" and record the body height of each participant. In this way, we obtain an unlabeled dataset $D \subseteq (0, \infty)$, compare Remark 1.2. The picture shows the distribution of body heights.

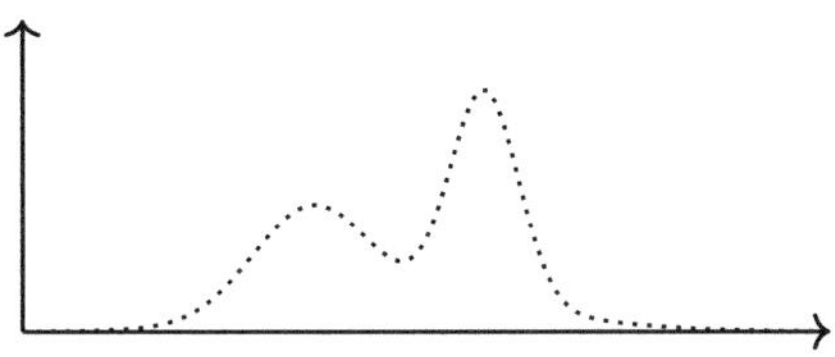

If a labeled dataset $D \subseteq X \times Y$ is given, a central task is to assign a function $f \colon X \to Y$ to the dataset. One can take three perspectives, which are not sharply separated, but merge into each other:

1. There exists an unknown "true" function $f_0 \colon X \to Y$ and the dataset is of the form $D = \{(x_1, f_0(x_1)), \ldots, (x_n, f_0(x_n))\}$ with $x_i \in X$. We want to approximate f_0 by f, thus achieving $f(x) \approx f_0(x)$ for all $x \in X$. This is called the *approximation perspective*.

2. The dataset is generated by a random perturbation of some function $f_0 \colon X \to Y$, e.g., via $y_i = f_0(x_i) + \varepsilon_i$, where $\varepsilon_i \in \mathbb{R}$ are realizations of a normally distributed random variable. We are looking for that function f from which the data "most likely" originates. This is called the *probability perspective*.

3. We do not assume that there is a true function, but rather adjust f by minimizing a cost function $\phi = \phi(f, D)$ to fit the data. This is called the *optimization perspective*.

Depending on context and application, we call the function f a *regressor*, *classifier*, *predictor*, or *approximant*. The dataset used to determine f is then called the *training data*, and the process of determining f is referred to as *supervised learning*.

Let us say that in Example 1.3(i) an affine linear regression is performed. Then one can take the point of view that there is indeed a "causal relationship" of the form

$$\text{exam result} = a_1 \cdot \text{preparation time} + a_2 \cdot \text{social media time} + b$$

and that the true constants a_1, a_2, and b are determined using the data. In Example 1.3(ii), it is rather difficult to come up with an idea for an underlying function; instead, one can minimize a suitably defined cost, or error, function over a suitable class of potential approximants and then hope that a new handwritten letter will be recognized correctly. If continuous labels come from physical measurements, then it might be a natural assumption that there is a normally distributed measurement error involved.

If an unlabeled dataset $D \subseteq X$ is given, e.g., as in Example 1.3(iv), it is also a natural task to assign a function to the data. In the aforementioned example, it is plausible to assume that the sketched distribution arises from the superposition of two normal distributions (height of male and female participants):

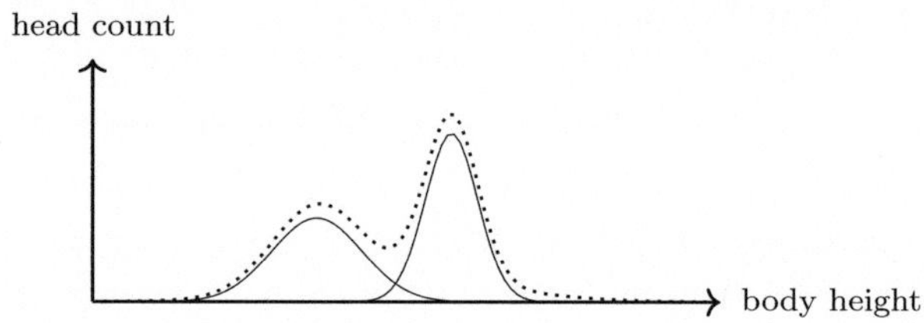

The goal is then to find a function $f : D \to \{\mathrm{m}, \mathrm{f}\}$ that assigns a gender to a body height, i.e., separates the two "clusters". Once this is done, another task is to estimate mean and variance of the two individual normal distributions.

Processes in which unlabeled datasets are given and patterns within the data are recognized or predictions are made based on the unlabeled data are referred to as *unsupervised learning*.

We provide further examples.

Example 1.4. (i) Let a social network be illustrated by the following picture, where a connecting line between names indicates that these users are friends. A natural task is to find the "clusters". In the picture below, the latter can be easily recognized, but in a larger social network this is of course not the case.

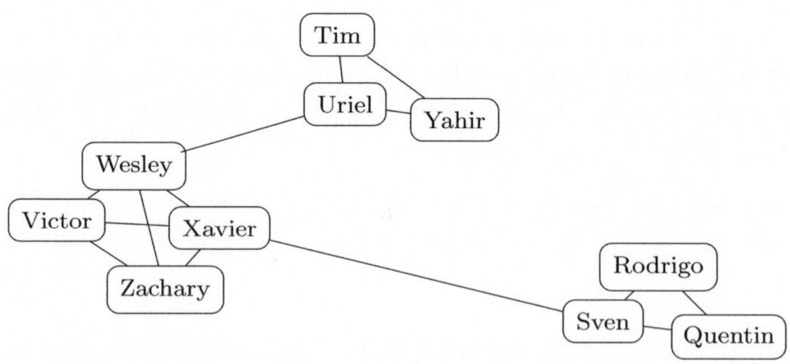

(ii) The following table (taken from [LRU12]) contains ratings from 0–5 for five movies by seven reviewers. Natural tasks include the recognition of patterns, such as which reviewers have similar movie tastes, and the prediction of ratings when new reviewers or new movies come into play.

	Alien	Casablanca	Star Wars	Titanic	Matrix
Abbie	0	2	0	2	1
Bailey	1	0	1	0	1
Catherine	5	0	5	0	5
Darlene	0	4	0	4	2
Elena	3	0	3	0	3
Fatima	0	5	0	5	0
Gladys	4	0	4	0	4

(iii) We consider the following grayscale image, which can be represented by a (320×240)-matrix of numbers between zero and one. A natural task here is data compression, where we are looking for a method with which the image can be reconstructed or approximated from fewer than $(320 \cdot 240)$-many real numbers.

We note that a priori none of the Examples 1.4(i)–(iii) are covered by our initial Definition 1.1 of a dataset. In the case of the social network, however, we can number the users with $1, 2, 3, \ldots, d$ and then represent the i-th user by a vector $a_i := (a_{i1}, \ldots, a_{id}) \in \mathbb{R}^d$, where a_{ij} is one if the i-th user is friends with the j-th user, and zero otherwise. In this case it is however more convenient to use the symmetric *data matrix*

$$A = \begin{bmatrix} a_{11} & \cdots & a_{1d} \\ \vdots & & \vdots \\ a_{d1} & \cdots & a_{dd} \end{bmatrix}$$

instead of the dataset $D = \{a_1, \ldots, a_d\}$.

In the case of the rating table in Example 1.4(ii), the data comes already in matrix form, but we could also model the rows or columns as data points. Note that this leads to two different datasets: Either a set of reviewers, represented by elements of $\mathbb{R}^5$, or a set of movies, which are then given by elements of $\mathbb{R}^7$.

In the case of the grayscale image in Example 1.4(iii), the data also comes as a matrix, namely as an element of $\mathbb{R}^{320\times 240}$, and it seems rather unnatural to interpret, for example, the rows or columns as a set of individual data points. If one treats the matrix as a whole, the mentioned compression can be achieved, for example, by approximation through a matrix of lower rank. The following images show two such approximations.

rk = 5

rk = 15

If one now considers the rows of the image in Example 1.4(iii) as data points in $\mathbb{R}^{320}$, the latter images correspond projections of these 240 data points onto a 5- and a 15-dimensional subspace of $\mathbb{R}^{320}$, respectively. These two spaces were of course selected in a sophisticated way.

Applications that reduce the *dimensionality* of a dataset in this sense are particularly important because many datasets naturally live in a space with very high dimension. In some cases, the dimension of the space is even significantly larger than the number of given data points: Consider, for example, again the participants of a course "Mathematical Introduction to Data Science", represented by all of their social media posts, i.e., each participant is given by a feature vector that contains all the texts, images, and videos that this participant has ever posted.

We conclude this chapter with the following remark.

Remark 1.5. If one assigns a function $f\colon X \to Y$ to a labeled or unlabeled dataset D via an algorithm, this is often referred to as *machine learning* and one says that the function is *learned* from the data. We advise the reader to be cautious with this notion. In fact, in the context of Example 1.3(ii), we mentioned already that the "learning" of handwritten letters corresponds to the numerical minimization of a cost function, and we will see that "learning" the taste of the reviewers in Example 1.4(ii) will be achieved by determining eigenvalues.

2

Affine Linear, Polynomial and Logistic Regression

We will begin our study of data science with the classic topic of regression methods. This may seem unspectacular at first glance, but in fact, many ideas, concepts, and technical tricks coming into play here will prove fundamental throughout the whole book. For example, we will encounter the minimization of a convex cost function in the context of support vector machines and we will use the maximum likelihood method for parameter estimation in the context of high-dimensional Gaussians. In the subchapter on polynomial regression, as later with the kernel method, we will reduce a nonlinear problem to a linear one. Finally, the perceptron, and thus the central building block of neural networks, is closely related to the method of logistic regression, which we will discuss at the end of the current chapter.

In due course we will repeatedly point out concrete connections with other chapters and reference accordingly. At some points, we will also give forward references to later chapters, but we do not need anything deep — but only results on the existence and uniqueness of minimizers of convex functions.

2.1 Affine Linear Regression in One Dimension

Let a dataset

$$D := \{(x_i, y_i) \in \mathbb{R}^2 \mid i = 1, \ldots, n\} \tag{2.1}$$

be given, where each data point consists of a real feature x_i and a real label y_i. Our goal is to find an affine linear function $f\colon \mathbb{R} \to \mathbb{R}$, such that the sum

$$\sum_{i=1}^{n} (f(x_i) - y_i)^2$$

S. A. Wegner, *Mathematical Introduction to Data Science*,
https://doi.org/10.1007/978-3-662-69426-8_2

of the squared distances between function value and label over all data points is minimal. Intuitively, this means that in the following picture we are looking for that solid line for which the sum of the squares of the lengths of the dotted lines takes the smallest possible value.

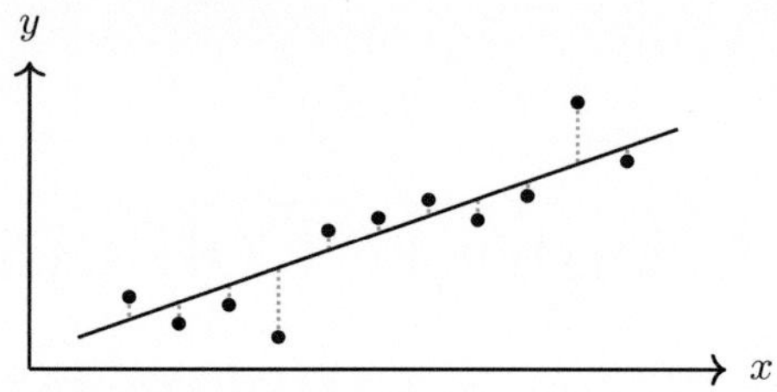

A priori, it is of course not clear whether such a line, or affine linear function, exists at all and, if so, whether it is uniquely determined. Finally, the question arises whether f can be explicitly calculated from the data. All three questions are answered by the following theorem.

Theorem 2.1. (on simple affine linear regression) *Let a dataset $D := \{(x_i, y_i) \mid i = 1, \dots, n\} \subseteq \mathbb{R}^2$ be given such that not all x_i coincide. Then there is precisely one affine linear function*

$$f^* = \operatorname{argmin}\Big\{ \sum_{i=1}^{n} (f(x_i) - y_i)^2 \mid f\colon \mathbb{R} \to \mathbb{R} \textit{ affine linear} \Big\}$$

and it holds that $f^*(x) = a^*x + b^*$ for $x \in \mathbb{R}$ with

$$a^* = \frac{\sum_{i=1}^{n} (x_i - \overline{x})(y_i - \overline{y})}{\sum_{i=1}^{n} (x_i - \overline{x})^2} = \frac{\big(\sum_{i=1}^{n} x_i y_i\big) - n\overline{x}\,\overline{y}}{\big(\sum_{i=1}^{n} x_i^2\big) - n\overline{x}^2} \quad \textit{and} \quad b^* = \overline{y} - a^*\overline{x}.$$

Here, $\overline{x} = (x_1 + \cdots + x_n)/n$ and $\overline{y} = (y_1 + \cdots + y_n)/n$ are the averages of the features and the labels, respectively.

Proof. ① We first show that the two expressions for a^* coincide. Using $x_1 + \cdots + x_n = n\overline{x}$, and the corresponding formula for $\overline{y}$, we get for the numerator

$$\begin{aligned}\sum_{i=1}^{n}(x_i - \overline{x})(y_i - \overline{y}) &= \sum_{i=1}^{n} x_i y_i - \overline{x}\sum_{i=1}^{n} x_i - \overline{x}\sum_{i=1}^{n} y_i + n\overline{x}\,\overline{y} \\ &= \Big(\sum_{i=1}^{n} x_i y_i\Big) - n\overline{x}\,\overline{y}.\end{aligned}$$

The equality of the denominators follows by specializing $y_i = x_i$. In particular, none of the denominators are equal to zero, since the x_i's are by assumption not all equal. Next, we need to show that the function

$$\phi\colon \{f\colon \mathbb{R} \to \mathbb{R} \mid f \text{ affine linear}\} \to \mathbb{R},\ \phi(f) := \sum_{i=1}^{n} (f(x_i) - y_i)^2$$

has a minimizer. Since affine linear functions are uniquely determined by two real parameters a and b, we can read the aforementioned ϕ as a function of these parameters and minimize

$$\phi\colon \mathbb{R}^2 \to \mathbb{R},\ \phi(a, b) = \sum_{i=1}^{n} (ax_i + b - y_i)^2.$$

By construction, $\phi \geqslant 0$ holds and ϕ is a continuous function. To guarantee the existence of a minimizer, it suffices to show that the limit inferior of $\phi(a, b)$ for $\|(a, b)\| \to \infty$ is equal to infinity. Assuming the latter does not hold, there is a sequence $(a^{(k)}, b^{(k)})_{k\in\mathbb{N}}$ with

$$\|(a^{(k)}, b^{(k)})\| \to \infty \quad \text{and} \quad \phi(a^{(k)}, b^{(k)}) = \sum_{i=1}^{n} (a^{(k)} x_i + b^{(k)} - y_i)^2 \to c \in \mathbb{R}.$$

We choose i and j such that $x_i \neq x_j$ holds. Since only nonnegative terms are added in the sum on the right, all summands must be bounded. This implies that both $(a^{(k)} x_i + b^{(k)} - y_i)_{k\in\mathbb{N}}$ and $(a^{(k)} x_j + b^{(k)} - y_j)_{k\in\mathbb{N}}$ are bounded, and thus the difference $(a^{(k)}(x_i - x_j) - y_i + y_j)_{k\in\mathbb{N}}$ is also bounded. Since $x_i - x_j \neq 0$, the latter means that $(a^{(k)})_{k\in\mathbb{N}}$ is bounded and therefore $|b^{(k)}| \to \infty$ holds. Together, this implies $\phi(a^{(k)}, b^{(k)}) \to \infty$ contradicting the assumption. The limit inferior of $\phi(a, b)$ for $\|(a, b)\| \to \infty$ is thus equal to ∞ and we obtain the existence of a minimizer.

② Next, we consider

$$\nabla\phi(a, b) = \begin{bmatrix} \frac{\partial\phi}{\partial a} \\ \frac{\partial\phi}{\partial b} \end{bmatrix} = \begin{bmatrix} \sum_{i=1}^{n} 2(ax_i + b - y_i)x_i \\ \sum_{i=1}^{n} 2(ax_i + b - y_i)1 \end{bmatrix} \stackrel{!}{=} 0.$$

The second equation is equivalent to

$$0 = \sum_{i=1}^{n} (ax_i + b - y_i) = a \sum_{i=1}^{n} x_i + nb - \sum_{i=1}^{n} y_i = an\overline{x} + n\overline{b} - n\overline{y},$$

from which we get $b = \overline{y} - a\overline{x}$. Subbing this into the first equation yields

$$0 = \sum_{i=1}^{n} (ax_i + \overline{y} - a\overline{x} - y_i)x_i = a \sum_{i=1}^{n} x_i^2 + \overline{y} \sum_{i=1}^{n} x_i - a\overline{x} \sum_{i=1}^{n} x_i - \sum_{i=1}^{n} y_i x_i$$

$$= a\sum_{i=1}^{n} x_i^2 + \overline{y}n\overline{x} - a\overline{x}n\overline{x} - \sum_{i=1}^{n} x_i y_i$$
$$= a\Big(\sum_{i=1}^{n} x_i^2 - n\overline{x}^2\Big) - \Big(\sum_{i=1}^{n} x_i y_i - n\overline{x}\overline{y}\Big),$$

which can be solved for a. The derivative of the everywhere differentiable function ϕ thus vanishes at precisely one point, namely at (a^*, b^*), where a^* and b^* are given by the formulas noted in the theorem. Together with the first part of the proof, it follows that (a^*, b^*), or in other words $f^*\colon \mathbb{R} \to \mathbb{R}$, $x \mapsto a^*x + b^*$, is the uniquely determined minimizer of ϕ. □

Definition 2.2. Let D be a dataset as in Theorem 2.1. Then the function $f^*\colon \mathbb{R} \to \mathbb{R}$, $f^*(x) = a^*x + b^*$ as in Theorem 2.1 is called the *affine linear regressor* for D and its graph is called the *regression line* for D.

Remark 2.3. (i) Under the assumptions of Theorem 2.1, $f^*(\overline{x}) = \overline{y}$ holds, i.e., the regression line always passes through the point $(\overline{x}, \overline{y})$.

(ii) The function ϕ used in the proof of Theorem 2.1 is called *error function*, *cost function*, *loss function* or *objective function*. The squares included in it ensure that ϕ becomes differentiable. If one uses, for example,

$$\psi(f) := \sum_{i=1}^{n} |f(x_i) - y_i|$$

instead, one can still show that a minimizer exists, but it is in general not unique, see Exercise 2.2, and may differ from the affine linear regressor.

The above approach to defining a regressor is an example of the optimization perspective explained in Chapter 1, where a chosen error function is minimized over a chosen class of functions.

Using the error function ϕ as in Theorem 2.1 is the so-called *method of least squares*. Besides the mentioned effect that the squares make ϕ differentiable, their use may seem somewhat arbitrary at first glance. While ψ in Remark 2.3(ii) treats all deviations equally, ϕ "punishes" deviations greater than one more harshly than deviations less than one. On second glance, however, it will now be shown that the squares are very natural.

We adopt the probability perspective already mentioned in Chapter 1, i.e., we assume that there actually is an affine linear function $f_0\colon \mathbb{R} \to \mathbb{R}$, $f_0(x) = ax + b$, whose parameters $a, b \in \mathbb{R}$ are unknown to us. We further assume that the dataset D arises in the following way: for a feature x_i the label is given by $y_i := f_0(x_i) + \varepsilon_i$, where the ε_i's are independent normally distributed perturbations. Given such a dataset D, we now want to determine an affine linear function $f^*\colon \mathbb{R} \to \mathbb{R}$, $f^*(x) = a^*x + b^*$, such that the likelihood for the data coming from f^* is maximal compared to all other affine linear functions from which the data could also have originated. To make this more precise, we consider the *likelihood function*

$$L\colon \{f\colon \mathbb{R}\to\mathbb{R}\mid f \text{ affine linear}\}\to\mathbb{R},\ L(f):=\prod_{i=1}^{n}\frac{1}{\sqrt{2\pi}\sigma}\,\mathrm{e}^{-\frac{(f(x_i)-y_i)^2}{2\sigma^2}} \tag{2.2}$$

for a dataset D and $\sigma>0$. If $(\Omega,\Sigma,\mathrm{P})$ is a probability space, $\mathcal{E}_1,\dots,\mathcal{E}_n\colon\Omega\to\mathbb{R}$ are independent random variables with $\mathcal{E}_i\sim\mathcal{N}(0,\sigma^2)$, and $Y_i(f):=f(x_i)+\mathcal{E}_i$ for a given affine linear f, then we have $Y_i(f)\sim\mathcal{N}(f(x_i),\sigma^2)$, compare Proposition A.19. Consequently, for fixed $y_i\in\mathbb{R}$ and for $\delta>0$ we get

$$\mathrm{P}\big[|Y_i(f)-y_i|<\delta\big]=\int_{x_i-\delta}^{x_i+\delta}\frac{1}{\sqrt{2\pi}\sigma}\,\mathrm{e}^{-\frac{(f(t)-y_i)^2}{2\sigma^2}}\,\mathrm{d}t.$$

By independence, it follows

$$\mathrm{P}\big[|Y_i(f)-y_i|<\delta \text{ for all } i\big]=\int_{Q_\delta(x)}\prod_{i=1}^{n}\frac{1}{\sqrt{2\pi}\sigma}\,\mathrm{e}^{-\frac{(f(t_i)-y_i)^2}{2\sigma^2}}\,\mathrm{d}(t_1,\dots,t_n),$$

if we abbreviate the cube around $x=(x_1,\dots,x_n)\in\mathbb{R}^n$ with edge length 2δ by $Q_\delta(x)=[x_1-\delta,x_1+\delta]\times\cdots\times[x_n-\delta,x_n+\delta]$. We now think of $\delta>0$ to be a very small but fixed constant. Then, by the above, the likelihood that $f(x_i)+\mathcal{E}_i\approx y_i$ holds for all i is large when $L(f)$ is large. For this reason, we call a maximizer f^* of L a *maximum likelihood estimator*. The following theorem will show that the latter exists and relate it to the method of least squares.

Theorem 2.4. *Let $D:=\{(x_i,y_i)\mid i=1,\dots,n\}\subseteq\mathbb{R}^2$ be a dataset such that not all x_i coincide. Then for each $\sigma>0$ there is precisely one maximizer*

$$f^*=\operatorname{argmax}\big\{L(f)\mid f \textit{ affine linear}\big\}$$

of the likelihood function L from (2.2) and $f^*\colon\mathbb{R}\to\mathbb{R}$, $f^*(x)=a^*x+b^*$ coincides with the affine linear regressor from Theorem 2.1.

Proof. Instead of maximizing L, we maximize the so-called *log-likelihood function* $\ell:=\log\circ L$, for which we get

$$\begin{aligned}\ell(f)&=\log\Big(\prod_{i=1}^{n}\frac{1}{\sqrt{2\pi}\sigma}\,\mathrm{e}^{-\frac{(f(x_i)-y_i)^2}{2\sigma^2}}\Big)\\&=n\log\big(\tfrac{1}{\sqrt{2\pi}\sigma}\big)-\tfrac{1}{2\sigma^2}\sum_{i=1}^{n}(f(x_i)-y_i)^2\\&=c_1+c_2\,\phi(f),\end{aligned} \tag{2.3}$$

where c_1, $c_2\in\mathbb{R}$ are constants independent of f, and ϕ is the error function from Theorem 2.1. Since $c_2 < 0$, the conclusion follows immediately from Theorem 2.1. □

Definition 2.5. We denote by

$$\mathrm{cov}\colon \mathbb{R}^n \times \mathbb{R}^n \to \mathbb{R},\ \mathrm{cov}(x,y) := \tfrac{1}{n}\sum_{i=1}^{n}(x_i - \overline{x})(y_i - \overline{y}) \quad \text{and}$$
$$\mathrm{var}\colon \mathbb{R}^n \to \mathbb{R},\ \mathrm{var}(x) := \mathrm{cov}(x,x)$$

the *covariance* and the *variance* functions. The proof of Theorem 2.1 has shown that $\mathrm{cov}(x,y) = \overline{xy} - \overline{x}\,\overline{y}$ and $\mathrm{var}(x) = \overline{x^2} - \overline{x}^2$ hold, where we use the abbreviation $\overline{xy} = \frac{1}{n}(x_1y_1 + \cdots + x_ny_n)$.[1] If we evaluate the (co)variance on a sample of a random variable, we refer to it as *sample covariance* and *sample variance*, respectively.

Using variance and covariance, the parameters of the affine linear function f^* in Theorems 2.1 and 2.4 can be written as follows:

$$a^* = \frac{\mathrm{cov}(x,y)}{\mathrm{var}(x)} \quad \text{and} \quad b^* = \overline{y} - a^*\,\overline{x}.$$

We point out that many authors define the (co)variance with a factor $\frac{1}{n-1}$ instead of $\frac{1}{n}$. In the above formula for a^* this makes however no difference and for large n it also does not matter too much for variance and covariance individually. Nevertheless, there are good reasons to use the factor $\frac{1}{n-1}$, compare Exercise 2.14.

The assumption that a dataset stems from an unknown affine linear function with a normally distributed perturbation leads to the question of whether for large samples the estimated parameters a^* and b^* are close to the parameters a and b of the "true" function. The following theorem answers this question asymptotically.

Theorem 2.6. *Let $(x_i)_{i\in\mathbb{N}} \subseteq \mathbb{R}$ be a bounded sequence, let $(\Omega, \Sigma, \mathrm{P})$ be a probability space, let $\mathcal{E}_i\colon \Omega \to \mathbb{R}$ be independent random variables with $\mathcal{E}_i \sim \mathcal{N}(0,\sigma^2)$ for $i \in \mathbb{N}$ and let $Y_i := ax_i + b + \mathcal{E}_i$, where $a, b \in \mathbb{R}$ and $\sigma > 0$. For $n \in \mathbb{N}$, we assume that $x^{(n)} := (x_1, \ldots, x_n)$ is not constant, that the sequence of means $(\overline{x^{(n)}})_{n\in\mathbb{N}}$ converges and the sequence of variances $(\mathrm{var}\, x^{(n)})_{n\in\mathbb{N}}$ converges to $v \neq 0$. With the abbreviation $Y^{(n)} := (Y_1, \ldots, Y_n)$, we find that the sequences of random variables*

$$A^{(n)} := \frac{\mathrm{cov}(x^{(n)}, Y^{(n)})}{\mathrm{var}(x^{(n)})} \xrightarrow{n\to\infty} a \quad \text{and} \quad B^{(n)} := \overline{Y^{(n)}} - A^{(n)}\,\overline{x^{(n)}} \xrightarrow{n\to\infty} b$$

[1] Formally, we are taking the mean of the Hadamard product $\overline{x \odot y}$ of the vectors x and y, compare Definition 16.31.

converge in probability. Explicitly, this means that

$$\lim_{n\to\infty} \mathrm{P}\big[|A^{(n)} - a| > \varepsilon\big] = 0 \quad \text{and} \quad \lim_{n\to\infty} \mathrm{P}\big[|B^{(n)} - b| > \varepsilon\big] = 0$$

hold for every $\varepsilon > 0$.

Proof. Using the formulas given in Definition 2.5 and the abbreviation $\mathcal{E}^{(n)} = (\mathcal{E}_1, \dots, \mathcal{E}_n)$, we obtain

$$\begin{aligned}
\operatorname{cov}(x^{(n)}, Y^{(n)}) &= \overline{x^{(n)}Y^{(n)}} - \overline{x^{(n)}}\,\overline{Y^{(n)}} \\
&= \tfrac{1}{n}\sum_{i=1}^{n} x_i(ax_i + b + \mathcal{E}_i) - \overline{x^{(n)}}\big(\tfrac{1}{n}\sum_{i=1}^{n} ax_i + b + \mathcal{E}_i\big) \\
&= a\overline{x^{(n)2}} + \overline{x^{(n)}}b + \overline{(x\mathcal{E})^{(n)}} - a\overline{x^{(n)}}^{2} - \overline{x^{(n)}}b - \overline{x^{(n)}}\,\overline{\mathcal{E}^{(n)}} \\
&= a\ \operatorname{var}(x^{(n)}) + \overline{(x\mathcal{E})^{(n)}} - \overline{x^{(n)}}\,\overline{\mathcal{E}^{(n)}}.
\end{aligned}$$

The first term converges pointwise to the constant random variable $v \neq 0$ by assumption. Since we have $\mathcal{E}_i \sim \mathcal{N}(0, \sigma^2)$, the law of large numbers, see Theorem A.14, implies that $\overline{\mathcal{E}^{(n)}} \to 0$ in probability. As the sequence of means $(\overline{x^{(n)}})_{n\in\mathbb{N}}$ is bounded by assumption, the last term therefore converges to zero in probability. It remains to consider the middle term

$$\overline{(x\mathcal{E})^{(n)}} = \tfrac{1}{n}\sum_{i=1}^{n} x_i\mathcal{E}_i,$$

for which one immediately sees $\mathrm{E}(\overline{(x\mathcal{E})^{(n)}}) = 0$. For the variance it follows that

$$\begin{aligned}
\mathrm{V}\overline{(x\mathcal{E})^{(n)}} &= \tfrac{1}{n^2}\,\mathrm{E}\big((x_1\mathcal{E}_1 + \cdots + x_n\mathcal{E}_n)^2\big) \\
&= \tfrac{1}{n^2}\sum_{i=1}^{n} x_i^2\,\mathrm{E}(\mathcal{E}_i^2) + 2\sum_{i<j} x_i x_j\,\mathrm{E}(\mathcal{E}_i)\,\mathrm{E}(\mathcal{E}_j) \\
&= \tfrac{1}{n}\sigma^2\overline{x^{(n)2}}.
\end{aligned}$$

The Chebyshev inequality provides for $\varepsilon > 0$

$$\mathrm{P}\big[|\overline{(x\mathcal{E})^{(n)}} - 0| \geqslant \varepsilon\big] \leqslant \frac{\mathrm{V}\overline{(x\mathcal{E})^{(n)}}}{\varepsilon^2} = \frac{\sigma^2}{n\varepsilon^2}\overline{x^{(n)2}} \xrightarrow{n\to\infty} 0,$$

since $\overline{x^{(n)2}} = \operatorname{var}(x^{(n)}) + \overline{x^{(n)}}^{2}$ as the sum of two convergent sequences is itself convergent. It now immediately follows that $A^{(n)} \to av/v = a$ in probability. From this, because $\overline{\mathcal{E}^{(n)}} \to 0$ and because $(\overline{x^{(n)}})_{n\in\mathbb{N}}$ converges by assumption, we also get

$$B^{(n)} = a\overline{x^{(n)}} + b + \overline{\mathcal{E}^{(n)}} - A^{(n)}\overline{x^{(n)}} \to b$$

in probability. □

More colloquially, Theorem 2.6 states that for a dataset D as in (2.1), where the y_i are generated from the x_i by a normally distributed additive perturbation of an affine linear function $f\colon \mathbb{R} \to \mathbb{R}$, $f(x) = ax + b$, the probability that the estimated parameters a^* and b^* are arbitrarily close to the "true" parameters a and b can be kept arbitrarily close to one, if the size of the dataset is chosen large enough. We refer to Exercise 2.9 for an experimental treatment.

After this probabilistic detour, we now return to the optimization perspective. As already demonstrated in Theorem 2.1, we can approximate any data from the class of affine linear functions. However, it may happen, beyond the situation discussed in Theorem 2.6, that a good approximation by an affine linear function is not possible at all. The following notion allows to quantify "how well" the data can be described by an affine linear function.

Definition 2.7. Let $D := \{(x_i, y_i) \mid i = 1, \ldots, n\} \subseteq \mathbb{R}^2$ be a dataset, in which not all x_i and not all y_i coincide. Then we call

$$r_{xy} = \frac{\operatorname{cov}(x,y)}{\operatorname{var}(x)^{1/2}\operatorname{var}(y)^{1/2}} = \frac{\sum\limits_{i=1}^{n}(x_i - \overline{x})(y_i - \overline{y})}{\Big(\sum\limits_{i=1}^{n}(x_i - \overline{x})^2\Big)^{1/2}\Big(\sum\limits_{i=1}^{n}(y_i - \overline{y})^2\Big)^{1/2}}$$

the (*affine linear*) *regression coefficient* of D.

Theorem 2.8. *Let $D := \{(x_i, y_i) \mid i = 1, \ldots, n\} \subseteq \mathbb{R}^2$ be a dataset as in Definition 2.7. Then $r_{xy} \in [-1, 1]$ and $r_{xy} = \pm 1$ holds if and only if all (x_i, y_i) lie on a straight line with $\pm$ve slope.*

Proof. ① With $u := (x_i - \overline{x})_{i=1,\ldots,n}$, $v := (y_i - \overline{y})_{i=1,\ldots,n} \in \mathbb{R}^n$ the assertion follows from the Cauchy-Schwarz-Bunyakovsky inequality $|\langle u, v\rangle| \leqslant \|u\|\|v\|$ in $\mathbb{R}^n$, because the second formula in Definition 2.7 shows exactly

$$r_{xy} = \frac{\langle u, v\rangle}{\|u\|\|v\|}.$$

② In CSB, equality $|\langle u, v\rangle| = \|u\|\|v\|$ holds if and only if u and v are linearly dependent. Since $u \neq 0 \neq v$, this is equivalent to the existence of a $\lambda \in \mathbb{R}$ with $v = \lambda u$. Explicitly, this means

$$\forall\, i = 1, \ldots, n\colon y_i = \lambda x_i - \lambda\overline{x} + \overline{y}$$

and implies that all (x_i, y_i) lie on a straight line with slope λ if and only if $r_{xy} \in \{+1, -1\}$. If the latter is the case, then $\operatorname{sgn}(r_{xy}) = \operatorname{sgn}(\langle a, b\rangle) = \operatorname{sgn}(\lambda)$, from which the statements about the slope follow. □

Theorem 2.8 suggests that data can be better approximated by an affine linear function the closer r_{xy} is to $+1$ or -1. We will confirm this heuristic experimentally in Exercise 2.9.

Now we want to examine the case that we excluded in Theorem 2.1, namely that all x_i in our dataset $D := \{(x_i, y_i) \mid i = 1, \dots, n\}$ are equal. In this case, there is no uniquely determined minimizer $f^* \colon \mathbb{R}_x \to \mathbb{R}_y,\ x \mapsto f^*(x)$.

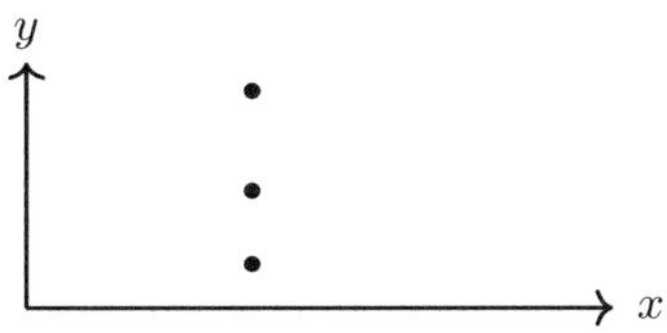

On the other hand, now all data points lie on a straight line by assumption, and the latter can easily be represented by an affine linear function if we only swap the axes, i.e., use the transposed dataset $\check{D} := \{(y_i, x_i) \mid i = 1, \dots, n\}$ instead of D. For this to work out, we need that not all y_i coincide and thus we can treat all datasets that contain at least two different points with Theorem 2.1.

On the other hand, if we assume that not all x_i and also not all y_i are equal, Theorem 2.1 can be applied to D and to $\check{D}$. We then call the line belonging to D the *first regression line*, and the one belonging to $\check{D}$ the *second regression line*. The latter is represented by $g \colon \mathbb{R}_y \to \mathbb{R}_x,\ g(y) = \check{a}y + \check{b}$ with

$$\check{a} = \frac{\operatorname{cov}(x,y)}{\operatorname{var}(y)} \quad \text{and} \quad \check{b} = \overline{x} - \check{a}\overline{y}$$

and we see that g, just as the first regression line, passes through the point $(\overline{x}, \overline{y})$ when we plot both lines in the xy-coordinate system. The slope of the second regression line is here equal to the slope of g^{-1}, i.e., $1/\check{a}$, if $\check{a} \neq 0$. In general, both lines are not equal, see Exercise 2.1. They coincide precisely when $r_{xy} = \pm 1$, see Exercise 2.4.

2.2 Multidimensional Affine Linear Regression

The method discussed so far is often referred to as *simple linear regression*, in contrast to *multivariable* and *multivariate linear regression*, which deal with multidimensional features and multidimensional labels, respectively. We start with the multivariable case, i.e., we consider a dataset

$$D = \{(x_i, y_i) \in \mathbb{R}^d \times \mathbb{R} \mid i = 1, \dots, n\},$$

which we want to approximate by an affine linear function $f\colon \mathbb{R}^d \to \mathbb{R}$, $f(x) = \langle a, x\rangle + b$, with $a \in \mathbb{R}^d$ and $b \in \mathbb{R}$. For $d = 2$ this corresponds to the following picture in which the data points are approximated by a plane.

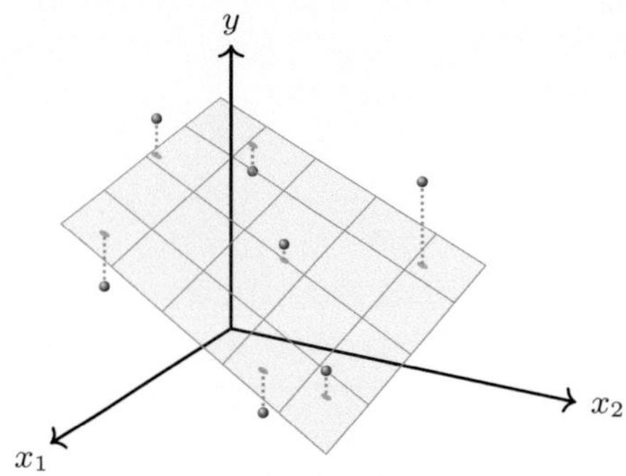

For arbitrary d, we obtain the following extension of Theorem 2.1.

Theorem 2.9. (on multivariable affine linear regression) *Let* $D = \big\{(x_i, y_i) \in \mathbb{R}^d \times \mathbb{R} \mid i = 1, \ldots, n\big\}$ *be a dataset with* $x_i = (x_{i1}, x_{i2}, \ldots, x_{id})$ *for* $i = 1, \ldots, n$. *We assume that the matrix*

$$X := \begin{bmatrix} 1 & x_{11} & \cdots & x_{1d} \\ \vdots & \vdots & & \vdots \\ 1 & x_{n1} & \cdots & x_{nd} \end{bmatrix} \in \mathbb{R}^{n\times(d+1)}$$

has rank $d+1$. *Then there exists precisely one affine linear function* $f^*\colon \mathbb{R}^d \to \mathbb{R}$, *i.e.,* $f(x) = \langle a^*, x\rangle + b^*$ *with unique* $a^* = (a_1^*, \ldots, a_d^*) \in \mathbb{R}^d$ *and* $b^* \in \mathbb{R}$, *such that*

$$f^* = \operatorname{argmin}\Big\{\phi(f) := \sum_{i=1}^{n}(f(x_i) - y_i)^2 \mid f\colon \mathbb{R}^d \to \mathbb{R} \textit{ affine linear}\Big\}$$

holds. The function f^* *is given by the parameters*

$$\begin{bmatrix} b^* \\ a_1^* \\ \vdots \\ a_d^* \end{bmatrix} = (X^\intercal X)^{-1} X^\intercal \begin{bmatrix} y_1 \\ \vdots \\ y_n \end{bmatrix}.$$

Proof. As the formula for a^* and b^* in the theorem suggests, it is convenient to combine a^* and b^* into one $(d+1)$-dimensional vector. In this sense, we write an affine linear function $f\colon \mathbb{R}^d \to \mathbb{R}$ as

$$f(z) = \langle a, z\rangle + b = \sum_{i=1}^{n} a_i z_i + b \cdot 1 = \left\langle \begin{bmatrix} b \\ a_1 \\ \vdots \\ a_d \end{bmatrix}, \begin{bmatrix} 1 \\ z_1 \\ \vdots \\ z_d \end{bmatrix} \right\rangle$$

where $z = (1, z_1, \ldots, z_d) \in \mathbb{R}^d$. Thus, we can minimize over functions of the form

$$f\colon \mathbb{R}^{d+1} \to \mathbb{R},\ f = \langle \tilde{a}, \cdot \rangle$$

or the vectors $\tilde{a} = (b, a_1, \ldots, a_d) \in \mathbb{R}^{d+1}$ that uniquely determine them, if we extend all data points by a new first entry equal to 1. The objective function then becomes

$$\begin{aligned}\phi(\tilde{a}) &= \sum_{i=1}^{n}\Bigg(\Bigg\langle \begin{bmatrix} b \\ a_1 \\ \vdots \\ a_d \end{bmatrix}, \begin{bmatrix} 1 \\ x_{i1} \\ \vdots \\ x_{id} \end{bmatrix} \Bigg\rangle - y_i\Bigg)^2 = \sum_{i=1}^{n}\big((X\tilde{a})_i - y_i\big)^2 \\ &= \langle X\tilde{a} - y, X\tilde{a} - y\rangle = \langle X\tilde{a}, X\tilde{a}\rangle - \langle X\tilde{a}, y\rangle - \langle y, X\tilde{a}\rangle + \langle y, y\rangle \\ &= \langle \tilde{a}, X^\intercal X\tilde{a}\rangle - 2\langle \tilde{a}, X^\intercal y\rangle + \|y\|^2,\end{aligned}$$

where X denotes the matrix defined in the theorem. We now claim

$$\phi'(\tilde{a})h = 2\langle h, X^\intercal X\tilde{a}\rangle - 2\langle h, X^\intercal y\rangle.$$

The second term stems from the fact that $\tilde{a} \mapsto 2\langle \tilde{a}, X^\intercal y\rangle$ is linear. For the first term we calculate

$$\begin{aligned}\tfrac{|\phi(\tilde{a}+h)-\phi(\tilde{a})-2\langle h, X^\intercal X\tilde{a}\rangle|}{\|h\|} &= \tfrac{1}{\|h\|}\big|\langle \tilde{a}, X^\intercal X\tilde{a}\rangle + \langle \tilde{a}, X^\intercal Xh\rangle + \langle h, X^T X\tilde{a}\rangle \\ &\quad + \langle h, X^T Xh\rangle - \langle \tilde{a}, X^\intercal X\tilde{a}\rangle - 2\langle h, X^\intercal X\tilde{a}\rangle\big| \\ &= \frac{|\langle Xh, Xh\rangle|}{\|h\|} \leqslant \|X\|_{\text{op}}^2\|h\| \xrightarrow{h\to 0} 0,\end{aligned}$$

where $\|X\|_{\text{op}}$ denotes the operator norm of X. This establishes the claim. Using the formula, we see that $\phi'(\tilde{a}) = 0$ holds if and only if $2\langle h, X^\intercal X\tilde{a}\rangle - 2\langle h, X^\intercal y\rangle = 0$ for all $h \in \mathbb{R}^{d+1}$. This is equivalent to

$$X^\intercal X\tilde{a} - X^\intercal y = 0.$$

To solve this last equation for $\tilde{a}$, we need to show that the matrix $X^\intercal X \in \mathbb{R}^{(d+1)\times(d+1)}$ is invertible. To this end, let $X^\intercal Xv = 0$ with some $v \in \mathbb{R}^{d+1}$. This implies $\langle v, X^\intercal Xv\rangle = \langle Xv, Xv\rangle = \|Xv\|^2 = 0$, thus $Xv = 0$, which implies $v = 0$ due to $\operatorname{rk} X = d + 1$. Therefore, $X^\intercal X$ is invertible.

From what we proved so far it follows that $\tilde{a} = (X^\intercal X)^{-1}X^\intercal y$ is the only critical point of $\phi\colon \mathbb{R}^{d+1} \to \mathbb{R}$. In contrast to the one-dimensional proof, it is not easy to calculate $\liminf_{\|\tilde{a}\|\to\infty} \phi(\tilde{a})$ explicitly. Instead, we will now make a forward reference to the theory of convex functions that we are going to develop in Chapter 17. First, we observe that the objective function

$$\phi(\tilde{a}) = \sum_{i=1}^{n} \big((X\tilde{a})_i - y_i\big)^2 = \|X\tilde{a} - y\|^2$$

is the composition of an affine linear function and the Euclidean norm. As we will show in Example 17.15 and Lemma 17.16, $\phi\colon \mathbb{R}^{d+1} \to \mathbb{R}$ is therefore convex and for a convex function a critical point is automatically a minimizer, see Corollary 17.5. This finishes the proof. □

If we specialize $d = 1$ in Theorem 2.9, the matrix X simplifies to

$$X := \begin{bmatrix} 1 & x_1 \\ \vdots & \vdots \\ 1 & x_n \end{bmatrix} \in \mathbb{R}^{n\times 2}$$

and we see that the rank condition $\operatorname{rk} X = 2$ from Theorem 2.9 is precisely satisfied when the assumptions from Theorem 2.1 apply, i.e., if not all $x_1, \ldots, x_n$ are equal. From the respective uniqueness statements of the theorems, it immediately follows that both provide the parameters of the same affine linear regressor. If the reader wants to skip the now somewhat redundant Theorem 2.1 entirely, they may also verify directly that the formula

$$\begin{bmatrix} b^* \\ a^* \end{bmatrix} = (X^\intercal X)^{-1} X^\intercal \begin{bmatrix} y_1 \\ \vdots \\ y_n \end{bmatrix}$$

from Theorem 2.9 specializes for $d = 1$ to the formulas for a^* and b^* given in Theorem 2.1. We leave this as Exercise 2.5. The special case $d = 2$ of Theorem 2.9 provides an approximation of points in $\mathbb{R}^3$ by a plane, as shown in the picture on page 16. As above, the three parameters a_1^*, a_2^* and b^* can be explicitly given by using the sample variance and covariance of the data points, see Exercise 2.6.

Next, we investigate the question of how to measure the quality of approximation by an affine linear regressor in the multivariable case. To this end, we generalize the regression coefficient as follows.

Definition 2.10. Let $D = \big\{(x_i, y_i) \in \mathbb{R}^d \times \mathbb{R} \,|\, i = 1, \ldots, n\big\}$ be a dataset with $x_i = (x_{i1}, x_{i2}, \ldots, x_{id})$ for $i = 1, \ldots, n$. We assume that the matrix

$$X := \begin{bmatrix} 1 & x_{11} & \cdots & x_{1d} \\ \vdots & \vdots & & \vdots \\ 1 & x_{n1} & \cdots & x_{nd} \end{bmatrix} \in \mathbb{R}^{n\times(d+1)}$$

has rank $d + 1$ and, furthermore, that not all y_i coincide. Let $f^*\colon \mathbb{R}^d \to \mathbb{R}$ be the affine linear regressor for D from Theorem 2.9. Then

$$R^2 := \frac{\|f^*(x) - \overline{y}\|^2}{\|y - \overline{y}\|^2}$$

is called the *coefficient of determination* of D, where we understand

$$\overline{y} = (\overline{y}, \ldots, \overline{y}) \text{ and } f^*(x) = (f^*(x_1), \ldots, f^*(x_n))$$

as vectors in $\mathbb{R}^n$ and, in both numerator and denominator, the Euclidean norm on $\mathbb{R}^n$ is taken.

We first note that the coefficient of determination in the case $d = 1$ coincides with the square of the regression coefficient: $R^2 = r_{xy}^2$, see Exercise 2.8. In contrast to the one-dimensional case, we only define R^2 and not R itself for $d > 1$. In the multivariable case, we cannot expect a statement analogous to the last part of Theorem 2.8. However, we note that R^2 by definition compares the deviation of the approximant from the mean with the deviation of the data from the mean, so it is to be expected that R^2 is close to one if and only if the data is well approximated by the affine linear regressor. This is confirmed by the following theorem.

Theorem 2.11. *Let D, X and f^* be as in Definition 2.10.*

(i) We have $0 \leqslant R^2 = 1 - \dfrac{\|y - f^*(x)\|^2}{\|y - \overline{y}\|^2} \leqslant 1.$

(ii) We have $R^2 = 1$ if and only if $f^*(x_i) = y_i$ holds for all $i = 1, \ldots, n$, i.e., if and only if all (x_i, y_i) lie in a hyperplane.

Proof. (i) It suffices to show the formula for R^2. The two estimates then follow immediately. We use the abbreviations

$$\hat{y} := f^*(x) \underset{\substack{\uparrow \\ \text{Thm} \\ 2.9}}{=} X(X^\intercal X)^{-1}X^\intercal y =: Py \quad \text{and} \quad Q := I - P.$$

Direct calculation shows that $P^2 = P$ and $Q^\intercal = Q$ hold, so P is idempotent and Q is symmetric, from which one derives easily $QP = 0$. Using this, we first show several identities.

① By subbing in, one immediately verifies $\langle \overline{y}, \overline{y} \rangle = n\overline{y}^2$ and $\langle y, \overline{y} \rangle = n\overline{y}^2$.

② Moreover, $\langle y, \hat{y} \rangle = \langle y - \hat{y} + \hat{y}, \hat{y} \rangle = \langle Qy, Py \rangle + \langle \hat{y}, \hat{y} \rangle \underset{\substack{\uparrow \\ Q=Q^\intercal}}{=} \langle y, QPy \rangle + \langle \hat{y}, \hat{y} \rangle \underset{\substack{\uparrow \\ QP=0}}{=} \langle \hat{y}, \hat{y} \rangle.$

③ Finally, we claim $\overline{y} = \overline{\hat{y}}$. First, it holds

$$X^\intercal Qy = X^\intercal(y - Py) = X^\intercal y - X^\intercal X(X^\intercal X)^{-1}X^\intercal y = X^\intercal y - X^\intercal y = 0,$$

which is why the matrix product of each row of $X^\intercal$ with Qy is zero. But this means that the scalar product of each column of X with Qy is zero. This is particularly true for the first column of X and it follows $\langle \mathbb{1}, Qy\rangle = 0$ with the abbreviation $\mathbb{1} = (1, \dots, 1)$. This yields

$$\overline{y} = \tfrac{1}{n}\sum_{i=1}^{n} y_i = \tfrac{1}{n}\langle \mathbb{1}, y\rangle = \tfrac{1}{n}\langle \mathbb{1}, y - \hat{y} + \hat{y}\rangle = \tfrac{1}{n}\langle \mathbb{1}, Qy\rangle + \tfrac{1}{n}\langle \mathbb{1}, \hat{y}\rangle = \overline{\hat{y}}.$$

Now we show the claimed equation. We start with

$$\begin{aligned} 1 - \frac{\|y - f^*(x)\|^2}{\|y - \overline{y}\|^2} &= \frac{\|y - \overline{y}\|^2 - \|y - \hat{y}\|^2}{\|y - \overline{y}\|^2} \\ &= \frac{-2\langle y, \overline{y}\rangle + \langle \overline{y}, \overline{y}\rangle + 2\langle y, \hat{y}\rangle - \langle \hat{y}, \hat{y}\rangle}{\|y - \overline{y}\|^2} =: (\circ) \end{aligned}$$

and further calculate the numerator. We apply

$$2\langle y, \overline{y}\rangle \overset{①}{=} 2\langle \overline{y}, \overline{y}\rangle \overset{③}{=} 2\langle \overline{\hat{y}}, \overline{\hat{y}}\rangle \overset{①}{=} 2\langle \hat{y}, \overline{\hat{y}}\rangle \overset{③}{=} 2\langle \hat{y}, \overline{y}\rangle \quad \text{and} \quad 2\langle y, \hat{y}\rangle - \langle \hat{y}, \hat{y}\rangle \overset{②}{=} \langle \hat{y}, \hat{y}\rangle$$

and obtain

$$(\circ) = \frac{\langle \hat{y}, \hat{y}\rangle - 2\langle \hat{y}, \overline{y}\rangle + \langle \overline{y}, \overline{y}\rangle}{\|y - \overline{y}\|^2} = \frac{\|\hat{y} - \overline{y}\|^2}{\|y - \overline{y}\|^2} = \frac{\|f^*(x) - \overline{y}\|^2}{\|y - \overline{y}\|^2} = R^2.$$

(ii) According to (i), $R^2 = 1$ holds if and only if $y - f^*(x) = 0$. □

Remark 2.12. If a dataset with both multidimensional features and multidimensional labels

$$D = \{(x_i, y_i) \in \mathbb{R}^d \times \mathbb{R}^k \mid i = 1, \dots, n\}$$

is given, it can be approximated by an affine linear function $f^* \colon \mathbb{R}^d \to \mathbb{R}^k$, $f^*(x) = A^*x + b^*$, with $A^* \in \mathbb{R}^{d\times k}$, $b^* \in \mathbb{R}^k$, by minimizing the error function

$$\phi(f) = \sum_{i=1}^{n} \|f(x_i) - y_i\|^2$$

provided that the same rank condition as in Theorem 2.9, namely $\operatorname{rk} X = d+1$, holds. By design of ϕ, Theorem 2.9 can then be applied term-wise to obtain the coordinate functions of f^*. This provides the rows of the matrix A^* and the entries of b^*.

Before we move on to the next topic, let us note that inverting the matrix $X^\intercal X$ in high dimensions cannot be done efficiently. Therefore, although the above results guarantee that there is a uniquely determined regressor, they do not provide a practical way to calculate it. On the other hand, our proof method, i.e., minimizing a cost function, shows how f^* can be calculated, namely, in the notation of Theorem 2.9, by numerical approximation of

$$\underset{\tilde{a}\in\mathbb{R}^{d+1}}{\operatorname{argmin}} \|X\tilde{a} - y\|^2.$$

A very popular and intuitive procedure to do so is the so-called *gradient descent method*, which we will discuss in Chapter 17.

2.3 Polynomial Regression

Next, we consider datasets that cannot be well described by an affine linear function, but instead by a polynomial.

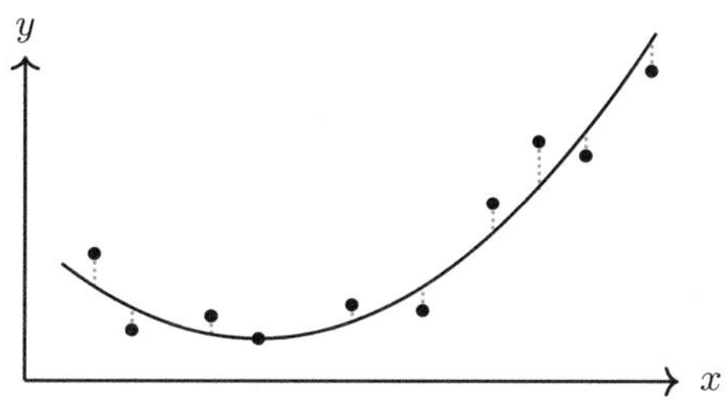

We stick with the method of least squares and obtain the following theorem, in which we limit ourselves to the one-dimensional case.

Theorem 2.13. (on polynomial regression) *Let* $D = \{(x_i, y_i) \in \mathbb{R} \times \mathbb{R} \mid i = 1, \ldots, n\}$ *be a dataset. We assume that the matrix*

$$X := \begin{bmatrix} 1 & x_1^1 & \cdots & x_1^d \\ \vdots & \vdots & & \vdots \\ 1 & x_n^1 & \cdots & x_n^d \end{bmatrix} \in \mathbb{R}^{n\times(d+1)}$$

has rank $d+1$*. Then there is precisely one polynomial* $P^* \in \mathbb{R}[X]$ *with* $\operatorname{rk} P^* \leqslant d$*, i.e.,* $P^* = a_0^* + a_1^* X^1 + \cdots + a_d^* X^d$ *with unique* $a^* = (a_0^*, \ldots, a_d^*) \in \mathbb{R}^{d+1}$*, such that*

$$P^* = \operatorname{argmin}\Big\{\phi(P) := \sum_{i=1}^{n}(P(x_i) - y_i)^2 \mid P \in \mathbb{R}[X] \text{ with } \operatorname{rk} P \leqslant d\Big\}$$

holds. The polynomial P^* *is called the* polynomial regressor *for the dataset* D*. Its coefficients are given by*

$$\begin{bmatrix} a_0^* \\ \vdots \\ a_d^* \end{bmatrix} = (X^\intercal X)^{-1} X^\intercal \begin{bmatrix} y_1 \\ \vdots \\ y_n \end{bmatrix}.$$

Proof. We define the mapping $\psi\colon \mathbb{R} \to \mathbb{R}^{d+1}$, $x \mapsto (1, x, x^2, \ldots, x^d)$, and consider the dataset

$$\hat{D} := \{(\psi(x_i), y_i) \mid i = 1, \ldots, n\} \subseteq \mathbb{R}^{d+1} \times \mathbb{R}.$$

We then see that the data matrix X associated with $\hat{D}$ from Theorem 2.9 matches the one given in Theorem 2.13. Since we have assumed $\operatorname{rk} X = d+1$, we can apply Theorem 2.9 and obtain, in the notation of the current theorem, the uniquely determined minimizer $f^* = \langle (a_1^*, \ldots, a_d^*), \cdot \rangle + a_0^*$ of the objective function

$$\begin{aligned}
\phi(f) &= \sum_{i=1}^{n} (f(\psi(x_i)) - y_i)^2 \\
&= \sum_{i=1}^{n} \left(\left\langle \begin{bmatrix} a_1 \\ \vdots \\ a_d \end{bmatrix}, \begin{bmatrix} x_i^1 \\ \vdots \\ x_i^d \end{bmatrix} \right\rangle + a_0 - y_i \right)^2 \\
&= \sum_{i=0}^{n} ([a_0 + a_1 x_i^1 + \cdots + a_d x_i^d] - y_i)^2 \; = \; \phi(P),
\end{aligned}$$

where we first identified the function $f\colon \mathbb{R}^d \to \mathbb{R}$, $f = \langle (a_1, \ldots, a_d), \cdot \rangle + a_0$ with the parameters $a_0, \ldots, a_d$ and then identified these parameters again with the polynomial $P = a_0 + a_1 X + a_2 X^2 + \cdots + a_d X^d$. □

We note that the reduction of non-linear problems to linear ones by "embedding" the data into a space of higher dimension is a method that can also be applied in other contexts, see e.g., Chapter 15. We further mention that the above proof can easily be generalized to polynomials in several variables: We only need to replace the monomials in our "embedding" ψ by a basis of the corresponding space of polynomials. Finally we note that in case that $n \leqslant d+1$ holds the minimizer is equal to the interpolation polynomial.

2.4 Logistic Regression

We conclude this chapter by discussing the method of logistic regression. We start with a dataset

$$D = \{(x_i, y_i) \in \mathbb{R}^d \times \{0, 1\} \mid i = 1, \ldots, n\},$$

where, in contrast to everything so far, only the labels zero and one occur. If such a dataset is given, we are interested in determining a so-called *classifier* which for the moment shall mean a mapping $f\colon \mathbb{R}^d \to \{0,1\}$ such that $f(x_i) = y_i$ holds for as many $i \in \{1, \ldots, n\}$ as possible. In Chapters 3 and 13–15 we will treat classifiers more closely. At this point we show how such a classifier can be obtained using a regression method. As with affine linear and polynomial regression, we first need a class of functions from which we want to approximate. Here we choose the family of *logistic functions*, i.e., functions of the form

$$f\colon \mathbb{R}^d \to (0,1),\ f(z) = \frac{1}{1+\mathrm{e}^{-(w_1 z_1 + \cdots + w_d z_d + b)}},$$

where we vary $w_1, \ldots, w_d$ and $b \in \mathbb{R}$. In the case $d = 1$, one can substitute $s := 1/w_1$ and $\mu := -b/w_1$ to get $f(z) = 1/(1+\mathrm{e}^{-(z-\mu)/s})$ with *shift parameter* μ and *scaling parameter* s. The following picture shows two logistic functions with parameters $(b^{(1)}, w^{(1)})$ and $(b^{(2)}, w^{(2)})$, or $(\mu, s^{(1)})$ and $(\mu, s^{(2)})$, where $0 < s^{(1)} < s^{(2)}$.

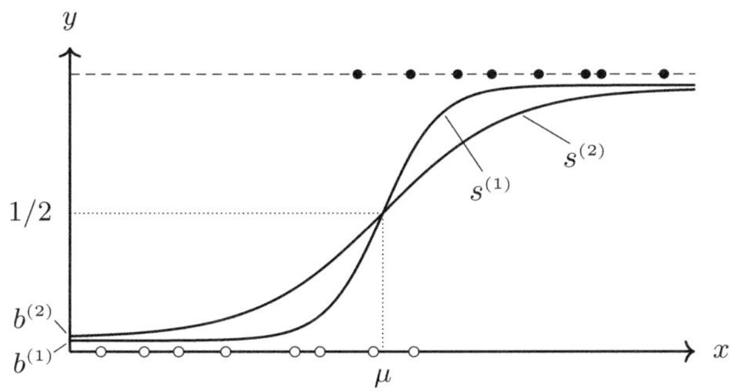

The plot suggests that we can adjust the function f to the data points by shifting and scaling. Since the data points shown in the picture "overlap", it seems likely that minimizing the sum of the (square) distances might again lead to a unique minimizer. Although this is not the strategy we will pursue below to find a *logistic regressor* f^* for the data, it still provides the right intuition. We mention already here, that once the regressor f^* has been found, we can easily get a $\{0,1\}$-valued classifier by rounding the values of f^*.

As it has already proven successful in affine linear regression, we again extend our data points with an additional entry of 1 — this time, however, we append it instead of putting it at the front[2]: For $x = (x_1, \ldots, x_d) \in \mathbb{R}^d$ we write $\widehat{x} := (x_1, \ldots, x_d, 1)$ for the *extended data point* and $(w, b) = (w_1, \ldots, a_d, b)$ for the *generalized weight vector*. With this we get

$$\langle w, z\rangle + b = \sum_{i=1}^{d} w_i z_i + b \cdot 1 \; = \left\langle \begin{bmatrix} w_1 \\ \vdots \\ w_d \\ b \end{bmatrix}, \begin{bmatrix} z_1 \\ \vdots \\ z_d \\ 1 \end{bmatrix} \right\rangle = \langle w, \widehat{z}\rangle$$

where we (abuse of notation!) denote the "normal" weight vector on the left

[2]We are changing the convention compared to Chapter 2.2, but in doing so we achieve consistency with Chapter 13 on the perceptron and Chapter 16 on neural networks, compare in particular Definition 13.9.

with w and the generalized weight vector on the right, also with w. If we now use the *sigmoid function*, essentially the prototype of a logistic function with shift parameter zero and scaling parameter one,

$$\mathrm{sig}\colon \mathbb{R} \to (0,1),\ \mathrm{sig}(t) := \frac{1}{1+\mathrm{e}^{-t}},$$

then any logistic function $f\colon \mathbb{R}^d \to (0,1)$ can be written as $f = \mathrm{sig}\big(\langle w, \cdot\rangle\big)$ with $w \in \mathbb{R}^{d+1}$.

Since the values of the logistic function can be interpreted as probabilities by rounding, it is natural to apply the maximum likelihood method to specify how the parameters $w \in \mathbb{R}^{d+1}$ should be adjusted to the dataset. To this end, we first consider the case where our data points actually come from a logistic function f in the following way: Given a feature x_i, we assume that the label y_i is 1 with probability $f(x_i)$, and 0 with probability $1 - f(x_i)$. Formally, we again consider a probability space $(\Omega, \Sigma, \mathrm{P})$ and independent random variables $Y_1, \ldots, Y_n\colon \Omega \to \mathbb{R}$, which are each Bernoulli distributed, i.e., $Y_i \sim \mathcal{B}(f(x_i))$ for fixed $x_1, \ldots, x_n$. Then we have

$$\mathrm{P}\big[Y_i(f) = y_i \text{ for all } i\big] = \prod_{i=1}^{n} f(x_i)^{y_i}(1 - f(x_i))^{1-y_i}$$

and accordingly, we define the likelihood function as

$$\begin{aligned} L\colon \big\{f\colon \mathbb{R} \to (0,1) \mid f \text{ logistic function}\big\} &\to \mathbb{R} \\ L(f) := \prod_{i=1}^{n} f(x_i)^{y_i}(1-f(x_i))^{1-y_i}. \end{aligned} \tag{2.4}$$

A maximizer f^* of L then maximizes the likelihood of the data in D coming from the logistic function f^* in the way described above. In contrast to affine linear regression, in the current setting we cannot expect that a maximizer always exists:

Example 2.14. Consider the dataset $D = \{(-1,0),(1,1)\} \subseteq \mathbb{R} \times \{0,1\}$ and the logistic functions $f\colon \mathbb{R} \to (0,1)$, $f(z) := (1+\mathrm{e}^{-(wz+b)})^{-1}$ with $w, b \in \mathbb{R}$. Then we have

$$\begin{aligned} L(f) &= f(-1)^0(1 - f(-1))^{1-(-1)} f(1)^1(1 - f(1))^{1-1} \\ &= \left(1 - \frac{1}{1+\mathrm{e}^{-(-w+b)}}\right)\left(\frac{1}{1+\mathrm{e}^{-(w+b)}}\right) < 1 \end{aligned}$$

for all f. For fixed b and $w \to \infty$, $L(f)$ converges to 1. Consequently, L cannot have a maximizer.

If one sketches the dataset from Example 2.14, one can see that the fact that the data does not "overlap" as in the picture on page 23 is responsible for the above effect. In fact, in Exercise 2.11 we will show that for the dataset $D := \{(-1,1),(0,0),(1,1)\}$ the likelihood function has a maximizer and that the latter can be calculated explicitly using analytical methods. In contrast to affine linear and polynomial regression, however, a general explicit formula for the parameters of f^* cannot be given. But, as we will show now, an "overlap condition" already hinted at above will guarantee the existence and, as in the case of multidimensional affine linear regression, a rank condition will guarantee uniqueness.

To show both, we now consider the likelihood function from (2.4) as a function $L = L(w)$ with $w \in \mathbb{R}^{d+1}$ and then consider the *negative log-likelihood function*, i.e., we define

$$\ell\colon \mathbb{R}^{d+1} \to \mathbb{R},\quad \ell(w) := -\log \prod_{i=1}^{n} \operatorname{sig}(\langle w, \widehat{x}_i\rangle)^{y_i}(1-\operatorname{sig}(\langle w, \widehat{x}_i\rangle))^{1-y_i} \tag{2.5}$$

and first note the following formula for ℓ.

Lemma 2.15. *Let $D = \{(x_i, y_i) \in \mathbb{R}^d \times \{0,1\} \mid i = 1,\dots,n\}$ be a dataset and ℓ the associated log-likelihood function as in* (2.5). *Then we have for $w \in \mathbb{R}^{d+1}$:*

$$\ell(w) = \sum_{i=1}^{n} -y_i\langle w, \widehat{x}_i\rangle + \log(1 + \mathrm{e}^{\langle w, \widehat{x}_i\rangle}).$$

Proof. Direct calculation shows

$$\begin{aligned}
\ell(w) &= -\sum_{i=1}^{n} y_i \log\Big(\frac{1}{1+\mathrm{e}^{-\langle w,\widehat{x}_i\rangle}}\Big) + (1-y_i)\log\Big(1 - \frac{1}{1+\mathrm{e}^{-\langle w,\widehat{x}_i\rangle}}\Big)\\
&= -\sum_{i=1}^{n} y_i \log\Big(\frac{1}{1+\mathrm{e}^{-\langle w,\widehat{x}_i\rangle}}\Big) + \log\Big(\frac{\mathrm{e}^{-\langle w,\widehat{x}_i\rangle}}{1+\mathrm{e}^{-\langle w,\widehat{x}_i\rangle}}\Big) - y_i\log\Big(\frac{\mathrm{e}^{-\langle w,\widehat{x}_i\rangle}}{1+\mathrm{e}^{-\langle w,\widehat{x}_i\rangle}}\Big)\\
&= -\sum_{i=1}^{n} y_i \log\Big(\frac{1}{1+\mathrm{e}^{-\langle w,\widehat{x}_i\rangle}}\cdot\frac{1+\mathrm{e}^{-\langle w,\widehat{x}_i\rangle}}{\mathrm{e}^{-\langle w,\widehat{x}_i\rangle}}\Big) + \log\Big(\frac{1}{1+\mathrm{e}^{\langle w,\widehat{x}_i\rangle}}\Big)\\
&= -\sum_{i=1}^{n} y_i\langle w,\widehat{x}_i\rangle - \log\big(1+\mathrm{e}^{\langle w,\widehat{x}_i\rangle}\big),
\end{aligned}$$

as claimed. □

With the above representation of the function ℓ, it is not difficult to determine its derivative. We leave this to the reader as Exercise 2.11 and only note the result here, namely

$$\nabla\ell(w) = \sum_{i=1}^{n}(\operatorname{sig}(\langle w,\widehat{x}_i\rangle) - y_i)\widehat{x}_i.$$

Let us assume for now that we have found a solution to the system of equations

$\nabla\ell(w) = 0$. In order to conclude that this solution is a minimizer, we next show that ℓ is convex. Note that the data matrix X is defined differently below than in Theorem 2.9, but that the rank condition there is equivalent to the following.

Lemma 2.16. *Let $D = \{(x_i, y_i) \in \mathbb{R}^d \times \{0,1\} \mid i = 1, \ldots, n\}$ be a dataset. Then the associated negative log-likelihood function ℓ as in (2.5) is convex. If additionally the data matrix*

$$X := \begin{bmatrix} x_{11} & \cdots & x_{1d} & 1 \\ \vdots & & \vdots & \vdots \\ x_{n1} & \cdots & x_{nd} & 1 \end{bmatrix} \in \mathbb{R}^{n\times(d+1)}$$

has rank $d+1$, then ℓ is even strictly convex.

Proof. We consider $g\colon \mathbb{R} \to \mathbb{R}$, $g(t) = -\log \operatorname{sig}(t) = \log(1 + \mathrm{e}^{-t})$ and see that $g'(t) = \operatorname{sig}(t) - 1$ strictly increases, from which it follows that g is strictly convex. We further consider $h\colon \mathbb{R} \to \mathbb{R}$, $h(t) = -\log(1 - \operatorname{sig}(t)) = g(t) + t$. Here $h'(t) = \operatorname{sig}(t) + 1$ and it follows that h is also strictly convex. Combining the summands in the proof of Lemma 2.15 differently we obtain

$$\begin{aligned} \ell(w) &= -\sum_{i=1}^{n} y_i \log\Big(\frac{1}{1+\mathrm{e}^{-\langle w,\widehat{x}_i\rangle}}\Big) + (1-y_i)\log\Big(1 - \frac{1}{1+\mathrm{e}^{-\langle w,\widehat{x}_i\rangle}}\Big) \\ &= \sum_{i=1}^{n} y_i\big(-\log\text{-}\langle w,\widehat{x}_i\rangle\big) + (1-y_i)\big(-\log(1-\operatorname{sig}\langle w,\widehat{x}_i\rangle)\big) \\ &= \sum_{i=1}^{n} y_i g(\langle w,\widehat{x}_i\rangle) + (1-y_i)h(\langle w,\widehat{x}_i\rangle). \end{aligned}$$

Since $y_i \in \{0,1\}$, we get that for each i either the term with g or the term with h remains while the other disappears due to y_i or $1 - y_i$ being zero. We can therefore write the function ℓ as

$$\ell(w) = \sum_{i=1}^{n} k_i(\langle w,\widehat{x}_i\rangle)$$

with $k_i \in \{g,h\}$ for $i = 1, \ldots, n$. To see the (strict) convexity, let $\lambda \in (0,1)$ and $v \neq w \in \mathbb{R}^{d+1}$. Then

$$\begin{aligned} \ell(\lambda w + (1-\lambda)v) &= \sum_{i=1}^{n} k_i(\lambda\langle w,\widehat{x}_i\rangle + (1-\lambda)\langle v,\widehat{x}_i\rangle) \\ &\leqslant \sum_{i=1}^{n} \lambda k_i(\langle w,\widehat{x}_i\rangle) + (1-\lambda)k_i(\langle v,\widehat{x}_i\rangle) \\ &= \lambda\ell(w) + (1-\lambda)\ell(v). \end{aligned}$$

Since all k_i are strictly convex, we get a strict estimate in the above calculation if only $\langle w,\widehat{x}_i\rangle \neq \langle v,\widehat{x}_i\rangle$ holds for at least one $i \in \{1, \ldots, n\}$. If this was not the case, then $\langle \widehat{x}_i, u \rangle = 0$ would hold for all i with $u := v - w \in \mathbb{R}^{d+1} \backslash \{0\}$, and we would get

$$Xu = \begin{bmatrix} x_{11} & \cdots & x_{1d} & 1 \\ \vdots & & \vdots & \vdots \\ x_{n1} & \cdots & x_{nd} & 1 \end{bmatrix} \begin{bmatrix} u_1 \\ \vdots \\ u_{d+1} \end{bmatrix} = \begin{bmatrix} \langle \widehat{x}_1, u \rangle \\ \vdots \\ \langle \widehat{x}_1, u \rangle \end{bmatrix} = 0$$

which contradicts $\operatorname{rk} X = d + 1$. □

If we have a solution w^* of $\nabla \ell(w) = 0$ and the rank condition in Lemma 2.16 is fulfilled, w^* is the unique minimizer of ℓ and thus provides the maximum likelihood estimator. For large datasets, however, it is not possible to solve $\nabla \ell(w) = 0$ analytically and a numerical method has to be applied instead. In Chapter 17, we will discuss the gradient descent method and show that this works particularly well when a convex function is to be minimized. This is, in addition to the question of uniqueness, another reason why we used the negative log-likelihood function above. Under the rank condition mentioned in Lemma 2.16, it is even guaranteed that there can be at most one minimizer. This follows by applying Proposition 17.14 from Chapter 17. In the last theorem of this chapter, we now want to address the question of existence. The necessary condition for this is the following.

Definition 2.17. Let $D = \{(x_i, y_i) \in \mathbb{R}^d \times \{0, 1\} \mid i = 1, \ldots, n\}$ be a dataset. We say that the data *overlaps* if for every $w \in \mathbb{R}^{d+1}$ there exists a k such that $y_k = 1$ and $\langle w, \widehat{x}_k \rangle < 0$, or $y_k = 0$ and $\langle w, \widehat{x}_k \rangle > 0$.

If we consider for $w \in \mathbb{R}^{d+1}$ the classifier

$$\mathbb{R}^d \to \{0, 1\}, \ x \mapsto \begin{cases} 1 & \text{if } \operatorname{sig}(\langle w, x \rangle) \geqslant 1/2, \\ 0 & \text{otherwise,} \end{cases}$$

i.e., the rounded logisitic regressor, Definition 2.17 implies that for every w at least one data point from D is misclassified. However, the condition is indeed stronger, consider, e.g., the one-dimensional dataset $D = \{(0, 0), (0, 1)\}$. In two dimensions, the following three pictures illustrate the situation.

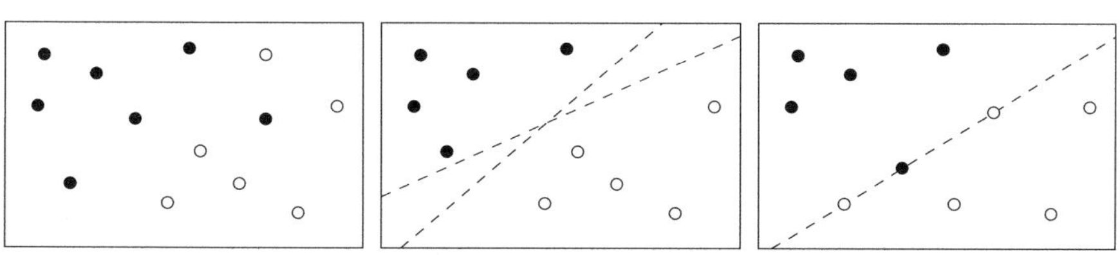

On the left, the data overlaps, in the middle and on the right it does not. However, in the middle picture there are even infinitely many w's that would lead to a correct classifier, while in the right picture there is no such w.

We will now show that in the case of overlapping data, i.e., as in the picture on the left, at least one maximum likelihood estimator in the sense of a maximizer of L from (2.4) exists. *Linearly separable datasets* as in the middle picture will be treated with other methods in later Chapters 13–14. There, we will also sketch how to handle the borderline case depicted in the right picture.

Theorem 2.18. (on logistic regresssion) *Let* $D = \{(x_i, y_i) \in \mathbb{R}^d \times \{0,1\} \mid i = 1,\dots,n\}$ *be an overlapping dataset and denote by* ℓ *be the associated negative log-likelihood function as defined in* (2.5). *Then there exists at least one minimizer*

$$w^* \in \operatorname*{argmin}_{w\in\mathbb{R}^{d+1}} \ell(w).$$

Proof. Since the function ℓ is continuous and defined on the whole space $\mathbb{R}^{d+1}$, it suffices to examine its behavior at infinity. Let $(w_k)_{k\in\mathbb{N}} \subseteq \mathbb{R}^{d+1}$ be a sequence with $\|w_k\| \to \infty$. We claim that $\ell(w_k) \to \infty$ holds for $k \to \infty$. In order to establish the claim we need several preparations.

① Our first goal is to show that the function

$$\ell^\infty \colon \mathbb{R}^{d+1}\backslash\{0\} \to (0,\infty),\;\; \ell^\infty(w) := \lim_{t\to\infty} \frac{\ell(tw)}{t}$$

is well-defined. Using the formula for ℓ from Lemma 2.15, we see that in the first term the t's cancel out, that the second term goes to zero provided $\langle w, \widehat{x}_i\rangle \leqslant 0$, and that it otherwise converges to the latter scalar product. It follows that the limit

$$\begin{aligned}\lim_{t\to\infty} \frac{\ell(tw)}{t} &= \lim_{t\to\infty} \sum_{i=1}^{n} -y_i \tfrac{1}{t}\langle tw, \widehat{x}_i\rangle + \tfrac{1}{t}\log(1+\mathrm{e}^{\langle tw,\widehat{x}_i\rangle})\\ &= \sum_{\substack{i=1\\ y_i=1}}^{n} -\langle w, \widehat{x}_i\rangle \;+\; \sum_{\substack{i=1\\ \langle w,\widehat{x}_i\rangle>0}}^{n} \langle w, \widehat{x}_i\rangle\end{aligned}$$

exists. Now let $k \in \{1,\dots,n\}$ be such that $y_k = 1$ and $\langle w, \widehat{x}_k\rangle < 0$ hold. We estimate

$$\begin{aligned}\ell(tw) &= -y_k t\langle w, \widehat{x}_k\rangle + \underbrace{\log(1+\mathrm{e}^{t\langle w,\widehat{x}_k\rangle})}_{\geqslant 0} + \sum_{i\neq k} -y_i t\langle w, \widehat{x}_i\rangle + \log(1+\mathrm{e}^{t\langle w,\widehat{x}_i\rangle})\\ &\geqslant -t\langle w, \widehat{x}_k\rangle + \sum_{\substack{i\neq k\\ y_i=0}} \underbrace{\log(1+\mathrm{e}^{t\langle w,\widehat{x}_i\rangle})}_{\geqslant 0} + \sum_{\substack{i\neq k\\ y_i=1}} -t\langle w, \widehat{x}_i\rangle + \underbrace{\log(1+\mathrm{e}^{t\langle w,\widehat{x}_i\rangle})}_{\geqslant t\langle w,\widehat{x}_i\rangle}\\ &\geqslant -t\langle w, \widehat{x}_k\rangle\end{aligned}$$

and deduce that our limit is strictly positive for each $w \neq 0$. If $k \in \{1,\dots,n\}$

is such that $y_k = 0$ and $\langle w, \widehat{x}_i \rangle > 0$, we get

$$\ell(tw) = 0 + \underbrace{\log(1+\mathrm{e}^{t\langle w,\widehat{x}_k\rangle})}_{\geqslant t\langle w,\widehat{x}_k\rangle} + \sum_{i\neq k} -y_i t\langle w,\widehat{x}_i\rangle + \log(1+\mathrm{e}^{t\langle w,\widehat{x}_i\rangle}) \geqslant t\langle w,\widehat{x}_k\rangle$$

where we see as above that the sum over $i \neq k$ is greater than or equal to zero.

② For $v,\, w \in \mathbb{R}^{d+1}$, we estimate next

$$\begin{aligned}
|\ell(v)-\ell(w)| &\leqslant \sum_{i=1}^{n}\big|-y_i\langle v,\widehat{x}_i\rangle + \log(1+\mathrm{e}^{\langle v,\widehat{x}_i\rangle}) + y_i\langle w,\widehat{x}_i\rangle - \log(1+\mathrm{e}^{\langle w,\widehat{x}_i\rangle})\big| \\
&\underset{\text{MVT}}{\overset{}{\leqslant}} \sum_{i=1}^{n}\big|\langle v-w,\widehat{x}_i\rangle\big| + \sup_{s\in\mathbb{R}}\big|\tfrac{\mathrm{d}}{\mathrm{d}s}\log(1+\mathrm{e}^{s})\big|\cdot\big|\langle v,\widehat{x}_i\rangle - \langle w,\widehat{x}_i\rangle\big| \\
&\leqslant \sum_{i=1}^{n} 2\|v-w\|\|\widehat{x}_i\| \\
&\leqslant L\|v-w\|
\end{aligned}$$

and we see that ℓ is Lipschitz continuous with constant $L := 2\max_{i=1,\dots,n}\|\widehat{x}_i\|$.

③ Let now $(w_k)_{k\in\mathbb{N}}$ be as at the beginning of the proof. By passing to a subsequence, we may assume without loss of generality that all w_k are non-zero. Then the sequence $(v_k)_{k\in\mathbb{N}} \subseteq \partial\mathrm{B}_1(0)$ with $v_k := w_k/\|w_k\|$ has a convergent subsequence with limit v in the unit sphere. Without loss of generality, we assume that $v_k \to v$ for $k\to\infty$ and claim

$$\lim_{k\to\infty}\frac{\ell(w_k)}{\|w_k\|} = \ell^{\infty}(v) \overset{①}{>} 0,$$

from which in particular $\ell(w_k)\to\infty$ follows. To show the equality above, we put $t_k := \|w_k\|$ and get $\ell(w_k)/\|w_k\| = \ell(t_kv_k)/t_k$. For $\varepsilon > 0$ we choose $k_0 \in \mathbb{N}$ such that for $k \geqslant k_0$ both $|\ell(t_kv)/t_k - \ell^{\infty}(v)| < \varepsilon/2$ and $|v_k - v| < \varepsilon/2$ hold. For $k \geqslant k_0$ we get

$$\begin{aligned}
\Big|\frac{\ell(w_k)}{\|w_k\|} - \ell^{\infty}(v)\Big| &\leqslant \Big|\frac{\ell(t_kv_k)}{t_k} - \frac{\ell(t_kv)}{t_k}\Big| + \Big|\frac{\ell(t_kv)}{t_k} - \ell^{\infty}(v)\Big| \\
&\overset{②}{\leqslant} \frac{L|t_kv_k - t_kv|}{t_k} + \frac{\varepsilon}{2} < \varepsilon
\end{aligned}$$

and we are done. □

At this point, we will not illustrate how to use the gradient descent method to compute a logistic regressor in a concrete example. Indeed, we will only explain the former method properly in Chapter 17, and moreover, there are ready-made packages in many programming languages which numerically determine the parameters $w_1^*, \dots, w_d^*$ and b^*, see Exercise 2.12. Once a w^* as in Theorem

2.16 has been found, we call

$$f^* : \mathbb{R}^d \to (0,1), \quad f^*(z) = \mathrm{sig}(\langle w^*, \widehat{z}\rangle)$$

a *logistic regressor* for the dataset D. We have already explained above that by rounding one can obtain a $\{0,1\}$-*valued classifier*.

Remark 2.19. Before we end this chapter, we would like to point out the following: By rearranging and taking the logarithm, the definition of a logistic function $f = \mathrm{sig}(\langle w, \cdot\,\rangle)$ with $w \equiv (w, b) \in \mathbb{R}^{d+1}$ yields the equation

$$\log \frac{f}{1-f} = w_1 z_1 + \cdots + w_d z_d + b$$

for $z = (z_1, \ldots, z_d) \in \mathbb{R}^d$. The left-hand side can be interpreted as the logarithm of the chance $\frac{f}{1-f}$ (f, that the label of z is 1, to $1-f$, that the label of z is 0). The right-hand side represents the latter as an affine linear function of the features. However, the attempt to transform the dataset via the left function and then apply "ordinary" affine linear regression fails here — unlike for polynomial regression in Chapter 2.3 — as the obviously not well-defined dataset

$$\hat{D} := \left\{(x_i, \log \frac{y_i}{1-y_i}) \mid i = 1, \ldots, n\right\}$$

would have to be used. If, on the other hand, a dataset (that we interpret categorically) is given, whose labels are actually in $(0,1)$, the above method may be applied, compare Exercise 2.13.

References

The standard results on linear regression presented at the beginning of this chapter can be found in any book. The term "second regression line" is taken from the manuscript [Rin08]. The consistency statement in Theorem 2.6 follows [Sha15, Lecture 3]. The section on polynomial regression is based on [SSBD14, Section 9.2.2]. A good reference for existence and uniqueness statements of the logistic regressor is [AA84]. There, three cases are considered: overlap, separation, and quasi-separation. Our proof in Theorem 2.18 is based on the concept of the recession function, see e.g., [Giu03]. The author would like to thank Thomas Schmidt for several helpful discussions in the latter context.

In connection with the maximum likelihood functions (2.2) and (2.4), and as a prerequisite in Theorem 2.6, we have considered finitely or countably many independent random variables with prescribed distributions on a fixed probability space $(\Omega, \Sigma, \mathrm{P})$. That such a space exists follows by using the product measure, see, e.g., [Shi16, Theorems 2.3.3 and 2.9.2]. A version that fits our purposes, taken from [Beh13, Chapter 4.5 on the "cloning theorem"], is given in the appendix as Theorem A.13.

Exercises

Exercise 2.1. Determine for $D = \{(1,2),(2,2),(2,3),(3,3),(4,3)\} \subseteq \mathbb{R}_x \times \mathbb{R}_y$ both regression lines, as well as the regression coefficients r_{xy} and r_{yx}, by hand (calculator or CAS is allowed, but, e.g., no Python package!). Sketch the data points and the regression lines.

Exercise 2.2. In this exercise, ϕ denotes the objective function from Theorem 2.1 and ψ the objective function from Remark 2.3(ii).

(i) Find a dataset D such that

$$\operatorname{argmin}\big\{\phi(f) \mid f\colon \mathbb{R} \to \mathbb{R} \text{ affine linear}\big\} \neq \operatorname{argmin}\big\{\psi(f) \mid f\colon \mathbb{R} \to \mathbb{R} \text{ affine linear}\big\}$$

holds.

(ii) Find a dataset where not all x-values are the same and yet there is more than one function f^* with

$$\psi(f^*) = \min\big\{\psi(f) \mid f\colon \mathbb{R} \to \mathbb{R} \text{ affine linear}\big\}.$$

Exercise 2.3. Carry out the method of linear regression for the following dataset. Below, the exam preparation time is the x-variable and the exam result is the y-variable. Draw all data points and also the regression line.

Student	Exam preparation in h	Exam result in %
1	21.0	82
2	18.0	69
3	15.0	29
4	8.0	41
5	8.0	44
6	1.5	8
7	0.0	10

Exercise 2.4. Let $D = \{(x_i, y_i) \in \mathbb{R}^2 \mid i = 1, \ldots, n\}$ be a dataset, where neither all x_i nor all y_i are equal. Show that the first and second regression lines coincide if and only if $r_{xy} = \pm 1$.

Exercise 2.5. Let D be a dataset as in Theorem 2.1 and X defined as in Theorem 2.9. Show *by direct calculation* that

$$\begin{bmatrix} b^* \\ a^* \end{bmatrix} := (X^\intercal X)^{-1} X^\intercal \begin{bmatrix} y_1 \\ \vdots \\ y_n \end{bmatrix}$$

provides precisely the formulas for a^* and b^* that were given in Theorem 2.1.

Exercise 2.6. Show that under the assumptions of Theorem 2.9, but in the special case $N = 2$, the linear regressor is given by $f\colon \mathbb{R}^2 \to \mathbb{R}$, $f = \langle a, \cdot \rangle + b$ with

$$b = \overline{y} - a_1\overline{x}_1 - a_2\overline{x}_2$$

$$a_1 = \frac{\operatorname{cov}(x_1, y)\operatorname{cov}(x_2, x_2) - \operatorname{cov}(x_2, y)\operatorname{cov}(x_1, x_2)}{\operatorname{cov}(x_1, x_1)\operatorname{cov}(x_2, x_2) - \operatorname{cov}(x_1, x_2)^2}$$

$$a_2 = \frac{\operatorname{cov}(x_2, y)\operatorname{cov}(x_1, x_1) - \operatorname{cov}(x_1, y)\operatorname{cov}(x_1, x_2)}{\operatorname{cov}(x_1, x_1)\operatorname{cov}(x_2, x_2) - \operatorname{cov}(x_1, x_2)^2}$$

where $a = (a_1, a_2)$.
Hint: In the notation of Theorem 2.9, solve the system of linear equations $X^T X \left[\begin{smallmatrix} b \\ a \end{smallmatrix}\right] = X^T y$. In the process, to simplify notation, omit the summation index in occurring sums when no misunderstandings can arise, e.g., $\Sigma x_1 y := \sum_{i=1}^n x_{i1} y_i$. To express the result using the covariance, first show that $n \operatorname{cov}(r, s) = (\Sigma rs) - n\overline{r}\,\overline{s}$ holds for $r,\ s \in \mathbb{R}^n$.

Exercise 2.7. Implement the formulas from Exercise 2.6, e.g., in Python, and apply them to the following dataset, where exam preparation time and time on social media are the x-variables and the exam result is the y-variable. Create a plot of the data points and the regression plane.

Student	Exam preparation in h	Social media in h	Exam result in %
1	0.0	20.0	0.0
2	1.5	8.5	2.0
3	2.0	6.0	7.0
4	2.0	6.0	10.5
5	8.0	10.0	29.5
6	8.5	3.0	49.0
7	9.5	0.0	59.5
8	12.0	2.0	63.5
9	18.0	4.0	85.0
10	19.0	0.5	98.0

Note: If you are using Python, you can write your own `cov` function very efficiently using the `average` and `size` functions available in `numpy`. To double check, you can perform multivariable linear regression using `sklearn`. To use the latter package, the x-variables must be entered as a 10×2 matrix.

Exercise 2.8. Let $D = \{(x_i, y_i) \in \mathbb{R}^2 \mid i = 1, \dots, n\}$ be a dataset where not all x_i coincide, and let $f\colon \mathbb{R} \to \mathbb{R}$ be its affine linear regressor. Show:

$$r_{xy}^2 = 1 - \frac{\sum_{i=1}^n (y_i - f(x_i))^2}{\sum_{i=1}^n (y_i - \overline{y})^2} = \frac{\sum_{i=1}^n (f(x_i) - \overline{y})^2}{\sum_{i=1}^n (y_i - \overline{y})^2}.$$

Exercise 2.9. We consider the function $g\colon [-1, 1] \to \mathbb{R}$, $g(x) = 3x - 2$. Generate, for n to be specified later, in Python a dataset D_n, whose features are n-many randomly selected points in $[-1, 1]$. Assign to a feature x_i the label $y_i := g(x_i) + \varepsilon_i$, where the ε_i are independent samples of the normal distribution $\mathcal{N}(0, 1)$. Now perform affine linear regression on D_n for $n = 5, 10, 15, \dots$, e.g., by using the `sklearn` package.

Exercise 2.10. Find the polynomial $P \in \mathbb{R}[X]$ with $\operatorname{rk} P \leqslant 2$, which minimizes the quadratic distances to the following data points.

Data point	1	2	3	4	5
x	0.0	1.0	-1.5	3.5	4.0
y	2.0	2.2	0.0	2.5	3.0

Exercise 2.11. Let $D = \{(x_i, y_i) \in \mathbb{R}^d \times \{0,1\} \mid i = 1, \ldots, n\}$ be a dataset and ℓ the associated negative log-likelihood function from (2.5). Show that

$$\nabla \ell(w) = \sum_{i=1}^{n} \big(\text{sig}(\langle w, x_i \rangle) - y_i\big)\widehat{x}_i$$

holds. Furthermore, show that $\nabla \ell$ for the dataset $D = \{(-1,1), (0,0), (1,1)\}$ has precisely one zero $w^* \in \mathbb{R}^2$ and calculate its coordinates. Since ℓ is convex, $w^* \in \mathbb{R}^2$ provides the logistic regressor.

Exercise 2.12. The following dataset is shown in the picture on page 23:

$$\begin{aligned} D = \big\{&(0.5, 0), (0.35, 0), (0.5, 0), (0.35, 0), (0.1, 0), (0.72, 0), \\ &(0.8, 0), (0.24, 0), (1.10, 0), (0.97, 0), (1.5, 1), (1.9, 1), \\ &(1.65, 1), (1.35, 1), (1.7, 1), (1.24, 1), (1.09, 1), (0.92, 1)\big\}. \end{aligned}$$

Determine, for example using Python, the logistic regressor and plot it.

Exercise 2.13. To encourage customers to participate in a survey, different amounts of money are offered as compensation for their efforts. The table below lists the amounts offered and indicates how many customers accepted the respective offer.

USD	0.50	1	2	3	5	10	15	25	50
Acceptance	1/43	2/50	4/48	5/32	30/37	15/32	55/100	49/50	19/20

(i) Based on the table, create a dataset D_1, consisting of 412 data points with categorical labels, and determine its logistic regressor.

(ii) Create another dataset D_2, consisting of 9 data points with categorical labels, by applying majority vote for each amount. Determine the logistic regressor for D_2.

(iii) Now consider the dataset

$$D_3 := \big\{(x_i, \log \frac{y_i}{1 - y_i}) \mid i = 1, \ldots, 9\big\},$$

where the x_i are the amounts and the y_i are the unchanged acceptance rates from the table. Determine the affine linear regressor with parameters $(a, b) \in \mathbb{R}^2$ for D_3. Now compare the logistic function

$$f\colon \mathbb{R} \to (0,1),\ f(x) = \frac{1}{1 - \mathrm{e}^{-(ax+b)}}$$

with the logistic regressors from (i) and (ii).

Exercise 2.14. Let (Ω, Σ) be a measurable space and $X\colon \Omega \to \mathbb{R}$ a random variable with finite expectation μ and finite variance σ. Let $X_1, \ldots, X_n$ be independent copies of X. Define new random variables via

$$\mu_s := \frac{1}{n}\sum_{j=1}^{n} X_j \quad \text{and} \quad \sigma_s^2 := \frac{1}{n}\sum_{j=1}^{n}(X_j - \mu_s)^2$$

and show that $\mathrm{E}(\sigma_s^2) = \frac{n-1}{n}\sigma^2$ holds. Compare with our remarks after Definition 2.5.

3

k-Nearest Neighbors

At the end of the previous Chapter 2 we used logistic regression in order to assigns a classifier to a binary labeled dataset. This method has *provable* properties, such as producing a maximum likelihood estimator, or providing a unique classifier for overlapping data. In Chapters 13–14 we will see that a rigorous theory can also be developed for binary labeled and linearly separable datasets.

In this chapter, we want to make almost no assumptions on the datasets and demonstrate methods that lead to classifiers and regressors with rather simple mathematical tools. This has the advantage of universal applicability and provides in many cases a "first approximation" that can then be compared with mathematically more sophisticated methods. The disadvantage of this approach is of course that we cannot expect a deep provable theory.

3.1 k-NN Classifier

Let from now on $D \subseteq X \times Y$ be a categorically labeled dataset, where X is a metric space and Y is a finite set. We start with an example.

Example 3.1. In the following picture, $X \subseteq \mathbb{R}^2$ is a cuboid and $Y = \{1, 2, 3\}$.

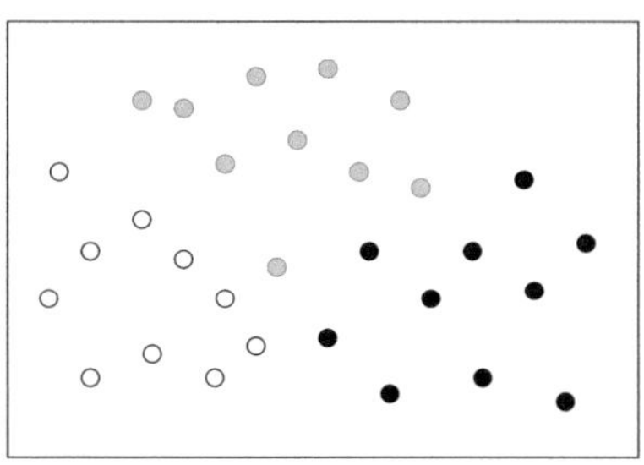

S. A. Wegner, *Mathematical Introduction to Data Science*,
https://doi.org/10.1007/978-3-662-69426-8_3

We equip X with the Euclidean metric and understand white points as labeled with 1, gray points as labeled with 2, and black points as labeled with 3.

Our goal is to find a function $f\colon X \to Y$ that "fits the data as accurately as possible". In the simplest sense, the latter could mean that $f(x) = y$ should hold for all, or at least for as many data points as possible. Provided we have such a function f and are given further points $x \in X \backslash D$, whose labels are either unknown to us, or to which a label has not yet been assigned, we may *predict* respectively *assign* a label via $y := f(x)$. An obvious idea to get the function f is the following.

Definition 3.2. Let (X, ρ) be a metric space, Y a finite set, $D \subseteq X \times Y$ a dataset, and $k \in \mathbb{N}$. We define a function $f\colon X \to Y$ as follows. For a given $x \in X$, we first select $x_1, \ldots, x_k \in D_1 := \pi_1(D) = \{x \mid (x, y) \in D\}$ with

$$x_1 \in \operatorname*{argmin}_{z \in D_1} \rho(x, z) \quad \text{and} \quad x_j \in \operatorname*{argmin}_{z \in D_1 \backslash \{x_1, \ldots, x_{j-1}\}} \rho(x, z) \text{ for } j \geqslant 2$$

and call these points the *k-nearest neighbors* of x. Let $y_1, \ldots, y_k$ be the labels of $x_1, \ldots, x_k$. We consider the mapping

$$N\colon Y \to \mathbb{N}, \quad N(y) := \#\{i \mid y_i = y\},$$

which counts the number of occurrences of the label y among the previously determined k-nearest neighbors. Finally, we choose the function value

$$f(x) \in \operatorname*{argmax}_{y \in Y} N(y)$$

and call $f\colon X \to Y$ the *k-NN classifier with majority vote.*

In prose, Definition 3.2 can be put much less technically as follows: For $x \in X$, we take that point in D which is closest to x, then the second closest, and so on. Next we determine, among these k-nearest neighbors, which label appears most frequently and then assign the same label to the point x. If several points from D have the same distance to x, or if there is a tie among the neighbors' labels, we make an arbitrary choice.

The values of the function f can be determined by the following k-nearest neighbors algorithm.

Algorithm 3.3. *Let (X, ρ) be a metric space, Y a finite set, $D \subseteq X \times Y$ a dataset, and $k \in \mathbb{N}$. The following pseudocode outlines the k-*NN algorithm with majority vote.

function K-NN CLASSIFIER(D, k, x)
  $D' \leftarrow D,\ A \leftarrow \emptyset$
  for $j \leftarrow 1$ to k **do**
    $z^* \leftarrow \operatorname{argmin}_{z \in D'_1} \rho(x, z)$
    $A \leftarrow A \cup \{z^*\},\ D' \leftarrow D' \setminus \{(z^*, y^*)\}$
  for y in Y **do**
    $N(y) \leftarrow \#\{a \in A \mid \pi_2(a) = y\}$
  $\ell \leftarrow \operatorname{argmax}_{y \in Y} N(y)$
  return ℓ

Here, $\pi_2(x, y) = y$ denotes the projection onto the second entry of $(x, y) \in D$ and y^* is the label of z^*. In addition, we have replaced the explicit numbering of the k-nearest neighbors by using the set A.

Our Definition 3.2 guarantees that f is a map. If one implements the above, this is achieved almost automatically if we, for example, always run through D in a predetermined order and, in case that the argmin is not unique, always select the first minimizer that we find. The same can be done for the argmax when holding the majority vote.

Example 3.4. We now return to the dataset in Example 3.1 and visualize the k-NN classifier for two different k's. The colored areas each indicate the value of the function f in these areas.

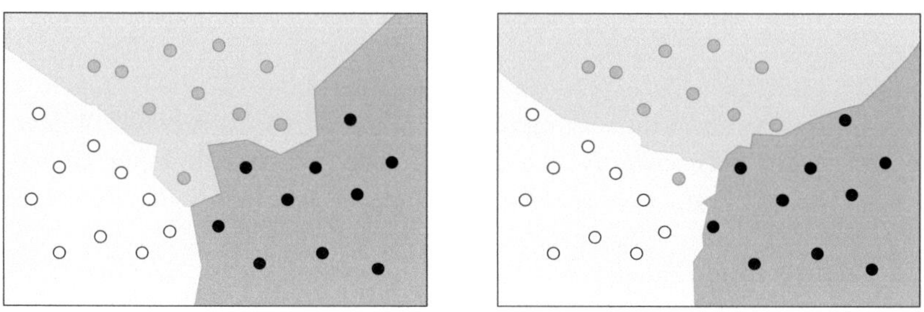

$k = 1$ $\qquad$ $k = 15$

When comparing the images, it is noticeable that the lines where the colors change are less jagged for $k = 15$ than for $k = 1$. In particular, for $k = 1$ there is a bulge around the gray point with the lowest ordinate. For $k = 15$ this bulge has disappeared, because the white and black points around the aforementioned gray point outnumber it in the majority vote. The price for removing the bulges, however, is that for the aforementioned gray point x with the lowest ordinate, we then have $f(x) \neq y$.

Definition 3.5. Let (X, ρ) be a metric space, Y a finite set, $D \subseteq X \times Y$ a dataset, $k \in \mathbb{N}$ and $f \colon X \to Y$ a k-NN classifier.

(i) The sets $X_y := \{x \in X \mid f(x) = y\}$ for $y \in Y$ are called *decision areas.*

(ii) The boundaries ∂X_y are called *decision boundaries.*

In Example 3.4, we observed that increasing k leads to less jagged decision boundaries and removes bulges, but can also lead to "misclassifications". At first glance, misclassification seems to be the greater evil and one might be inclined to prefer the classifier shown on the left in Example 3.4. Our goal is not necessarily to convince the reader of the opposite, but to make a case for the classifier on the right. In order to do so, we consider another example.

Example 3.6. Let $X \subseteq \mathbb{R}^2$ be again a cuboid and this time $Y = \{1, 2\}$. We consider a binary labeled set as in the following picture, which additionally shows the k-NN classifier for $k = 1$. We leave it to the reader to convince themselves that for $k \geqslant 3$ the dark gray island on the left will disappear.

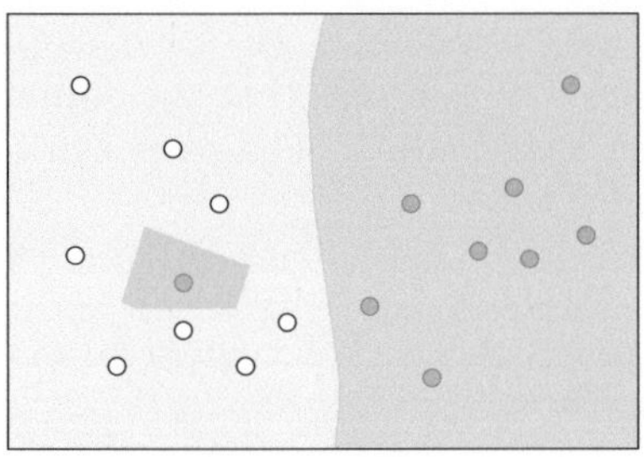

$k = 1$

Assuming that there is a true underlying function $g\colon X \to Y$ which we are trying to approximate by running k-NN on a sample D, it could of course be the case that g looks exactly like in the above picture—but it could also be the case that an error occurred in determining the "true" label of the point on the island and that the latter should actually be white! In this last case, the 1-NN algorithm adjusted the function f *too well* to the dataset, instead of *generalizing* from the data.

The following is not a formal mathematical notion, but it is of great importance in the context of machine learning. For this reason, we grant it a numbered definition.

Definition 3.7. Let (X, ρ) be a metric space, Y a finite set, $D \subseteq X \times Y$ a dataset and $k \in \mathbb{N}$. If $f\colon X \to Y$ is "too well" adapted to the dataset D we speak of *overfitting.*

In examples, one can check for overfitting using the following heuristic: Given a dataset D, we partition it into *training data* D_{train} and *test data* D_{test}. Then we determine a classifier $f\colon X \to Y$ based on D_{train}, and find out what percentage of points from D_{test} get correctly classified by f. If this percentage is rather small, this may indicate overfitting.

Next, we discuss the question of how complex, or "expensive", the algorithmic calculation of the k-NN classifier is, provided that a dataset with feature vectors in $\mathbb{R}^d$ is given and assuming that we use the Euclidean metric. We express this by counting the number of multiplications that need to be carried out when running Algorithm 3.3.

Theorem 3.8. (on the runtime of the k-NN algorithm) *Let $D \subseteq \mathbb{R}^d \times Y$ be given with $\#D = n$, $\#Y < \infty$ and $k \in \mathbb{N}$. Let $\mathbb{R}^d$ be equipped with the Euclidean metric and let $x \in \mathbb{R}^d$ be fixed. The calculation of the k-nearest neighbors of x can be implemented such that at most $(C \cdot n \cdot d \cdot k)$-many multiplications have to be carried out, where $C \in \mathbb{N}$ is a constant independent of d, k and n.*

Proof. The k-NN algorithm calculates $\big(n + (n-1) + \cdots + (n-k+1)\big)$-many distances between points in $\mathbb{R}^d$. In the Euclidean metric, it requires d-many multiplications to compute one distance if we omit the root, which we can do as it does not change the argmin. This leads to

$$d \cdot (n + (n-1) + \cdots + (n-k+1)) \leqslant C \cdot d \cdot k \cdot n$$

multiplications with a suitable $C \in \mathbb{N}$. □

The above result can be improved by computing the distances of all data points z, with $(z, y) \in D$, to the fixed x only once and then storing them for re-use. However, the dimension d still appears as a multiplicative factor in the runtime estimate. If we thus are given high-dimensional data, the runtime of the k-NN algorithm will be long — and this is the case even if the cardinality of D is rather small. We will return to this problem in Chapters 8–10.

Remark 3.9. Majority vote is just one way to determine $f(x)$ based on the k-nearest neighbors of $x \in X$. Alternatively, one may take into account the actual distances in the sense that a neighbor further away has less influence on the classification than a closer one. For example, in Definition 3.2, the function $N(y) := \#\{\, i \mid y_i = y\}$ could be replaced by

$$N(y) = \sum_{\substack{i=1 \\ y_i = y}}^{k} \frac{c}{1 + \rho(x_i, x)} \quad \text{or} \quad N(y) = \sum_{\substack{i=1 \\ y_i = y}}^{k} \mathrm{e}^{-c\rho(x_i,x)^2},$$

where $c > 0$ controls how quickly the contribution of a neighbor decreases if their distance from x increases. The second formula above is called *Gaussian weighting*.

3.2 k-NN Regressors

Let $D \subseteq X \times Y$ be a dataset, where (X, ρ) is a metric space and Y is a real vector space. For $k \in \mathbb{N}$ and $x \in X$, we choose k-nearest neighbors $x_1, \dots, x_n$ of x and denote their labels by $y_1, \dots, y_n$. We put

$$f\colon X \to Y,\ f(x) = \tfrac{1}{k} \sum_{i=1}^{k} y_i,$$

and call f the *k-NN regressor with arithmetic mean.*

Similar to Remark 3.9, one may take the actual distance into account also here, and for example use

$$f\colon X \to Y,\ f(x) = \frac{\sum_{i=1}^{n} w(x_i, x) \cdot y_i}{\sum_{i=1}^{n} w(x_i, x)}$$

with $w(z, x) = \frac{c}{1+\rho(z,x)}$, $w(z, x) = \mathrm{e}^{-c\rho(z,x)^2}$ or with other weight functions that are strictly positive and increase with decreasing distance.

3.3 Preprocessing

In the following, we consider a dataset $D \subseteq \mathbb{R}^d \times Y$ but focus on the set of feature vectors, which we denote for the moment with

$$F = \{x^{(1)}, \dots, x^{(n)}\} \subseteq \mathbb{R}^d.$$

If we use the Euclidean metric, or more generally the metric induced by $\|\cdot\|_p$ with $p > 0$, to calculate the k-NN predictor in the sense of the above explanations, all coordinates are treated equally. However, it is quite possible that the individual features in the dataset have very different orders of magnitude. If we know a certain pattern a priori, we may counteract this in an ad hoc fashion, e.g., by using a weighted metric

$$d\colon \mathbb{R}^d \times \mathbb{R}^d \to \mathbb{R},\ \rho_{p,w}(x, y) := \sum_{i=1}^{d} w_i |x_i - y_i|^p$$

for $p > 0$ and with suitably chosen $w_1, \dots, w_d > 0$.

However, if we have no knowledge about the meaning of the individual features, a natural approach is to treat all of the entries in a feature vector as equally as possible. To achieve this, one can try the following methods.

Definition 3.10. Let $F = \{x^{(1)}, \dots, x^{(n)}\} \subseteq \mathbb{R}^d$ be given, let $a < b$ be real numbers and let $\|\cdot\|$ be a norm on $\mathbb{R}^n$. We denote by $\overline{x_j^{(\cdot)}} = \frac{1}{n}(x_j^{(1)} + \dots + x_j^{(n)})$ the mean of the j-th entries over all data points and by $\sigma_j^2 = \operatorname{var}(x_j^{(\cdot)})$ the variance in the same sense. Then the set $\tilde{F} := \{\tilde{x}^{(1)}, \dots, \tilde{x}^{(n)}\}$ with

(i) $\tilde{x}^{(i)} = \Big(\dfrac{(x_1^{(i)} - \min_{j=1,\dots,n} x_1^{(j)})(b-a)}{\max_{j=1,\dots,n} x_1^{(j)} - \min_{j=1,\dots,n} x_1^{(j)}}, \dots\Big)$ is the *min-max normalization*,

(ii) $\tilde{x}^{(i)} = \Big(x_1^{(i)} - \frac{1}{n}\sum_{j=1}^{n} x_1^{(j)}, \dots\Big)$ is the *centering*,

(iii) $\tilde{x}^{(i)} = \Big(\dfrac{x_1^{(i)} - \overline{x_1}}{\sigma_1}, \dots\Big)$ is the *standardization*,

(iv) $\tilde{x}^{(i)} = \Big(\dfrac{x_1^{(i)}}{\|(x_1^{(1)}, \dots, x_1^{(n)})\|}, \dots\Big)$ is the *normalization* of F.

Here, in (i) we normalize *to* $[a, b]$ and in (iv) *with respect to* $\|\cdot\|$.

The above methods can be combined. For example, (iii) is a combination of (ii) and (iv), if we take the 2-norm on $\mathbb{R}^n$. But we could also first center and then normalize with $\|\cdot\|_\infty$. Then each entry would belong to $[-1, 1]$ and the coordinate-wise mean would be zero. Note, however, that this yields a different dataset $\tilde{F}$ than min-max normalization to $[-1, 1]$.

We will revisit normalization in particular in Exercise 3.5. Besides the "homogenization" of the features of a given dataset, centering in particular has another important application. To explain, let us consider data with feature vectors in $\mathbb{R}^d$ and real labels. Unlike before, we now assume that no data point

$$(x, y) = (x_1, \dots, x_d, y)$$

is known to us in its entirety, but instead a dataset is given in which only *a part* of the above $(d+1)$-many coordinates are available to us. A classic example is the following.

Example 3.11. Let a table be given that contains product ratings, where empty cells mean that the rating is unknown to us at this point.

	Product 1	Product 2	Product 3	Product 4	Product 5	Product 6	Product 7
Customer 1	★★★★☆	★★★★★	★★★★★	★☆☆☆☆		★★☆☆☆	
Customer 2		★☆☆☆☆	★★☆☆☆	★★★★★	★★★★★		★☆☆☆☆
Customer 3	★★☆☆☆	★★★☆☆	★★★☆☆	★★★☆☆		★★★☆☆	
Customer 4	★★★★★			★★☆☆☆	★★★☆☆	★★★★★	★★★★☆
Customer 5	★★★★★	★★★★☆		★☆☆☆☆	★★☆☆☆		★★★★☆

It is then a natural task to predict the missing ratings. In order to do so, e.g., for the rating of Product 7 by Customer 1, one is inclined to consider the customers as data points in $\mathbb{R}^6$, given by their ratings for Products 1–6, and the ratings for Product 7 as a label. On this basis, one could then determine a predictor $f\colon \mathbb{R}^6 \to \{1,\dots,5\}$, e.g., via the k-NN Algorithm 3.3, *if* all entries except the one containing the rating of Customer 1 for Product 7 were available.

In order to be able to make predictions despite missing coordinates and labels, the unknown entries must first be somehow filled in. One possibility is to center the feature vectors first and then to fill the empty entries with zeros. We demonstrate this procedure using the above rating matrix.

Example 3.12. (i) We consider the table in Example 3.11, divided into features and labels as described above. First centering the features and then filling in the blanks with zeros leads to the dataset $D = \{(x^{(i)}, y^{(i)}) \mid i = 2, 4, 5\}$ and unlabeled feature vectors $x^{(1)}$, $x^{(3)}$, as indicated in the following table.

	Feature 1	Feature 2	Feature 3	Feature 4	Feature 5	Feature 6	Label
$x^{(1)}$	0.60	1.60	1.60	-2.40	0.00	-1.40	
$(x^{(2)}, y^{(2)})$	0.00	-2.25	-1.25	1.75	1.75	0.00	1
$x^{(3)}$	-0.80	0.20	0.20	0.20	0.00	0.20	
$(x^{(4)}, y^{(4)})$	1.25	0.00	0.00	-1.75	-0.75	1.25	4
$(x^{(5)}, y^{(5)})$	2.00	1.00	0.00	-2.00	-1.00	0.0	4

If we use the Euclidean metric ρ on $\mathbb{R}^6$, we get

$$\rho(x^{(1)}, x^{(2)}) = 6.749, \quad \rho(x^{(1)}, x^{(4)}) = 3.361 \quad \text{and} \quad \rho(x^{(1)}, x^{(5)}) = 2.828,$$

from which the k-nearest neighbors can be seen. We ask the reader to look at the original table again in light of the distances just calculated! Using now majority vote, we obtain four stars for the rating of Product 7 by Customer 1, independent of $k = 1, 2, 3$. If we use the arithmetic mean, followed by rounding, we get four stars if $k = 1, 2$, and three stars if $k = 3$.

(ii) The above method disregards any information contained in the ratings of Customer 3, as their rating for Product 7 is unavailable to us. If we however want to incorporate this information as well, we can center over the entire rows instead of just their first six entries. This leads to $D = \{(x^{(i)}, y^{(i)}) \,|\, i = 1, \dots, 5\}$ as follows.

	Feature 1	Feature 2	Feature 3	Feature 4	Feature 5	Feature 6	Label
$(x^{(1)}, y^{(1)})$	0.60	1.60	1.60	-2.40	0.00	-1.40	0.00
$(x^{(2)}, y^{(2)})$	0.00	-1.80	-0.80	2.20	2.20	0.00	-1.80
$(x^{(3)}, y^{(3)})$	-0.80	0.20	0.20	0.20	0.00	0.20	0.00
$(x^{(4)}, y^{(4)})$	1.20	0.00	0.00	-1.80	-0.80	1.20	0.20
$(x^{(5)}, y^{(5)})$	1.80	0.80	0.00	-2.20	-1.20	0.00	0.80

Now, we again calculate the distances of the rows with respect to the Euclidean metric on $\mathbb{R}^7$ and obtain

$$\rho\left(\begin{bmatrix} x^{(1)} \\ y^{(1)} \end{bmatrix}, \begin{bmatrix} x^{(2)} \\ y^{(2)} \end{bmatrix}\right) = 6.991, \quad \rho\left(\begin{bmatrix} x^{(1)} \\ y^{(1)} \end{bmatrix}, \begin{bmatrix} x^{(3)} \\ y^{(3)} \end{bmatrix}\right) = 3.899,$$

$$\rho\big(\big[\begin{smallmatrix} x^{(1)} \\ y^{(1)} \end{smallmatrix}\big], \big[\begin{smallmatrix} x^{(4)} \\ y^{(4)} \end{smallmatrix}\big]\big) = 3.644, \quad \rho\big(\big[\begin{smallmatrix} x^{(1)} \\ y^{(1)} \end{smallmatrix}\big], \big[\begin{smallmatrix} x^{(5)} \\ y^{(5)} \end{smallmatrix}\big]\big) = 2.953.$$

The k-nearest neighbors can be read from the above again. To apply majority vote and arithmetic mean, respectively, we use the original ratings in the case of Customers 2, 4 and 5, and for Customer 3, the centering process has to be reversed, i.e., here we put the rating to $(2+3+3+3+3)/5 \approx 3$ stars. For the rating of Product 7 by Customer 1, this results in four stars, independent of $k = 1, 2, 3$ and independent of choosing majority vote or arithmetic mean and rounding to compute the predictor.

Remark 3.13. (i) First, we note that once the table in Example 3.12(ii) is centered, all other previously empty cells could now also be filled. We thus obtain a first method that allows us to fill a "holey" data matrix. In Chapter 7.3, we will discuss much more sophisticated techniques to do this.

(ii) The above methods for prediction are collaborative in the sense that the prediction of the rating for Product 7 by Customer 1 is based not only on the other ratings by Customer 1, but on the comparison of ratings of Customer 1 with those of all other customers. This is referred to as *user-to-user collaborative filtering*.

(iii) Instead of considering the customers as points in $\mathbb{R}^7$, we could also have considered the products as points in $\mathbb{R}^5$. If one does this, a column-wise centering is appropriate, compare Exercise 3.6(ii)–(iii).

(iv) Instead of the Euclidean metric, one may also use other metrics, or even so-called distance measures. We will deal with this in the following subchapter.

3.4 Cosine Similarity

In Chapters 3.1–3.3, we have always worked with an underlying metric space (X, ρ). In this chapter, we want to generalize this. First, we note that both k-NN classifiers and k-NN regressors, in the sense of the previous subchapters, can still be defined even if ρ is not definite and does not satisfy the triangle inequality.

Definition 3.14. Let X be a non-empty set. A function $\rho\colon X \times X \to \mathbb{R}$ is called a *distance measure* if, for all $x, y \in X$, the following holds:

(DM1) $\rho(x, y) \geqslant 0$, (positivity)

(DM2) $\rho(x, x) = 0$, ("weak" definiteness)

(DM3) $\rho(x, y) = \rho(y, x)$. (symmetry)

For distance measures, the heuristic is that small values mean that points have similar properties. Note, that for the cosine similarity defined below things are different.

Definition 3.15. Let $\mathbb{R}^d$ be equipped with the standard scalar product and the induced Euclidean norm. For $x, y \in \mathbb{R}^d \backslash \{0\}$, we then define

$$\operatorname{cossim}(x, y) := \frac{\langle x, y \rangle}{\|x\| \|y\|} = \frac{\sum_{i=1}^{d} x_i y_i}{\left(\sum_{i=1}^{d} x_i^2\right)^{1/2} \left(\sum_{i=1}^{d} y_i^2\right)^{1/2}} \in [-1, 1]$$

to be the *cosine similarity* of x and y. If we denote by $\theta = \measuredangle(x, y)$ the angle

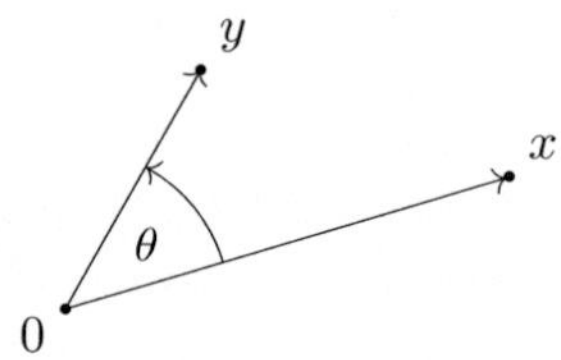

between x and y, we get $\langle x, y \rangle = \|x\| \|y\| \cos(\theta)$. It follows $\operatorname{cossim}(x, y) = \cos(\theta)$, which justifies the term "cosine similarity".

In contrast to a distance measure, the heuristic underlying the cosine similarity is that x and y are similar if $\operatorname{cossim}(x, y)$ is close to 1. If $\operatorname{cossim}(x, y)$ is close to 0, then x and y are rather dissimilar. The case of negative cosine similarity expresses something that cannot be described by a distance measure at all.

If we restrict ourselves to a subset of $\mathbb{R}^d \backslash \{0\}$, the cosine similarity induces a distance measure in the following way.

Lemma 3.16. *Let*

$$\mathbb{R}^d_{>0} := \{(x_1, \ldots, x_d) \in \mathbb{R}^d \backslash \{0\} \mid x_i \geqslant 0 \text{ for all } i = 1, \ldots, d\}.$$

Then $\operatorname{cosdist}\colon \mathbb{R}^d_{>0} \times \mathbb{R}^d_{>0} \to \mathbb{R}$, $\operatorname{cosdist}(x, y) := 1 - \operatorname{cossim}(x, y)$, *is a distance measure, which we refer to as the* cosine distance.

Proof. All three properties to be checked follow directly from properties of the (standard) scalar product and the (Euclidean) norm. □

We leave it to the reader as Exercise 3.7 to verify that ρ as in Lemma 3.16 is not positive definite and does not satisfy the triangle inequality.

A typical application of cosine similarity is text mining.

Example 3.17. We consider the following three texts.

On November 22, 1963, American President John Fitzgerald Kennedy was assassinated in Dallas. Kennedy had been in office for two years at this point. The president was shot with a rifle during a ride through Dallas in an open car. His wife Jackie Kennedy was also in the car, but remained unharmed. John Fitzgerald Kennedy was buried in Arlington on November 25, 1963.

On November 22, 1963, Lee Harvey Oswald assassinated then US President John Fitzgerald Kennedy in Dallas. Oswald shot Kennedy with a rifle while he was driving through Dallas in an open car with his wife Jackie Kennedy. Oswald was shot on November 24, 1963, during his transfer to the Dallas County Jail by nightclub owner Jack Ruby in an underground garage.

At John Fitzgerald Kennedy Airport in Dallas, there is a wide selection of restaurants. The Porterhouse Steak and the T-bone Steak, each served with grilled tomatoes at the 5ive Steak restaurant, are particularly recommended. But vegetarians also get their money's worth at John Fitzgerald Kennedy Airport: 5ive Steak, for example, offers a Black Bean Burger. At Jikji Cafe, there are vegetarian dishes from Korean cuisine.

We now assign to each text a vector in $\mathbb{R}^d$ in the following way. First, we consider the set of all words occurring in all three texts

$$W := \{\mathsf{On}, \mathsf{November}, \mathsf{22}, \mathsf{1963}, \mathsf{American}, \mathsf{President}, \dots\},$$

put $d := \#W$ and choose a bijection $\{1, \dots, d\} \to W$. Via this bijection, we can write vectors in $\mathbb{R}^d$ as $(x_w)_{w \in W}$. In particular, we can define the *vectorizations* for the texts no. 1–3 from above via

$$\begin{aligned} x^{(i)} &:= (x_w^{(i)})_{w \in W} \text{ where} \\ x_w^{(i)} &:= \text{number of occurrences of word } w \text{ in text } i \end{aligned}$$

for $i = 1, 2, 3$. By construction, we get $x^{(1)}, x^{(2)}, x^{(3)} \in \mathbb{R}^d_{\geqslant 0}$. In the above example one can calculate the pairwise cosine similarities

$$\operatorname{cossim}(x^{(1)}, x^{(2)}) = 0.69, \ \operatorname{cossim}(x^{(1)}, x^{(3)}) = 0.31, \ \operatorname{cossim}(x^{(2)}, x^{(3)}) = 0.22$$

with a simple program, compare Exercise 3.8. We thus see that texts no. 1 and text no. 2 are significantly more cosine similar than text no. 1 and text no. 2 or text no. 1 and text no. 3. If we use now the cosine distances

$$\operatorname{cosdist}(x^{(1)}, x^{(2)}) = 0.31, \ \operatorname{cosdist}(x^{(1)}, x^{(3)}) = 0.69, \ \operatorname{cosdist}(x^{(2)}, x^{(3)}) = 0.78,$$

and, for example, binary labels "👍" and "👎" are assigned by some reader to a suitable number of texts, we may use the k-NN Algorithm 3.3 to predict, for a new text, if this particular reader will enjoy it or not.

Remark 3.18. (i) Beyond the *raw count* used above, there are other ways to vectorize which, for example, compensate for the fact that the raw count grows with the text length; that words like "the", "and", "that" etc. occur often in every text; or that, e.g., "drive", "drove" and "driven" are treated as different words. We discuss some of these methods in Exercise 3.8.

(ii) Text mining is a good example of how one very naturally and very quickly has to deal with high-dimensional data: The three vectorizations of our short example texts are already elements of $\mathbb{R}^{108}$.

(iii) Instead of the cosine distance, one could also use the Euclidean metric for text comparison. In general, however, the cosine distance is better suited here. This becomes clear when considering a word that already occurs often in one text. The Euclidean distance to another text is then minimal if this word occurs in it exactly as often as in the first text. The cosine distance, on the other hand, becomes smaller and smaller the more often the word appears in the second text. We also refer to the artificial, but instructive, example in Exercise 3.9.

Also for rating predictions it can make sense to use the cosine similarity. Of course this leads to different results compared to using the Euclidean metric.

Example 3.19. We consider the product reviews from Example 3.11 and recall below the original rating table. Next to it we centered the table row-wise and then filled the empty cells with zeros. We denote the entries on the left with $x_j^{(i)}$, those on the right with $\tilde{x}_j^{(i)}$. In particular we refrain in this example from designating some coordinate as a label.

	x_1	x_2	x_3	x_4	x_5	x_6	x_7
$x^{(1)}$	4	5	5	1		2	
$x^{(2)}$		1	2	5	5		1
$x^{(3)}$	2	3	3	3		3	
$x^{(4)}$	5			2	3	5	4
$x^{(5)}$	5	4		1	2		4

	$\tilde{x}_1$	$\tilde{x}_2$	$\tilde{x}_3$	$\tilde{x}_4$	$\tilde{x}_5$	$\tilde{x}_6$	$\tilde{x}_7$
$\tilde{x}^{(1)}$	0.6	1.6	1.6	-2.4	0.0	-1.4	0.0
$\tilde{x}^{(2)}$	0.0	-1.8	-0.8	2.2	2.2	0.0	-1.8
$\tilde{x}^{(3)}$	-0.8	0.2	0.2	0.2	0.0	0.2	0.0
$\tilde{x}^{(4)}$	1.2	0.0	0.0	-1.8	-0.8	1.2	0.2
$\tilde{x}^{(5)}$	1.8	0.8	0.0	-2.2	-1.2	0.0	0.8

The cosine similarities of $\tilde{x}^{(1)}$ and the remaining centered data points amount to

$$\operatorname{cossim}(\tilde{x}^{(1)}, \tilde{x}^{(2)}) = -0.634, \ \operatorname{cossim}(\tilde{x}^{(1)}, \tilde{x}^{(3)}) = -0.185,$$

$$\operatorname{cossim}(\tilde{x}^{(1)}, \tilde{x}^{(4)}) = 0.355, \ \operatorname{cossim}(\tilde{x}^{(1)}, \tilde{x}^{(5)}) = 0.640$$

and we encourage the reader to compare the cosine similarities with the Euclidean distances in Example 3.12(ii). A straightforward idea to generate a prediction might be to weight the ratings with their respective cosine similarities. As an example, we again take the rating of Customer 1 for Product 7, and use the two most cosine similar customers, or in other words the 2-nearest customers in the sense of cosine distance. This leads to

$$\underset{\text{Prediction}}{\underset{\uparrow}{\hat{x}_7^{(1)}}} := \frac{\operatorname{cossim}(\tilde{x}^{(1)}, \tilde{x}^{(5)}) \cdot x_7^{(5)} + \operatorname{cossim}(\tilde{x}^{(1)}, \tilde{x}^{(4)}) \cdot x_7^{(4)}}{\operatorname{cossim}(\tilde{x}^{(1)}, \tilde{x}^{(5)}) + \operatorname{cossim}(\tilde{x}^{(1)}, \tilde{x}^{(4)})}$$

$$= \frac{0.640 \cdot 4 + 0.355 \cdot 4}{0.640 + 0.355} = 4.$$

However, if we use the three most cosine similar customers who put in a rating for Product 7, we get

$$\hat{x}_7^{(1)} \underset{\substack{\uparrow \\ \text{Prediction}}}{:=} \frac{0.640 \cdot 4 + 0.355 \cdot 4 - 0.634 \cdot 1}{0.640 + 0.355 - 0.634} \approx 9.24.$$

The fact that the negative cosine similarity of $x^{(1)}$ and $x^{(2)}$ and the low rating $x_7^{(1)}$ result in an *increase* in the prediction may seem reasonabe, but on the other hand, it clearly overshoots the mark here. Before this last result scares off the reader too much, we point out that with a large amount of data and not too large k, it is to be expected that for each prediction only customers with a high similarity will be included in the formula, and we will therefore rarely be in the last described situation.

The above example shows that the potentially negative values of the cosine similarity should be treated with caution. On the other hand, they also offer the possibility in certain cases to distinguish not only between "similar" and "not similar", but to add something like "opposite" to the scale. We also mention that some authors divide in the prediction formula of Example 3.19 by the sum of the *absolute values* of the cosine similarities. If we do this above, however, the prediction *decreases*.

References

Due to its simplicity and universal applicability, the k-NN algorithm is very popular and is covered in many books and lecture notes on data science or machine learning. The same applies to the preprocessing methods that we discussed. Our main source for this chapter was [LRU12]. More details on collaborative filtering can be found in [SKKR01] and [Agg16].

Exercises

Exercise 3.1. We consider the dataset consisting of the first six images below with labels as indicated. Classify the 7th image using 1-NN (with majority vote). Use the metric induced by the 1-norm on $\mathbb{R}^{35}$.

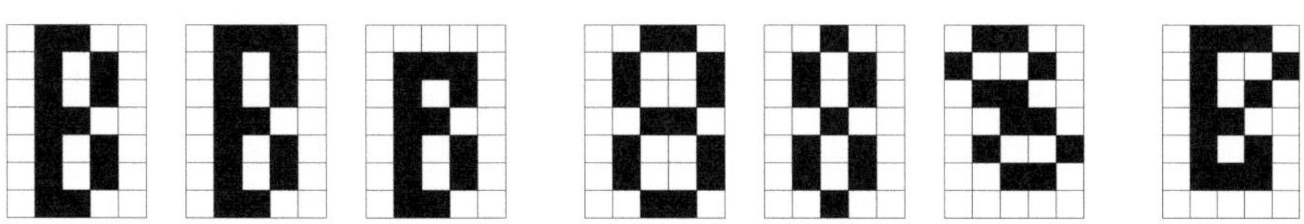

Label B Label B Label B Label 8 Label 8 Label 8 Label ?

Exercise 3.2. We consider the following dataset $D \subseteq \mathbb{R}^2 \times \{1, 2\}$, where white points have label 1 and black points have label 2. The label of the point marked with a cross is unknown to us.

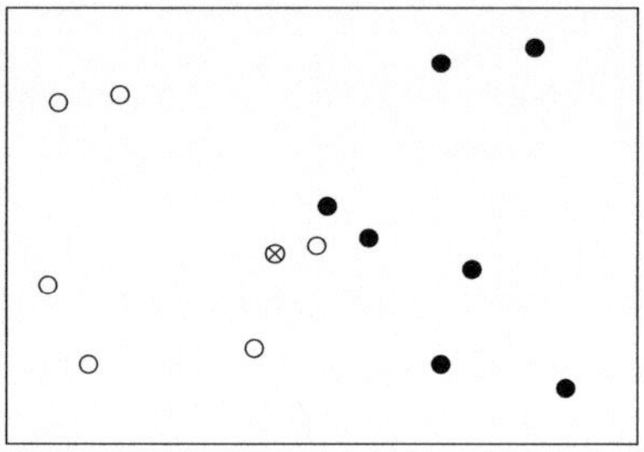

Determine (a rough estimate is sufficient!) for the point marked with a cross the label according to k-NN with $k = 1, \ldots, 7$ and majority vote.

Exercise 3.3. Sketch the decision areas and decision boundaries with respect to 1-NN, provided that we use the metric induced by the 2-norm, the 1-norm, or the ∞-norm, respectively.

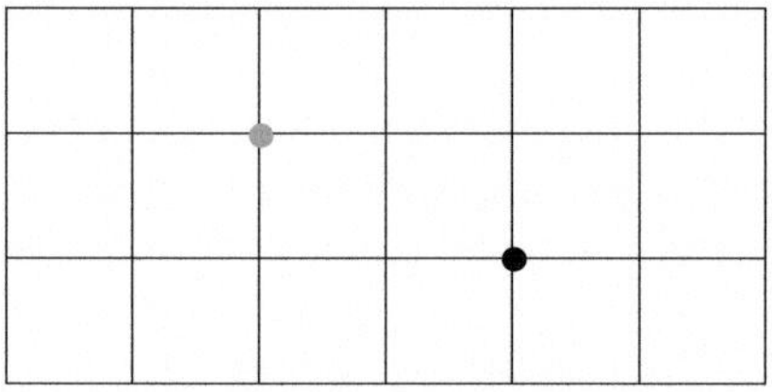

Exercise 3.4. Consider the dataset from Exercise 2.3 and apply 2-NN with arithmetic mean. Sketch the corresponding regressor and the data points. Compare the former with the affine linear regressor from Exercise 2.3. Do you find that 2-NN makes sense here?

Exercise 3.5. Consider the following 2-dimensional dataset with labels in $\{-1, +1\}$.

Data point	Feature 1	Feature 2	Label
1	111	0.3	+1
2	82	0.4	+1
3	110	0.4	+1
4	148	0.7	+1
5	91	0.5	+1
6	133	0.9	+1
7	71	1.2	−1
8	111	1.1	−1
9	123	1.6	−1
10	151	1.9	−1
11	89	1.3	−1
12	99	2.0	−1

(i) Implement k-NN with Euclidean metric and majority vote, such that your program gets $k \in 2\mathbb{N} - 1$ and a point $x \in \mathbb{R}^2$ as input, and outputs whether x should get label $+1$ or -1.

(ii) Do you notice any problem? If not, then apply k-NN to $x = (112, 1.9)$ and generate a plot where ordinate and abscissa are scaled equally.

(iii) Min-max normalize data and x to $[0, 1]$. Then run k-NN with majority vote again. Explain the (surprising?) result.

Exercise 3.6. We consider the following table of product ratings, where a score between 0 and 10 can be given, and where several ratings are unknown to us.

	Product 1	Product 2	Product 3	Product 4
Customer 1	10	7	□	4
Customer 2	□	□	5	3
Customer 3	3	0	3	6
Customer 4	7	9	□	5
Customer 5	10	9	8	□

(i) What ratings for Product 4 by Customer 5 arise from the method explained in Example 3.12 and which from the method explained in Example 3.19?

(ii) In (i) we employed user-to-user collaborative filtering. How do the results of (i) change if we use item-to-item collaborative filtering instead?

(iii) What are advantages and disadvantages of the user-to-user method and the item-to-item method, respectively?

Exercise 3.7. Show that the cosine distance on $\mathbb{R}^2_{\geqslant 0}$ is neither positive definite nor satisfies the triangle inequality.

Exercise 3.8. Vectorize the texts from Example 3.17 in a programming language of your choice using suitable packages, in Python e.g., `Re`, and verify the cosine similarities noted in 3.17. Also calculate the Euclidean distances. The texts, already processed with `Re`, can be found at `https://mathematicaldatascience.github.io/Kennedy_en.html`.

Exercise 3.9. The following three lists are vectorizations of texts via raw count:

```
T1 = [  12,  0,  4,  1,  0,  1,  12,  10  ]
T2 = [  11,  1,  0,  2,  0,  2,  11,  10  ]
T3 = [  24,  0,  8,  2,  0,  2,  24,  20  ]
```

What is the 1-NN of `T1` with respect to Euclidean distance? Which text is most cosine similar to `T1`?

Exercise 3.10. Let several text documents $D_1, \ldots, D_n$ as in Example 3.17 be given and let W be the numbered set of all words. If D is one of the texts and $w \in W$, we denote by $\operatorname{rc}(w, D)$ the number of occurrences of the word w in the text D. We further define the *term frequency* and the *inverse document frequency* by

$$\operatorname{tf}(w, D) := \frac{\operatorname{rc}(w, D)}{\max\{\operatorname{rc}(v, D) \mid v \in D\}} \quad \text{and} \quad \operatorname{idf}(w, D_1, \ldots, D_n) := \log\Big(\frac{n}{\#\{i \mid w \in D_i\}}\Big).$$

The product of term frequency and inverse document frequency is denoted by

$$\text{tf-idf}(w, D) := \text{tf}(w, D) \cdot \text{idf}(w, D_1, \ldots, D_m),$$

where $D \in \{D_1, \ldots, D_n\}$. Explain the following heuristics for comparing the vectorization of D by $(\text{rc}(w, D))_{w \in W}$, $(\text{tf}(w, D))_{w \in W}$ or $(\text{tf-idf}(w, D))_{w \in W}$:

(i) Compared to the raw count, the term frequency compensates for the fact that in longer texts, all words tend to occur more frequently than in shorter texts.

(ii) By multiplying with the inverse document frequency, one can additionally address the fact that words irrelevant for text comparison, such as "the", "and", "or", etc, occur very frequently in practically every long text.

4

Clustering

In the preceding chapters, we dealt with classifiers and regressors. In doing so, we always started out with a labeled dataset, based on which a predictor for new and unlabeled data points was to be determined. In this chapter, we consider unlabeled datasets $D \subseteq (X, \rho)$, where $\rho\colon X \times X \to \mathbb{R}$ is a distance measure in the sense of Definition 3.14. Our goal is to find a partition, i.e., sets $C_1, \ldots, C_k \subseteq D$ such that

$$D = C_1 \,\dot{\cup}\, \cdots \,\dot{\cup}\, C_k,$$

where, ideally, data points $x, y \in D$ for which $\rho(x, y)$ is small should belong to the same set C_i. If the latter was the case, we are inclined to call the C_i the *clusters* of D. Instead of trying to formalize this idea in an abstract way, let us look at the following very illustrative example (taken from [SSBD14, p. 265]).

Example 4.1. We consider the set D shown in the following picture as a subset of $\mathbb{R}^2$ and equip $\mathbb{R}^2$ with the Euclidean metric ρ.

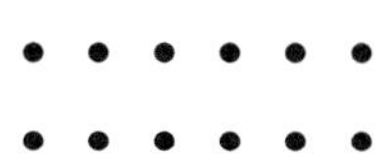

The set D is now to be divided into two clusters, i.e., $X = C_1 \,\dot{\cup}\, C_2$, where points that are close to each other should belong to the same cluster. Given this task, one quickly comes up with two obvious, but indeed quite different, solutions:

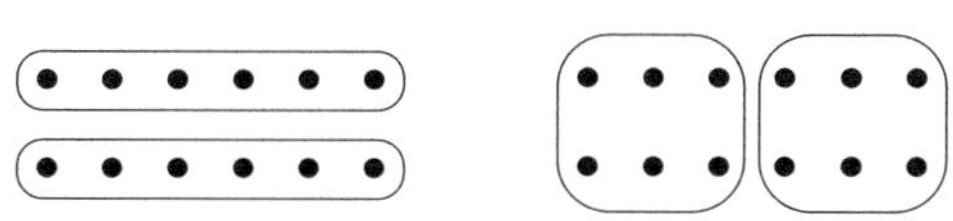

Since the vertical distances between the points are slightly larger than the horizontal distances, it can be seen in the left picture that for each point, its 1-nearest neighbor is in the same cluster. However, this results in distances within the cluster that are much larger than those between clusters. In the

S. A. Wegner, *Mathematical Introduction to Data Science*,
https://doi.org/10.1007/978-3-662-69426-8_4

right picture, the diameters of the clusters are significantly smaller. However, this comes at the cost that there are now points whose 1-nearest neighbor is not in the same cluster, while the 2-nearest is.

Example 4.1 makes it clear that the initial heuristic that "points which are close together should belong to the same cluster" must be in some way restricted. We now present two different ways to do this.

4.1 Linkage-Based Clustering

The first option, which corresponds to the left picture in Example 4.1, is *linkage-based clustering*. For this, we first recall the following definition.

Definition 4.2. Let X be a set and $\rho\colon X \times X \to \mathbb{R}$ be a distance measure. For $A, B \subseteq X$ we define

$$\rho(A,B) = \min_{\substack{a\in A\\ b\in B}} \rho(a,b)$$

and call $\rho(A,B)$ the *distance* between A and B.

The idea behind linkage-based clustering is now very simple. Let a dataset

$$D = \{x_1, \ldots, x_n\} \subseteq (X, \rho)$$

be given. Then we start with the *discrete clustering* $D = \{x_1\} \cup \cdots \cup \{x_n\}$. From this, we step-by-step merge those clusters with minimum distance. This procedure thus defines a sequence of clusterings, until in the end the *trivial clustering* $D = \{x_1, \ldots, x_n\}$ appears. The sequence of clusterings can be represented by a so-called *dendrogram*, which is illustrated below (very similar to [SSBD14, p. 267]).

Example 4.3. Let $X \subseteq \mathbb{R}^d$ be a cuboid, ρ the Euclidean metric on X, and D be the dataset sketched below, consisting of six points.

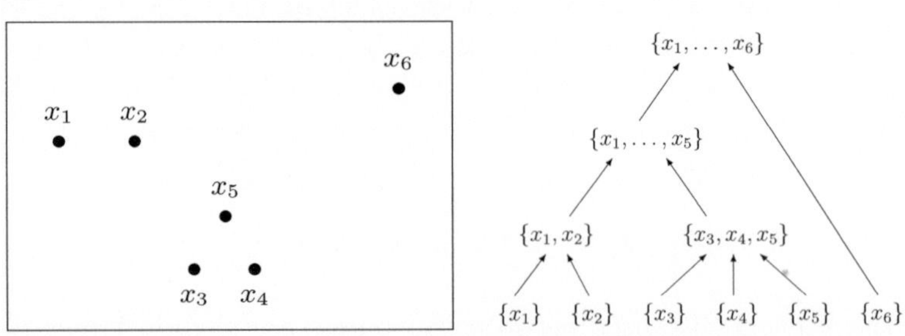

We point out that in the dendrogram, the rows do not correspond to the "rounds" of the algorithm; for example, in the first round we merged $\{x_3, x_4, x_5\}$

while $\{x_1, x_2\}$ got merged only in the round after that. Apart from that, in the case of $\{x_3\}$, $\{x_4\}$, $\{x_5\}$ with pairwise equal distance, one could first merge only two sets and then add the third point only in the next round. Here, one would have to determine in some way which points to start with, compare the explanations after Definition 3.2 for determining the k-nearest neighbors in the case of pairwise equal distances.

If one is not interested in the entire dendrogram, one must stop the aforementioned process at an appropriate point. For this, one can limit the number of rounds in the above explained procedure, or one can limit the number of clusters. A third alternative is to put a bound on the minimum of the pairwise distances between the clusters

$$\min_{i \neq j} \rho(C_i, C_j),$$

which increases in each round. In the following pseudocode, we present this variant and leave it as Exercise 4.1 to formulate a termination criterion via the number of clusters in pseudocode.

Algorithm 4.4. *Let $D \subseteq (X, \rho)$. The following pseudocode outlines linkage-based clustering, where the process is stopped before the minimum distance between the clusters in the next round would exceed a given value $\delta > 0$ for the first time.*

1: **function** LINKAGE-BASED CLUSTERING (X, ρ, D, δ)
2:     $n \leftarrow \#D$
3:     **for** $i \leftarrow 1$ to n **do**
4:         $C_i \leftarrow \{x_i\}$
5:     **while** $\min_{i \neq j} \rho(C_i, C_j) \leqslant \delta$ **do**
6:         $m \leftarrow 0$
7:         $(i^*, j^*) \leftarrow \operatorname{argmin}_{i \neq j} \rho(C_i, C_j)$
8:         **for** $k \leftarrow 1$ to n **do**
9:             **if** $k = \min(i^*, j^*)$ **then**
10:                 $C_k \leftarrow C_{i^*} \cup C_{j^*}$
11:             **if** $k = \max(i^*, j^*)$ **then**
12:                 $m \leftarrow 1$
13:             **else**
14:                 $C_k \leftarrow C_{k+m}$
15:         $n \leftarrow n - 1$
16:     **return** $C_1, \ldots, C_k$

If the argmin in line 7 is not unique, we choose an arbitrary minimizer $(i^*, j^*) \in \operatorname{argmin}_{i \neq j} \rho(C_i, C_j)$. Moreover, we understand $\min_{i \neq j} \rho(C_i, C_j) = \infty$ in the case that there is only one cluster left.

Remark 4.5. Instead of the usual distance between subsets, as in Definition 4.2, one may choose to use any of the following functions, which will of course lead to different clusterings.

(i) $\rho_1(A,B) := \frac{1}{|A||B|} \sum_{\substack{a\in A\\ b\in B}} \rho(a,b)$ (average of all pairwise distances)

(ii) $\rho_2(A,B) := \max_{\substack{a\in A\\ b\in B}} \rho(a,b)$ (maximum of all pairwise distances)

We will return to this in Exercise 4.2.

4.2 Cost-Minimizing Clustering

In contrast to linkage-based clustering, we now consider a *cost function* or *objective function* on the set of all possible clusterings. We assign to partitions $X = C_1 \dot{\cup} \cdots \dot{\cup} C_k$ with undesirable properties, such as too large diameter, high costs, and then arrive at clusterings that have the desired properties through a minimization process.

Let $D \subseteq X$ be a dataset, $\rho\colon X \times X \to \mathbb{R}$ a distance measure, and $k \in \mathbb{N}$. We denote by

$$\mathcal{C}_k := \big\{(C_1,\dots,C_k) \in \mathcal{P}(D)^k \mid D = C_1 \dot{\cup} \cdots \dot{\cup} C_k\big\}$$

the set of all k*-clusterings* of D. If $Z\colon \mathcal{C}_k \to \mathbb{R}$ is a given cost function, we call a minimizer $(C_1,\dots,C_k) \in \operatorname{argmin}_{C\in\mathcal{C}_k} Z(C)$ of Z a *cost-minimizing clustering* of D. The most popular example of such a cost function is the following.

Definition 4.6. Let X be a set and ρ be a distance measure on X such that for every finite set $A \subseteq X$ a *centroid*

$$\mu(A) \in \operatorname*{argmin}_{\mu\in X} \sum_{x\in A} \rho(x,\mu)^2$$

exists in X. Let $D \subseteq X$ be a dataset, let $k \in \mathbb{N}$ and let $\mathcal{C}_k$ be the set of all k-clusterings of D. For $k \in \mathbb{N}$ we call

$$K\colon \mathcal{C} \to \mathbb{R},\ K(C_1,\dots,C_k) := \sum_{i=1}^{k} \sum_{x\in C_i} \rho(x,\mu(C_i))^2$$

the k*-means cost function* on (X,ρ).

Remark 4.7. (i) If $X = \mathbb{R}^d$ is equipped with the Euclidean metric, then $\mu(A)$ exists for every finite $A \subseteq \mathbb{R}^d$, compare Remark 4.10 and the following example (left picture). If $X = \mathbb{R}^d$ is equipped with the norm induced by the $\|\cdot\|_\infty$, centroids always exist, but in general they are not unique (right picture).

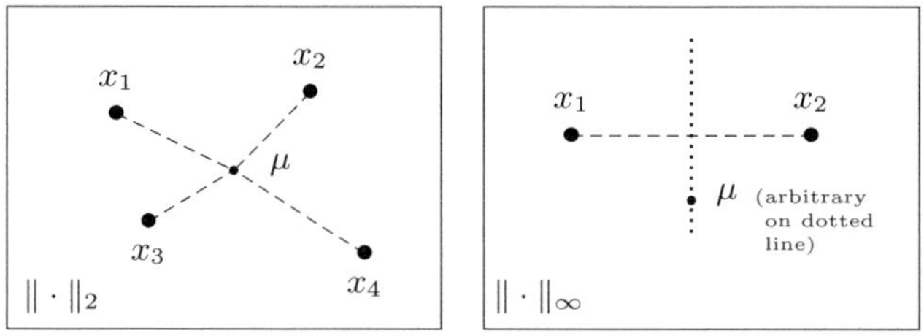

(ii) In the situation of Definition 4.6 we always have

$$K(C_1,\dots,C_k) \;=\; \min_{\mu_1,\dots,\mu_k\in X} \sum_{i=1}^{k}\sum_{x\in C_i} \rho(x,\mu_i)^2.$$

Here, "$\geqslant$" is clear and for "$\leqslant$" one first swaps the minimum and the sum over i and then observes that on the left-hand side we evaluate on a minimizer.

(iii) Given seven points and two different clusterings (C_1, C_2) and (C_1', C_2'), the following picture shows the corresponding centroids with respect to the Euclidean metric.

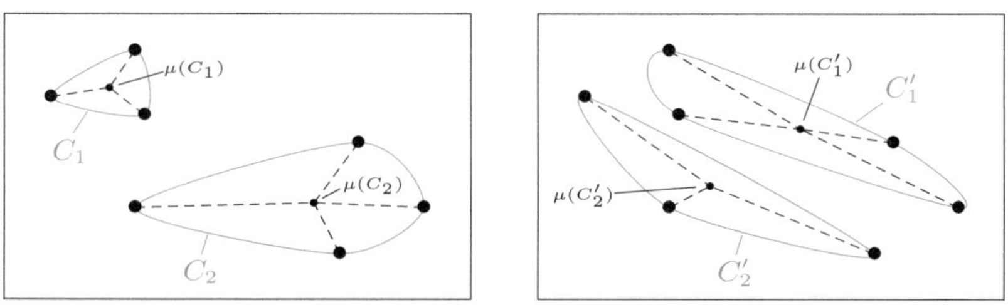

The 2-means cost function K calculates the sum of the squared distances of the data points from the respective centroid of the cluster. In the example above, we immediately see that $K(C_1, C_2) < K(C_1', C_2')$ holds.

To find a minimizer for K, we use the following *k-means algorithm*. We initially select pairwise different $\mu_1,\dots,\mu_k \in X$ (these are often chosen in D to exclude pathological situations). Then we define the clusters by assigning a data point $x \in D$ to the i-th cluster C_i, if x is closest to μ_i; in case this is true for several i, we choose among these. Then we re-calculate the centroids of the clusters and repeat the previous step.

Example 4.8. We start with the following example of a dataset with five elements in $\mathbb{R}^2$ with Euclidean metric and $k = 2$.

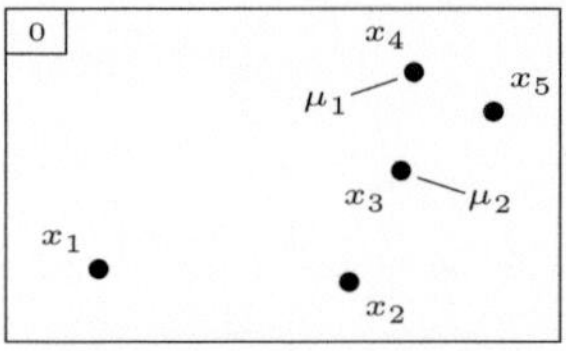

At the beginning, $\mu_1 \neq \mu_2$ is chosen arbitrarily from $\{x_1, \ldots, x_5\}$; in the example above $\mu_1 = x_4$ and $\mu_2 = x_3$.

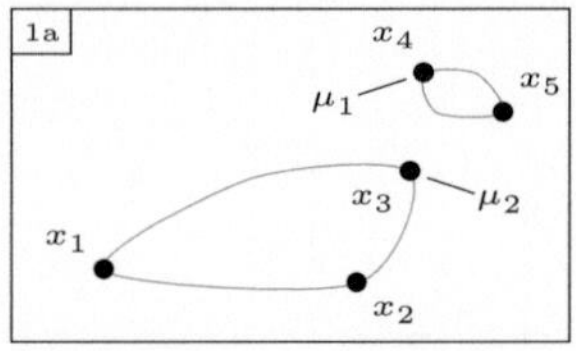

Then those points that are closer to μ_1 than to μ_2 are assigned to the cluster C_1 and those that are closer to μ_2 than to μ_1 are added to the cluster C_2. This yields $C_1 = \{x_4, x_5\}$ and $C_2 = \{x_1, x_2, x_3\}$.

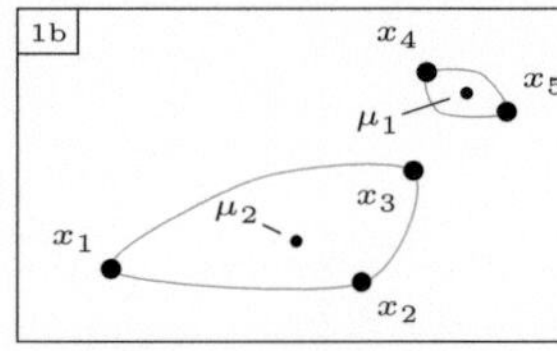

Next, the centroids of C_1 and C_2 are calculated and the μ_1, μ_2 are updated accordingly.

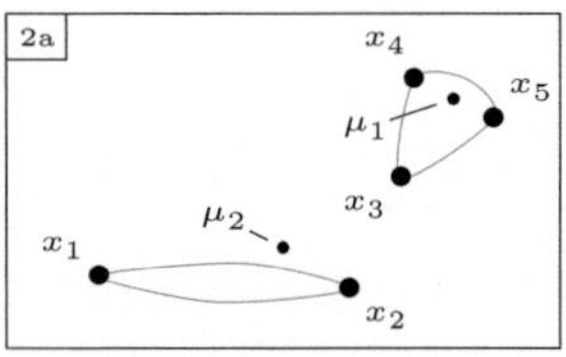

Now the points are re-assigned according to the same principle as before, but with respect to the new centroids. This leads to an update for the clusters $C_1 = \{x_3, x_4, x_5\}$ and $C_2 = \{x_1, x_2\}$.

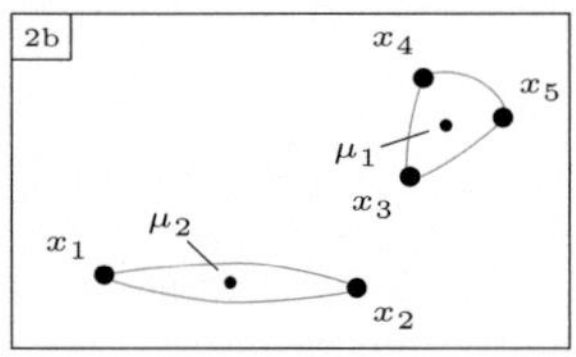

Next the centroids of C_1 and C_2 are calculated again and μ_1, μ_2 are updated once more.

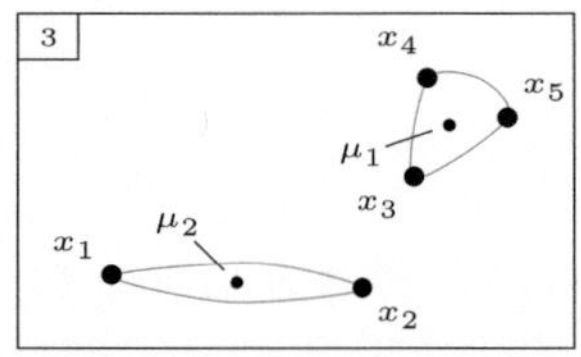

Now it would be time to reorganize the clusters according to the new centroids. We see however that we end up with the same clusters as in the step before. Therefore, we abort the process at this point and take the last clustering $C_1 = \{x_3, x_4, x_5\}$ und $C_2 = \{x_1, x_2\}$ as our final result.

We now formulate the algorithm in pseudocode.

Algorithm 4.9. *Let X be a set and ρ a distance measure on X, such that for every finite set $A \subseteq X$ a centroid $\mu(A) \in X$ exists. Let $D \subseteq X$ be a dataset and let $k \in \mathbb{N}$. The following pseudocode delivers a minimizer of the k-means cost function.*

function K-MEANS (D, k, X, ρ)
    $\mu_1, \ldots, \mu_k \leftarrow$ pairwise different points from X
    for $i \leftarrow 1$ to k **do**
        $C_i \leftarrow \{x \in D \mid i \in \operatorname{argmin}_{j=1,\ldots,k} \rho(x, \mu_j)\}$
    $U \leftarrow$ True
    while $U =$ True **do**
        $U \leftarrow$ False
        for $i \leftarrow 1$ to k **do**
            $C_i' \leftarrow \{x \in D \mid i \in \operatorname{argmin}_{j=1,\ldots,k} \rho(x, \mu_j)\}$
            $\mu_i \leftarrow \mu(C_i)$
            if $C_i' \neq C_i$ **then**
                $C_i \leftarrow C_i'$
                $U \leftarrow$ True
    return $C_1, \ldots, C_k$

Remark 4.10. (i) If $X = \mathbb{R}^d$ and ρ is the Euclidean metric, then the centroid $\mu(A)$ of every finite set $A \subseteq X$ exists, is uniquely determined, and equal to

$$\mu(A) = \frac{1}{|A|} \sum_{a \in A} a,$$

see Exercise 4.5, from whence the term "k-means" stems. The k-means algorithm in this case thus calculates k-many means. Instead of centroids, one can optionally also use the *medoid* or the *geometric median*

$$\mu(A) \in \operatorname*{argmin}_{\mu \in X} \sum_{x \in A} \rho(x,\mu)^2, \quad \text{respectively} \quad \mu(A) \in \operatorname*{argmin}_{\mu \in \mathcal{X}} \sum_{x \in A} \rho(x,\mu).$$

The corresponding algorithms are then called k-medoid or k-median algorithm.

(ii) Cost functions do not have to be based on centroids, medians, or versions thereof; for example, one may also require that the sum of all pairwise distances between points of a cluster is minimized. This would correspond to the cost function

$$Z(C_1,\dots,C_k) = \sum_{i=1}^{k} \sum_{x,y \in C_i} \rho(x,y).$$

From the formulation of Algorithm 4.9 it is not a priori clear that it will terminate, i.e., that from some round, no further update of the clusters occurs. In general, this cannot be expected, compare Remark 4.12. However, the following always holds.

Theorem 4.11. *Let X be a set and ρ a distance measure on X such that for every finite set $A \subseteq X$ a centroid $\mu(A) \in X$ exists. Let $D \subseteq X$ be a dataset and let $k \in \mathbb{N}$. Then the sequence $(K(C_1^{(j)},\dots,C_k^{(j)}))_{j\in\mathbb{N}}$ of evaluations of the k-means cost function on the clusterings produced by the k-means algorithm is monotonically decreasing.*

Proof. Let the algorithm be initialized with $\mu_1^{(0)},\dots,\mu_k^{(0)}$. For $j \geqslant 1$, we then denote by $(C_1^{(j)},\dots,C_k^{(j)})$ the clusters and by $\mu_1^{(j)} = \mu(C_1^{(j)})$, ..., $\mu_k^{(j)} = \mu(C_k^{(j)})$ their centroids in the j-th round of the algorithm. For $j \geqslant 2$ we have

$$\begin{aligned}
K(C_1^{(j)},\dots,C_k^{(j)}) &= \min_{\mu_1,\dots,\mu_k \in \mathcal{X}} \sum_{i=1}^{k} \sum_{x \in C_i^{(j)}} \rho(x,\mu_i)^2 \\
&\leqslant \sum_{i=1}^{k} \sum_{x \in C_i^{(j)}} \rho(x,\mu(C_i^{(j-1)}))^2 \\
&\leqslant \sum_{i=1}^{k} \sum_{x \in C_i^{(j-1)}} \rho(x,\mu(C_i^{(j-1)}))^2 \\
&= K(C_1^{(j-1)},\dots,C_k^{(j-1)}),
\end{aligned}$$

where we have used Definition 4.6 of the k-means cost function and Remark 4.7(ii) for the equalities. The first inequality is a specialization. For the second, we note that in line 9 of Algorithm 4.9, the set $C_i^{(j)}$ is defined in such a way that $\rho(x, \mu(C_i^{(j-1)}))$ is minimal for $x \in C_i^{(j)}$. Thus, the sum over the previous distances can only increase for any other clustering.

We advise the reader to look again at Example 4.8 for $j = 2$ to illustrate the last argument: There, Picture 2b corresponds to the penultimate line in the estimate and Picture 1b to the line above that. We then have

$$\sum_{j=3}^{5} \rho(x_j, \mu_1)^2 + \sum_{j=1}^{2} \rho(x_j, \mu_2)^2 \leqslant \sum_{j=4}^{5} \rho(x_j, \mu_1)^2 + \sum_{j=1}^{3} \rho(x_j, \mu_2)^2,$$

as we have just moved point x_3 from cluster C_2 in Figure 1b to cluster C_2 in Figure 2a, due to $\rho(x_3, \mu_1) < \rho(x_3, \mu_2)$. □

Remark 4.12. (i) Theorem 4.11 tells us that we cannot make things worse by running K-MEANS for a longer time, but it does not guarantee that the sequence of clusterings will eventually become stationary, see Exercise 4.7. To address this, one should limit the number of iterations of the while loop in addition to the termination condition in Algorithm 4.9.

(ii) It could happen that the sequence of clusterings $C_1^{(j)}, \ldots, C_k^{(j)}$ becomes stationary from a j_0 onwards, without $C_1^{(j_0)}, \ldots, C_k^{(j_0)}$ being a minimizer for the k-means objective function, compare Exercise 4.6.

(iii) If the initial values for the μ_i's are chosen unfavorably, it can happen that one or more C_i become empty. Therefore, it is recommended to run the k-means algorithm several times with different randomly chosen initializations for a given dataset and then to compare the results.

References

The clustering algorithms covered in this chapter are standard. Above, we have closely followed the excellent presentation [SSBD14, Chapter 22.1–22.2], but have added some details and examples. In particular, Exercise 4.6 matches [SSBD14, Exercise 22.2].

Exercises

Exercise 4.1. Outline a version of linkage-based clustering in pseudocode, where one can preselect the number of clusters, and then a clustering is computed that has precisely the specified number of clusters. Discuss possible pathologies that may occur.

Exercise 4.2. Perform linkage-based clustering twice for the following dataset and draw the dendrograms. Use the Euclidean metric ρ on $\mathbb{R}^2$ both times, but go once for the usual distance $\rho(A,B) = \min_{a\in A, b\in B} \rho(a,b)$ of subsets A, B, and then for the alternative distance $\rho_2(A,B) = \max_{a\in A, b\in B} \rho(a,b)$.

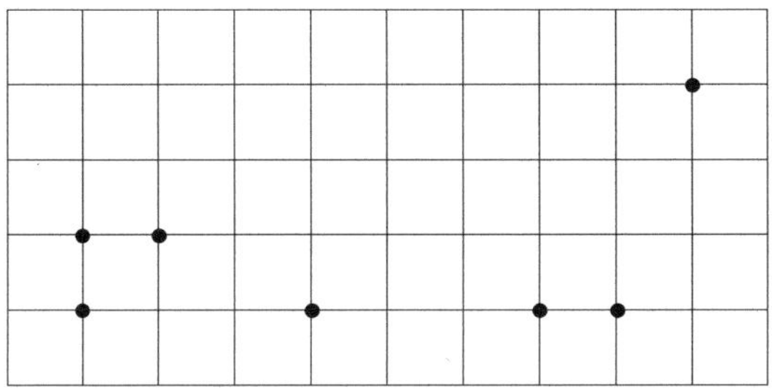

Exercise 4.3. Perform the 2-means algorithm on the following dataset. Use the Euclidean metric on $\mathbb{R}^2$ and initialize the centroids as $\mu_1 = x_1$ and $\mu_2 = x_6$. In each step indicate the clusterings (C_1, C_2) and the centroids μ_1 and μ_2 (rough estimates are sufficient!).

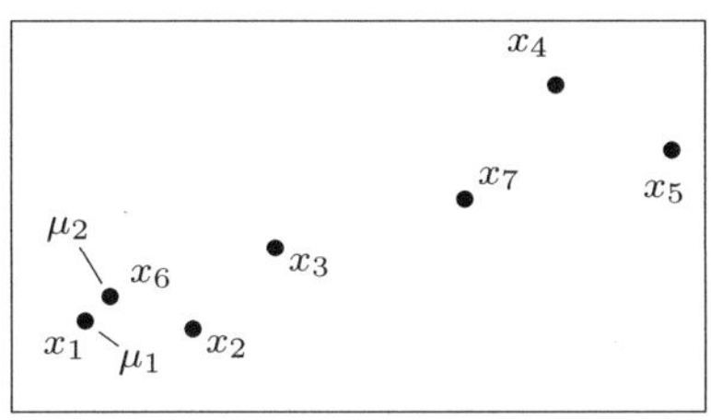

Exercise 4.4. Implement the k-means algorithm and use it to cluster the dataset given at `https://mathematicaldatascience.github.io/k-means_en.html` from the `sklearn` package: In addition to the existing list `L` of real labels, create a new list `N` with the labels determined by the algorithm. Compare the existing plot with a new plot in which the colors of the data points are determined by `N`. Try different k's and pay particular attention to the orange point in the purple cluster.

Exercise 4.5. Let $A \subseteq \mathbb{R}^d$ and denote by $\rho\colon \mathbb{R}^d \times \mathbb{R}^d \to \mathbb{R}$ the Euclidean metric. Show that

$$\tfrac{1}{|A|} \sum_{a\in A} a = \operatorname*{argmin}_{\mu\in\mathbb{R}^d} \sum_{a\in A} \rho(a,\mu)^2$$

holds for every finite set $A \subseteq \mathbb{R}^d$. In particular, the centroid is unique and consistent with the definition usually given in linear algebra courses.

Exercise 4.6. Let the dataset $\{1,2,3,4\} \subseteq \mathbb{R}$ be given, on which we apply the 2-means algorithm. We assume that we start with the centroids $\mu_1 = 2$ and $\mu_2 = 4$ and that, if $\operatorname{argmin}_{i=1,2} \|x - \mu_i\|$ is not a singleton, $i = 1$ is always chosen as the minimizer. Show that the sequence of clusterings $(C_1^{(j)}, C_2^{(j)})_{j\in\mathbb{N}}$ becomes stationary without that the stationary clustering minimizes the k-means cost function.

Exercise 4.7. Let

$$D = \{(-1,-1),(0,-1),(1,-1),(-1,1),(0,1),(1,1),(0,-0.5),(0,0.5)\} \subseteq \mathbb{R}^2.$$

We apply the 2-means algorithm to D, but this time with random initial values μ_1 and μ_2 and random selection of $i \in \operatorname{argmin}_{j=1,2} \|x - \mu_j\|$ in case the latter set contains more than one point.

(i) Show that after a finite number of iterations, the means $\mu_1 = (-0.5, 0)$ and $\mu_2 = (0.5, 0)$ *could* emerge.

(ii) Show, starting from (i), that it *could* happen that the sequence of clusterings $(C_1^{(j)}, C_2^{(j)})_{j \in \mathbb{N}}$ does not become stationary.

(iii) How likely is (ii)?

5

Graph Clustering

In Chapter 4, datasets D always came along as subsets of a space (X, ρ), where ρ was a distance measure, and we always assumed that it is possible to calculate the distance $\rho(x, y)$ between two data points directly, i.e., without any knowledge of other data points in D. In this chapter, we start from a completely different point, namely with datasets given by a so-called *graph*, see also Example 1.4(i).

Definition 5.1. A *graph* $G = (V, E)$ is a pair consisting of a finite set V, of so-called *vertices* or *nodes*, and a subset $E \subseteq \big\{\{v, w\} \,|\, v, w \in V,\, v \neq w\big\}$, whose elements we call *edges*.

Graphs as in Definition 5.1 are *undirected*, have no *loops* ( ) and no *multiple edges* ( ). Note that graphs are sometimes defined differently in the literature.

Example 5.2. The following example shows a graph with nineteen vertices. The edges are indicated by connecting lines.

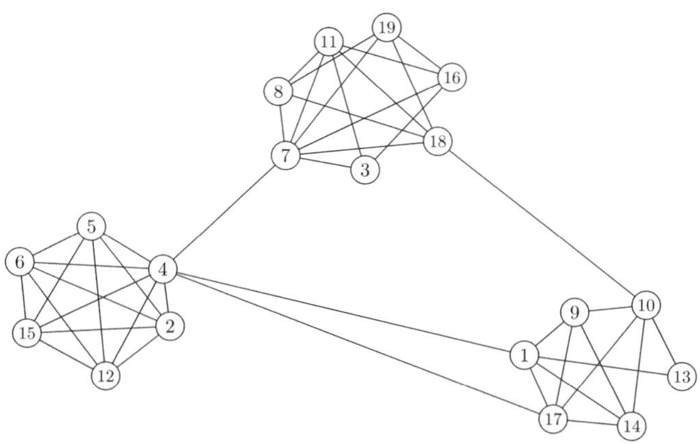

We consider the three subsets $C_1 = \{2, 4, 5, 6, 12, 15\}$, $C_2 = \{1, 9, 10, 13, 14, 17\}$ and $C_3 = \{3, 7, 8, 11, 16, 18, 19\}$ of V. Then one can immediately see in the

S. A. Wegner, *Mathematical Introduction to Data Science*,
https://doi.org/10.1007/978-3-662-69426-8_5

picture that each vertex has fewer edges connecting it to another of the subsets than it has edges to points within "its own" subset. It is therefore very natural to call C_1, C_2, C_3 the *clusters of the graph.*

Graphs that occur in applications are for example social networks. The users then correspond to the vertices and the edges to "being friends". Such very large graphs are then of course not given by a picture as above, but come as a list that specifies all users and all friendships. Formally, we introduce the following concept.

Definition 5.3. Let $G = (V, E)$ be a graph with $V = \{1, \dots, n\}$. Then

$$A = (a_{ij})_{i,j\in\mathbb{N}} \text{ with } a_{ij} = \begin{cases} 1, & \text{if } \{i,j\} \in E, \\ 0, & \text{otherwise} \end{cases}$$

is called the *adjacency matrix* of G and

$$L = (\ell_{ij})_{i,j\in\mathbb{N}} \text{ with } \ell_{ij} = \begin{cases} -1, & \text{if } \{i,j\} \in E, \\ \deg(i), & \text{if } i = j, \\ 0, & \text{otherwise} \end{cases}$$

is called the *Laplace matrix* or the *Laplacian* of G. Here, $\deg(i) = \#\{e \in E \mid \exists\, j \in V \colon \{i,j\} = e\}$ is the *degree* of $i \in V$, i.e., the number of edges that are *incident* with vertex i (e.g., the node  has degree 5).

We note the following simple properties.

Remark 5.4. (i) The matrices A and L are symmetric.

(ii) The i-th row sum/column sum of A is equal to $\deg(i)$.

(iii) The i-th row sum/column sum of L is equal to 0.

(iv) We have $L = D - A$ where $D = \operatorname{diag}(\deg(1), \dots, \deg(n))$.

Example 5.5. As a simple example, we consider the so-called complete graph with three vertices

$G =$ 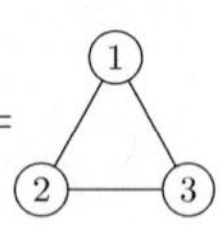

$$\text{with } A = \begin{bmatrix} 0 & 1 & 1 \\ 1 & 0 & 1 \\ 1 & 1 & 0 \end{bmatrix} \text{ and } L = \begin{bmatrix} 2 & -1 & -1 \\ -1 & 2 & -1 \\ -1 & -1 & 2 \end{bmatrix}.$$

We note the following property of Laplace matrices.

Proposition 5.6. *For every graph G, $\mathbb{1} = \begin{bmatrix} 1 \\ \vdots \\ 1 \end{bmatrix}$ is an eigenvector of the Laplacian with eigenvalue zero.*

Proof. $L\mathbb{1}$ has exactly the row sums of L as entries. Since all of these are zero, the equation $L\mathbb{1} = 0\mathbb{1}$ follows. □

From linear algebra, we know that for a symmetric matrix $M \in \mathbb{R}^{n \times n}$ all eigenvalues are real and that for each eigenvalue $\lambda \in \sigma(M)$ the geometric multiplicity ($= \dim \ker(M - \lambda)$) is equal to the algebraic multiplicity ($=$ number of times $x - \lambda$ occurs as a factor of the characteristic polynomial $\det(M - x)$). Counted with multiplicities, we have eigenvalues $\lambda_1 \leqslant \lambda_2 \leqslant \cdots \leqslant \lambda_n$ and can find an orthonormal basis of corresponding eigenvectors $v_1, \ldots, v_n$. We will now apply this to the Laplace matrices of graphs.

We start with two examples.

Example 5.7. We stick with the graph from Example 5.5 and the Laplace matrix noted there. From Proposition 5.6 we get the eigenvalue 0 with corresponding eigenvector $\mathbb{1}$. Calculation of the remaining eigenvalues yields, with multiplicities, $\lambda_1 = 0$, $\lambda_2 = 3$, $\lambda_3 = 3$ and a corresponding orthonormal basis

$$v_1 = \frac{1}{\sqrt{3}} \begin{bmatrix} 1 \\ 1 \\ 1 \end{bmatrix}, \quad v_2 = \frac{1}{\sqrt{6}} \begin{bmatrix} 1 \\ 1 \\ -2 \end{bmatrix}, \quad v_3 = \frac{1}{\sqrt{2}} \begin{bmatrix} -1 \\ 1 \\ 0 \end{bmatrix}.$$

Example 5.8. Next, we consider the following graph with six nodes

$G =$ 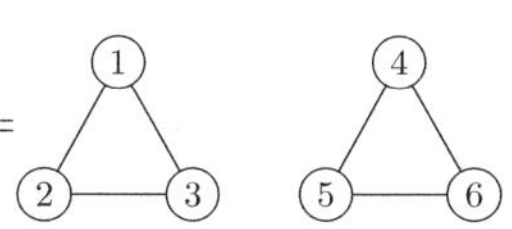

and get

$$L = \begin{bmatrix} 2 & -1 & -1 & & & \\ -1 & 2 & -1 & & 0 & \\ -1 & -1 & 2 & & & \\ & & & 2 & -1 & -1 \\ & 0 & & -1 & 2 & -1 \\ & & & -1 & -1 & 2 \end{bmatrix}.$$

From Example 5.7 we obtain the eigenvalues $\lambda_1 = 0$, $\lambda_2 = 0$, $\lambda_3 = 3$, $\lambda_4 = 3$, $\lambda_5 = 3$, $\lambda_6 = 3$ and eigenvectors

$$v_1 = \frac{1}{\sqrt{3}} \begin{bmatrix} 1 \\ 1 \\ 1 \\ 0 \\ 0 \\ 0 \end{bmatrix}, \quad v_2 = \frac{1}{\sqrt{3}} \begin{bmatrix} 0 \\ 0 \\ 0 \\ 1 \\ 1 \\ 1 \end{bmatrix}, \quad v_3 = \frac{1}{\sqrt{6}} \begin{bmatrix} 1 \\ 1 \\ -2 \\ 0 \\ 0 \\ 0 \end{bmatrix},$$

$$v_4 = \frac{1}{\sqrt{6}} \begin{bmatrix} 0 \\ 0 \\ 0 \\ 1 \\ 1 \\ -2 \end{bmatrix}, \quad v_5 = \frac{1}{\sqrt{2}} \begin{bmatrix} -1 \\ -1 \\ 0 \\ 0 \\ 0 \\ 0 \end{bmatrix}, \quad v_6 = \frac{1}{\sqrt{2}} \begin{bmatrix} 0 \\ 0 \\ 0 \\ -1 \\ -1 \\ 0 \end{bmatrix}.$$

Based on these examples, it is natural to conjecture that the Laplacian of any graph only has non-negative eigenvalues, and that furthermore the smallest eigenvalue λ_1 is always equal to zero. The second smallest eigenvalue λ_2 can be zero or non-zero and whether it is or is not seems to tell us something about the clusters: In Example 5.7, $\lambda_2 \neq 0$ and there are no clusters, in Example 5.8, $\lambda_2 = 0$ and there are apparently two clusters.

To prove that the above is actually true in general, we decompose the Laplacian into a sum of Laplacians of very simple graphs.

Definition 5.9. If $G = (V, E)$ is the graph with vertices $V = \{1, \ldots, n\}$ and precisely one edge $E = \{e\}$, we denote its Laplace matrix with L_e, for example:

$$L_{\{1,2\}} = \begin{bmatrix} 1 & -1 & 0 & \cdots & 0 \\ -1 & 1 & 0 & \cdots & 0 \\ 0 & \cdots & & \cdots & 0 \\ \vdots & & & & \vdots \\ 0 & \cdots & & \cdots & 0 \end{bmatrix}, \quad L_{\{1,3\}} = \begin{bmatrix} 1 & 0 & -1 & 0 & \cdots & 0 \\ 0 & 0 & 0 & 0 & \cdots & 0 \\ -1 & 0 & 1 & 0 & \cdots & 0 \\ 0 & \cdots & & & \cdots & 0 \\ \vdots & & & & & \vdots \\ 0 & \cdots & & & \cdots & 0 \end{bmatrix}.$$

Lemma 5.10. *Let G be a graph with Laplacian L. Then the following holds:*

(i) $L = \sum\limits_{e \in E} L_e$,

(ii) $\forall\, x \in \mathbb{R}^n \colon \langle x, Lx \rangle = \sum\limits_{\{i,j\} \in E} (x_i - x_j)^2 \geqslant 0$,

(iii) All eigenvalues of L are greater than or equal to zero.

Proof. (i) This is one of the rare cases where it is legitimate to do a "proof by example", as one can immediately see from the following that the same arguments also work for any other graph:

$L($ 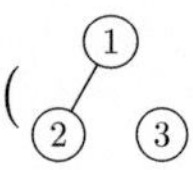 $) + L($ 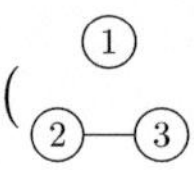 $) + L($ 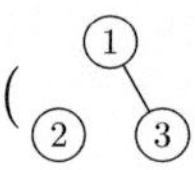 $) = \begin{bmatrix} 1 & -1 & 0 \\ -1 & 1 & 0 \\ 0 & 0 & 0 \end{bmatrix} + \begin{bmatrix} 0 & 0 & 0 \\ 0 & 1 & -1 \\ 0 & -1 & 1 \end{bmatrix} + \begin{bmatrix} 1 & 0 & -1 \\ 0 & 0 & 0 \\ -1 & 0 & 1 \end{bmatrix}$

$= \begin{bmatrix} 2 & -1 & -1 \\ -1 & 2 & -1 \\ -1 & -1 & 2 \end{bmatrix} = L($ 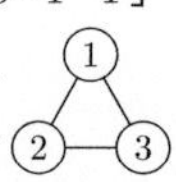 $).$

(ii) We assume without loss of generality that $V = \{1, \ldots, n\}$ holds. Using (i) and the definition of $L_{\{i,j\}}$ we get

$$\begin{aligned} \langle x, Lx \rangle &= \Big\langle x, \big(\sum_{e \in E} L_e\big) x \Big\rangle = \sum_{e \in E} \langle x, L_e x \rangle \\ &= \sum_{\{i,j\} \in E} \left\langle \begin{bmatrix} x_1 \\ \vdots \\ x_n \end{bmatrix}, \begin{bmatrix} 1 & -1 \\ -1 & 1 \end{bmatrix} \begin{bmatrix} x_1 \\ \vdots \\ x_n \end{bmatrix} \right\rangle \\ &= \sum_{\{i,j\} \in E} \left\langle \begin{bmatrix} x_1 \\ \vdots \\ x_n \end{bmatrix}, \begin{bmatrix} x_i - x_j \\ -x_i + x_j \end{bmatrix} \right\rangle \\ &= \sum_{\{i,j\} \in E} x_i(x_i - x_j) + x_j(-x_i + x_j) \\ &= \sum_{\{i,j\} \in E} (x_i - x_j)^2, \end{aligned}$$

where in the second row the entries indicated in the matrix are at the positions (i, i), (i, j), (j, i) and (j, j) and all other entries are zero.

(iii) Due to (ii), L is positive semidefinite and it follows that all eigenvalues are greater than or equal to zero. □

In the following, we denote by $0 = \lambda_1 \leqslant \lambda_2 \leqslant \cdots$ the eigenvalues of the Laplacian when a graph G is given and sometimes also call these the *eigenvalues of the graph*.

Definition 5.11. Let $G = (V, E)$ be a graph.

(i) We say that G is *connected* if for any $i \neq j \in V$ there are edges $\{i, k_1\}, \{k_1, k_2\}, \ldots, \{k_\ell, j\}$ that connect i and j.
(ii) A subset $V' \subseteq V$ is called a *connected component*, if $G' = (V', E')$ with $E' = \big\{\{i,j\} \in E \,|\, i, j \in V'\big\}$ is connected and maximal with this property under inclusion (that is, for all $V' \subseteq V'' \subseteq V$ with G'' connected, it follows $G' = G''$).

Obviously, belonging to a connected component is an equivalence relation and the connected components form a partition of V.

Theorem 5.12. *Let G be a graph. The number of connected components coincides with the multiplicity of the eigenvalue zero ($=\dim\ker L$) of the Laplacian. In particular, G is connected if and only if $\lambda_2 > 0$.*

Proof. "$\leqslant$" Analogous to Example 5.8, we obtain for each connected component C_i the eigenvector $\mathbb{1}_{C_i} = (\mathbb{1}_{C_i}(j))_{j=1,\ldots,n}$ corresponding to the eigenvalue zero, and these are orthogonal.

"$\geqslant$" Suppose, "$<$" holds. Let us say there are k-many connected components, but $\dim\ker L \geqslant k+1$. Then we can extend the linearly independent eigenvectors $v_1 = \mathbb{1}_{C_1}, \ldots, v_k = \mathbb{1}_{C_k}$ from above by $v_{k+1} \in \ker L$ to a linearly independent system $\{v_1, \ldots, v_{k+1}\}$. It follows that $Lv_{k+1} = 0$ and therefore

$$0 = \langle v_{k+1}, Lv_{k+1}\rangle \underset{\substack{\uparrow \\ \text{Lemma} \\ 5.10}}{=} \sum_{\{i,j\}\in E} \big((v_{k+1})_i - (v_{k+1})_j\big)^2.$$

Thus, $(v_{k+1})_i - (v_{k+1})_j = 0$ holds for all edges $\{i,j\}$. Thus, the mapping $V \to \mathbb{R}$, $i \mapsto (v_{k+1})_i$, is constant on connected components $C_1, \ldots, C_k \subseteq V$. Let the values there be $\alpha_1, \ldots, \alpha_k$. Then,

$$v_{k+1} = \alpha_1 \mathbb{1}_{C_1} + \cdots + \alpha_k \mathbb{1}_{C_k} = \alpha_1 v_1 + \cdots + \alpha_k v_k$$

contradicts the fact that $\{v_1, \ldots, v_{k+1}\}$ is linearly independent. □

With Theorem 5.12 we can identify connected components, which are in some sense the "best" but also the "trivial" clusters: Since we know their number, we can start somewhere and find all nodes that belong to the connected component by moving away from the starting vertex along edges, for example, by breadth-first search.

Clearly, the case where G is connected but still has clusters, e.g., as in Example 5.2, is more interesting. For this, we need the following tool from linear algebra.

Theorem 5.13. (Courant-Fischer formula) *Let $M \in \mathbb{R}^{n\times n}$ be symmetric, i.e., we have real eigenvalues $\lambda_1 \leqslant \lambda_2 \leqslant \cdots \leqslant \lambda_n$ with multiplicities and can choose an orthogonal system $\{v_1, \ldots, v_n\}$ of corresponding eigenvectors. Then for $k = 1, \ldots, n$*

$$\lambda_k = \min_{\substack{x\neq 0\\ \langle x,v_i\rangle=0\\ \textit{for } i=1,\ldots,k-1}} \frac{\langle x,Mx\rangle}{\langle x,x\rangle} \quad \textit{and} \quad v_k \in \operatorname*{argmin}_{\substack{x\neq 0\\ \langle x,v_i\rangle=0\\ \textit{for } i=1,\ldots,k-1}} \frac{\langle x,Mx\rangle}{\langle x,x\rangle}$$

hold, where we understand the scalar product condition as empty if $k = 1$.

Proof. Since $\langle x, v_i\rangle = 0$ holds if and only if $\langle x, v_i/\|v_i\|\rangle = 0$, we may assume without loss of generality that $\{v_1, \ldots, v_n\}$ is an orthonormal basis. Let $0 \neq x \in \mathbb{R}^n$ be arbitrary. Using the orthonormal basis, we write x as $x = \alpha_1 v_1 + \cdots + \alpha_n v_n$ with $\langle x, v_i\rangle = \langle \alpha_1 v_1 + \cdots + \alpha_n v_n, v_i\rangle = \alpha_1\langle v_1, v_i\rangle + \cdots + \alpha_n\langle v_n, v_i\rangle = \alpha_i$ for $i = 1, \ldots, n$. Let k be fixed. For $x \neq 0$ with $\langle x, v_i\rangle = 0$ for $i = 1, \ldots, k-1$ as in the theorem, we have

$$\begin{aligned}
\langle x, Mx\rangle &= \Big\langle \sum_{i=1}^n \alpha_i v_i, M\Big(\sum_{j=1}^n \alpha_j v_j\Big)\Big\rangle = \sum_{i,j=1}^n \alpha_i\alpha_j\langle v_i, Mv_j\rangle \\
&= \sum_{i,j=1}^n \alpha_i\alpha_j\langle v_i, \lambda_j v_j\rangle = \sum_{i,j=1}^n \alpha_i\alpha_j\lambda_j\langle v_i, v_j\rangle \\
&= \sum_{i=1}^n \alpha_i^2\lambda_i \underset{(*)}{\overset{}{=}} \sum_{i=k}^n \lambda_i\alpha_i^2 \geqslant \lambda_k \sum_{i=k}^n \alpha_i^2 \underset{(*)}{=} \lambda_k\sum_{i=1}^n \alpha_i^2,
\end{aligned}$$

where we used for $(*)$ that $\alpha_i = \langle x, v_i\rangle = 0$ for $i = 1, \ldots, k-1$. The estimate in between holds as the λ_i are sorted in ascending order. Analogously, we see

$$\langle x, x\rangle = \sum_{i=1}^n \alpha_i^2$$

and it thus follows $\frac{\langle x, Mx\rangle}{\langle x, x\rangle} \geqslant \lambda_k$ for all x that we considered above. That is, we have

$$\min_{\substack{x\neq 0\\ \langle x,v_i\rangle=0\\ \text{for } i=1,\ldots,k-1}} \frac{\langle x, Mx\rangle}{\langle x, x\rangle} \geqslant \lambda_k.$$

On the other hand, $\frac{\langle v_k, Mv_k\rangle}{\langle v_k, v_k\rangle} = \frac{\langle v_k, \lambda_k v_k\rangle}{\langle v_k, v_k\rangle} = \lambda_k$. It follows that above the equality holds and that v_k is a minimizer. □

Remark 5.14. (i) The fraction $\frac{\langle x, Mx\rangle}{\langle x,x\rangle}$ is often called the *Rayleigh quotient.*

(ii) Theorem 5.13 can be generalized as follows: For symmetric $M \in \mathbb{R}^{n\times n}$ with eigenvalues $\lambda_1 \leqslant \lambda_2 \leqslant \cdots \leqslant \lambda_n$ we have

$$\lambda_k = \max_{\substack{U\subseteq\mathbb{R}^n\\ \dim U=n-k+1}} \min_{\substack{x\in U\\ x\neq 0}} \frac{\langle x, Mx\rangle}{\langle x,x\rangle} = \min_{\substack{U\subseteq\mathbb{R}^n\\ \dim U=k}} \max_{\substack{x\in U\\ x\neq 0}} \frac{\langle x, Mx\rangle}{\langle x,x\rangle},$$

where the maximum respectively minimum is taken over all subspaces $U \subseteq \mathbb{R}^n$ of the given dimension. In fact, Theorem 5.13 shows the estimate "$\leqslant$" in the first equation. For the other direction, we consider for U with $\dim U = n-k+1$ the intersection $U \cap \operatorname{span}\{v_k, \ldots, v_n\}$ and choose $x = \alpha_1 v_1 + \cdots + \alpha_n v_n \neq 0$ in the latter, where the v_i form a basis of eigenvectors corresponding to the λ_i's. Then it follows

$$\frac{\langle x, Mx\rangle}{\langle x,x\rangle} = \frac{\sum_{i=k}^{n} \lambda_i \alpha_i^2}{\sum_{i=k}^{n} \alpha_i^2} \geqslant \frac{\lambda_k \sum_{i=k}^{n} \alpha_i^2}{\sum_{i=k}^{n} \alpha_i^2} = \lambda_k,$$

since the λ_i's are increasing. As $x \in U$ was arbitrary, $\max_{0\neq x\in U} \frac{\langle x, Mx\rangle}{\langle x,x\rangle} \geqslant \lambda_k$, and since U was arbitrary of the given dimension, we conclude

$$\max_{\substack{U\subseteq\mathbb{R}^n\\ \dim U=n-k+1}} \min_{\substack{x\in U\\ x\neq 0}} \frac{\langle x, Mx\rangle}{\langle x,x\rangle} \geqslant \lambda_k.$$

For the second equation, it is enough to apply the first one to $-M$.

(iii) Instead of minimizing/maximizing $\frac{\langle x, Mx\rangle}{\langle x,x\rangle}$ over all $x \neq 0$, one can also minimize/maximize only the numerator $\langle x, Mx\rangle$ over the unit sphere.

Applied to the Laplacian of a graph, we obtain the following corollary of the Courant-Fischer theorem.

Corollary 5.15. *Let G be a graph, let L be its Laplacian, and let $0 = \lambda_1 \leqslant \lambda_2 \leqslant \cdots \leqslant \lambda_n$ be its eigenvalues counted with multiplicities. We chose the eigenvalue $v_1 = \mathbb{1}$ corresponding to λ_1. Then the following holds for the second smallest eigenvalue λ_2 and any corresponding eigenvector v_2:*

$$\lambda_2 \underset{\substack{\text{Thm}\\ 5.13}}{\overset{\uparrow}{=}} \min_{\substack{x\neq 0\\ \langle x,\mathbb{1}\rangle=0}} \frac{\langle x, Lx\rangle}{\langle x,x\rangle} = \min_{\substack{\|x\|=1\\ \langle x,\mathbb{1}\rangle=0}} \langle x, Lx\rangle \underset{\substack{\text{Lem}\\ 5.10}}{\overset{\uparrow}{=}} \min_{\substack{\|x\|=1\\ \langle x,\mathbb{1}\rangle=0}} \sum_{\{i,j\}\in E} (x_i - x_j)^2$$

and

$$v_2 \in \operatorname*{argmin}_{\substack{\|x\|=1\\ \langle x,\mathbb{1}\rangle=0}} \langle x, Lx\rangle = \operatorname*{argmin}_{\substack{\|x\|=1\\ \langle x,\mathbb{1}\rangle=0}} \sum_{\{i,j\}\in E} (x_i - x_j)^2.$$ □

Corollary 5.15 leads to the following heuristic: Let G be a graph and let $v_2 = (x_1, \dots, x_n)$ be an eigenvector corresponding to the Laplacian of G's second smallest eigenvalue. We consider the vertices of G as points in $\mathbb{R}$ and use x_1 as the coordinate of vertex 1, x_2 as the coordinate of vertex 2, etc.:

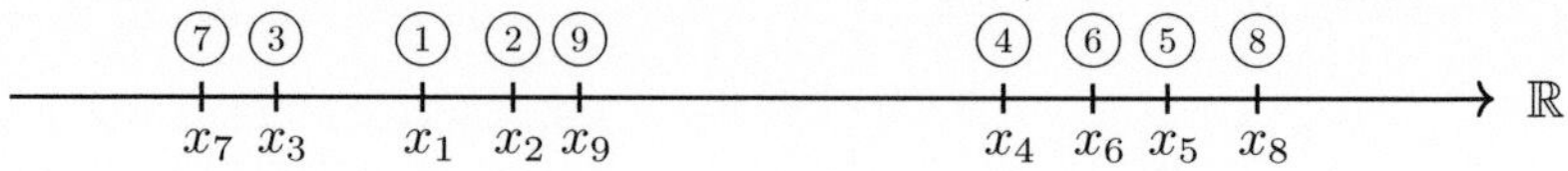

Then the equations in Corollary 5.15 mean that precisely there where the x_i's accumulate must be a cluster:

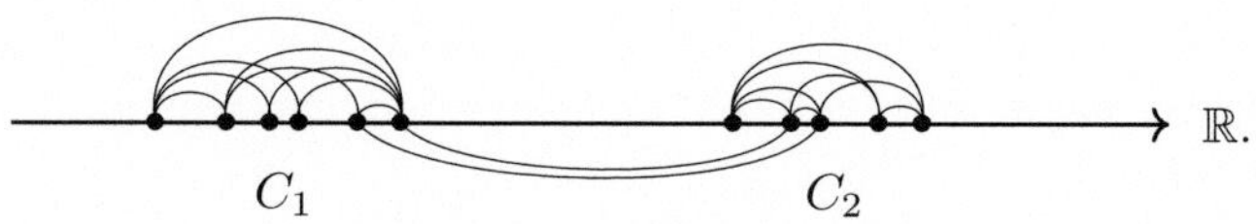

Indeed, if we would move a vertex from C_1 to C_2, we would get several more "long edges", and thus increase the value of $\sum_{\{i,j\}\in E}(x_i - x_j)^2$. On the other hand, C_1 and C_2 can also not be pushed closer together due to the condition $\|v_2\| = 1$. If $\lambda_2 \approx 0$, this means that an arrangement is possible where there are few connections from C_1 to C_2. This is then referred to as a *bottleneck*. Now let us formalize the above.

Definition 5.16. Let $G = (V, E)$ be a graph with $\deg(e) > 0$ for all $e \in E$.

(i) For $S \subseteq V$ we define its *volume* and *boundary* by

$$\operatorname{vol}(S) = \sum_{v \in S} \deg(v) \quad \text{and} \quad \partial S = \big\{\{i,j\} \in E \mid i \in S,\, j \in S^{c}\big\}.$$

(ii) For $\emptyset \neq S \subset V$ we define the *conductance* by

$$\phi(S) = \frac{\#\partial S}{\min(\operatorname{vol} S, \operatorname{vol} S^{c})}.$$

(iii) Finally, we define the *Cheeger constant* by

$$\operatorname{Cheeg}(G) := \min_{\emptyset \neq S \subset V} \phi(S).$$

Let $G = (V, E)$ be a graph. Then $\operatorname{Cheeg}(G)$ is small if there exists $S \subset V$ such that on the one hand $\#\partial S$ is small, i.e., in a decomposition of $V = S \cup S^{c}$ into two clusters, there are not too many edges between the clusters. On the other hand, $\min(\operatorname{vol} S, \operatorname{vol} S^{c})$ must be large; so if we would choose S very small in order to keep $\#\partial S$ small, the denominator of $\phi(S)$ would make the fraction large again.

One may consider the minimization of ϕ as looking for the *best cut through* G, where (a) as few edges as possible should be cut and at the same time (b) both resulting pieces should contain as many edges as possible.

As an instructive example, let us consider graphs consisting of two *cliques*. We compare two such graphs in the pictures below: In the graph on the left the two cliques are connected by one edge and on the right by three edges.

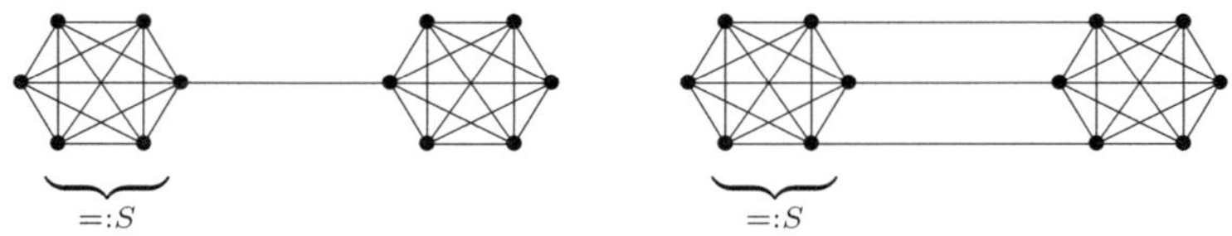

By counting, we see that (in the left graph)

$$\phi(S) = \tfrac{1}{5\cdot 5+1\cdot 6} \approx 0.03$$

holds and that any other choice of S leads to a larger fraction. The optimal cut is therefore the one that separates the two cliques and the Cheeger constant of the left graph is therefore 0.03. In the right graph, we get analogously

$$\operatorname{Cheeg}(G) = \phi(S) = \tfrac{3}{3\cdot 5+3\cdot 6} \approx 0.09.$$

The Cheeger constant is therefore three times smaller on the left. One may say that the "bottleneckedness" of the left graph is higher.

If we now want to find the clusters of a graph given by its adjacency matrix, we have to successively make the best possible cuts $S\cup S^{\mathrm{c}}$ as long as the Cheeger constant is small — and stop as soon as it becomes large. We will come back to the question of what precisely *large* and *small* mean in this context.

We note the following simple consequences of Definition 5.16.

Remark 5.17. Let $G=(V,E)$ be a graph.

(i) For $\emptyset\neq S\subset V$ we have $\phi(S)=\phi(S^{\mathrm{c}})$, $0\leqslant\phi(S)\leqslant 1$, thus in particular $0\leqslant\operatorname{Cheeg}(G)\leqslant 1$.

(ii) There exists $\emptyset\neq S\subset V$ with $\operatorname{Cheeg}(G)=\phi(S)=\frac{\#\partial S}{\operatorname{vol} S}$; moreover, if S is a minimizer, then also S^{c} is a minimizer as $\partial S=\partial(S^{\mathrm{c}})$ holds.

(iii) We have $\operatorname{vol}(V)=\operatorname{vol}(S)+\operatorname{vol}(S^{\mathrm{c}})$ for all $\emptyset\neq S\subset V$.

The Laplacian of a graph served us well when studying the connected components. For the main result of this chapter — an estimate of the Cheeger constants from above and below by eigenvalues — we have however to normalize the Laplacian first.

Definition 5.18. Let G be a graph with vertices $\{1,\dots,n\}$ and $\deg(i)>0$ for all $i=1,\dots,n$. Let L be its Laplacian and $D:=\operatorname{diag}(\deg(1),\dots,\deg(n))=:\operatorname{diag}(d_1,\dots,d_n)$. The matrix

$$\mathcal{L}:=D^{-1/2}LD^{-1/2}\quad\text{with}\quad D^{-1/2}=\operatorname{diag}(d_1^{-1/2},\dots,d_n^{-1/2})$$

is called the *normalized Laplacian* of G.

We first note an analogue to Corollary 5.15.

Lemma 5.19. *Let G be a graph with vertices $\{1,\dots,n\}$ and $\deg(i)>0$ for all $i=1,\dots,n$. We number the eigenvalues of the normalized Laplacian $\mathcal{L}$ in ascending order. Then $0=\lambda_1(\mathcal{L})\leqslant\lambda_2(\mathcal{L})\leqslant\cdots$ and it holds*

$$\lambda_2(\mathcal{L})=\min_{\substack{x\neq 0\\ Dx=0}}\frac{\sum\limits_{\{i,j\}\in E}(x_i-x_j)^2}{\sum\limits_{i=1}^{n}x_i^2d_i}.$$

Proof. For $x\in\mathbb{R}^n$ we have

$$\langle x,\mathcal{L}x\rangle=\langle x,D^{-1/2}LD^{-1/2}x\rangle=\langle D^{-1/2}x,LD^{-1/2}x\rangle=\langle y,Ly\rangle\underset{\substack{\uparrow\\ \text{Lemma}\\ 5.10}}{\geqslant}0$$

with $y:=D^{-1/2}x$, i.e., $\mathcal{L}$ is positive semidefinite and all eigenvalues are greater than or equal to zero. Furthermore, $\lambda_1(\mathcal{L})=0$ holds with eigenvector $D^{1/2}\mathbb{1}$ in view of

$$\mathcal{L}D^{1/2}\mathbb{1}=D^{-1/2}LD^{-1/2}D^{1/2}\mathbb{1}=D^{-1/2}L\mathbb{1}\underset{\substack{\uparrow\\ \text{Lemma}\\ 5.6}}{=}D^{-1/2}0\mathbb{1}=0D^{1/2}\mathbb{1}.$$

For the second smallest eigenvalue, we have

$$\begin{aligned}\lambda_2(\mathcal{L})&\overset{\substack{\text{Lemma}\\ 5.13}}{=}\min_{\substack{x\neq 0\\ \langle x,D^{1/2}\mathbb{1}\rangle=0}}\frac{\langle x,\mathcal{L}x\rangle}{\langle x,x\rangle}\overset{(*)}{=}\min_{\substack{y\neq 0\\ \langle D^{1/2}y,D^{1/2}\mathbb{1}\rangle=0}}\frac{\langle D^{1/2}y,\mathcal{L}D^{1/2}y\rangle}{\langle D^{1/2}y,D^{1/2}y\rangle}\\ &=\min_{\substack{y\neq 0\\ \langle y,D\mathbb{1}\rangle=0}}\frac{\langle y,Ly\rangle}{\langle y,Dy\rangle}=\min_{\substack{y\neq 0\\ Dy=0}}\frac{\sum_{\{i,j\}\in E}(y_i-y_j)^2}{\sum_{i=1}^{n}y_i^2d_i},\end{aligned}$$

where $(*)$ follows via the substitution $y:=D^{-1/2}x$ and applying the identity $L=D^{1/2}\mathcal{L}D^{1/2}$. □

We note that we have $\lambda_2(L)\neq\lambda_2(\mathcal{L})$ in general. However, we will see that $\lambda_2(\mathcal{L})>0$ characterizes the connectivity of G. If G is d-regular (that is,

$\deg(v) = d$ holds for all $v \in V$ with a some $d \in \mathbb{N}$) then $\lambda_2(\mathcal{L}) = \frac{1}{d}\lambda_2(L)$, compare Exercise 5.4.

For the proof of Theorem 5.21 we need the following technical lemma.

Lemma 5.20. *Let $G = (V, E)$ be a graph with $V = \{1, \ldots, n\}$. For $i = 1, \ldots, n$ we define $S_i := \{1, \ldots, i\}$ and $S_0 := \emptyset$. Let*

$$r := \max\{i \in \{1, \ldots, n\} \mid 2 \operatorname{vol} S_i \leqslant \operatorname{vol} V\}.$$

Then

(i) $\min(\operatorname{vol} S_k, \operatorname{vol} S_k^{\mathrm{c}}) = \operatorname{vol} S_k$ *and* $\operatorname{vol} S_k - \operatorname{vol} S_{k+1} = -d_{k+1}$ *hold whenever* $0 \leqslant k \leqslant r$*, and*

(ii) $\min(\operatorname{vol} S_k, \operatorname{vol} S_k^{\mathrm{c}}) = \operatorname{vol} S_k^{\mathrm{c}}$ *and* $\operatorname{vol} S_k^{\mathrm{c}} - \operatorname{vol} S_{k+1}^{\mathrm{c}} = d_{k+1}$ *hold whenever* $r \leqslant k \leqslant n$.

Proof. (i) Let $0 \leqslant k \leqslant r$. Then $2 \operatorname{vol} S_k \leqslant \operatorname{vol} V = \operatorname{vol} S_k + \operatorname{vol} S_k^{\mathrm{c}}$ holds by the definition of r. It follows $\operatorname{vol} S_k \leqslant \operatorname{vol} S_k^{\mathrm{c}}$. Furthermore, we have

$$\operatorname{vol} S_k - \operatorname{vol} S_{k+1} = \sum_{i=1}^{k} d_i - \sum_{i=1}^{k+1} d_i = -d_{k+1}.$$

(ii) Let $r \leqslant k \leqslant n$. Then $2 \operatorname{vol} S_k \geqslant \operatorname{vol} V = \operatorname{vol} S_k + \operatorname{vol} S_k^{\mathrm{c}}$ holds and consequently we obtain $\operatorname{vol} S_k \geqslant \operatorname{vol} S_k^{\mathrm{c}}$. Moreover, we get

$$\operatorname{vol} S_k^{\mathrm{c}} - \operatorname{vol} S_{k+1}^{\mathrm{c}} = \sum_{i=K+1}^{n} d_i - \sum_{i=k+2}^{n} d_i = d_{k+1}$$

as claimed. □

Now we get to the main result of this chapter. Notice that Theorem 5.21 implies that $\lambda_2 \approx 0$ holds if and only if $\operatorname{Cheeg}(G) \approx 0$ and that this means the following: It is then and only then possible to cut the graph into two parts such that few edges are cut in between the parts, compared to the number of edges within the smaller part, if $\lambda_2 \approx 0$ holds.

Theorem 5.21. (Cheeger's inequality) *Let $G = (V, E)$ be a graph with* $\deg(i) > 0$ *for all $i \in V$ and let $0 = \lambda_1 \leqslant \lambda_2 \leqslant \cdots$ be the eigenvalues of the normalized Laplacian of G. Then we have*

$$\frac{\lambda_2}{2} \leqslant \operatorname{Cheeg}(G) \leqslant \sqrt{2\lambda_2}.$$

Proof. As in earlier proofs, we assume without loss of generality that $V = \{1, \ldots, n\}$. We denote by $\mathbb{1}_S = (\mathbb{1}_{S,i})_{i=1,\ldots,n} \in \mathbb{R}^n$ the vector with entries

$\mathbb{1}_{S,i} = 1$ if $i \in S$ and $\mathbb{1}_{S,i} = 0$ if $i \notin S$. We will establish now the two estimates in the theorem seperately. We start with the left one.

① First part of Cheeger's inequality: $\lambda_2 \leqslant 2\,\mathrm{Cheeg}(G)$.

Let $\emptyset \neq S \subset V$ be a minimizer for the Cheeger constant. According to Remark 5.17(ii), we may assume without loss of generality that $\mathrm{Cheeg}(G) = \frac{\#\partial S}{\mathrm{vol}\, S}$. This means in particular $\min(\mathrm{vol}\, S, \mathrm{vol}\, S^{\mathrm{c}}) = \mathrm{vol}\, S$. We define the vector

$$x := \mathbb{1}_S - \frac{\mathrm{vol}\, S}{\mathrm{vol}\, V} \mathbb{1}_V \in \mathbb{R}^n.$$

Then $x \neq 0$,

$$Dx = \sum_{i=1}^{n} d_i x_i = \sum_{i=1}^{n} d_i \mathbb{1}_{S,i} - \tfrac{\mathrm{vol}\, S}{\mathrm{vol}\, V} \sum_{i=1}^{n} d_i \mathbb{1}_{V,i} = \mathrm{vol}\, S - \mathrm{vol}\, S = 0$$

and it follows with Lemma 5.19

$$\lambda_2 \leqslant \frac{\sum_{\{i,j\}\in E} (x_i - x_j)^2}{\sum_{i=1}^{n} x_i^2 d_i} =: \frac{Z}{N}.$$

We will now deal at first with the numerator. By the definition of x we have

$$x_i = \begin{cases} 1 - \frac{\mathrm{vol}\, S}{\mathrm{vol}\, V}, & \text{if } i \in S, \\ \quad - \frac{\mathrm{vol}\, S}{\mathrm{vol}\, V}, & \text{if } i \notin S \end{cases}$$

for $i = 1, \ldots, n$, and thus for $i \neq j$

$$|x_i - x_j| = \begin{cases} 0, & \text{if } i, j \in S \text{ or } i, j \notin S, \\ 1, & \text{otherwise.} \end{cases}$$

Consequently, we obtain

$$Z = \sum_{\{i,j\}\in E} (x_i - x_j)^2 = \#\big\{\{i,j\} \in E \mid i \in S \text{ and } j \notin S\big\} = \#\partial S.$$

Next, we estimate the denominator. Here, we get

$$\begin{aligned} N = \sum_{i=1}^{n} x_i^2 d_i &= \sum_{i=1}^{n} \big(\mathbb{1}_{S,i} - \tfrac{\mathrm{vol}\, S}{\mathrm{vol}\, V} \mathbb{1}_{V,i}\big)^2 d_i \\ &= \sum_{i=1}^{n} \mathbb{1}_{S,i}^2 d_i - 2\tfrac{\mathrm{vol}\, S}{\mathrm{vol}\, V} \sum_{i=1}^{n} \mathbb{1}_{S,i} \mathbb{1}_{V,i} d_i + \big(\tfrac{\mathrm{vol}\, S}{\mathrm{vol}\, V}\big)^2 \sum_{i=1}^{n} \mathbb{1}_{V,i}^2 d_i \\ &= \mathrm{vol}\, S - 2\frac{(\mathrm{vol}\, S)^2}{\mathrm{vol}\, V} + \frac{(\mathrm{vol}\, S)^2}{\mathrm{vol}\, V} \\ &= \mathrm{vol}\, S - \mathrm{vol}\, V \cdot \frac{\mathrm{vol}\, S}{\mathrm{vol}\, S + \mathrm{vol}\, S^{\mathrm{c}}} \end{aligned}$$

$$\begin{aligned}&\geqslant \operatorname{vol} S - \operatorname{vol} V \cdot \frac{\operatorname{vol} S}{2 \operatorname{vol} S},\\&= \frac{\operatorname{vol} S}{2},\end{aligned}$$

where we used $\operatorname{vol} S + \operatorname{vol} S^{\mathrm{c}} \geqslant 2\min(\operatorname{vol} S, \operatorname{vol} S^{\mathrm{c}}) = 2\operatorname{vol} S$. Combining the previous, we end up with the desired inequality:

$$\lambda_2 \leqslant \frac{Z}{N} \leqslant \frac{\#\partial S}{(\operatorname{vol} S)/2} = 2\frac{\#\partial S}{\operatorname{vol} S} = 2\operatorname{Cheeg}(G).$$

② Second part of Cheeger's inequality: $\lambda_2 \geqslant \frac{\operatorname{Cheeg}(G)^2}{2}$.

For $i = 1, \dots, n$ let $S_i := \{1, \dots, i\}$ and $S_0 := \emptyset$. Further, we define

$$\alpha := \min_{i=1,\dots,n} \phi(S_i) \geqslant \min_{\emptyset \neq S \subset V} \phi(S) = \operatorname{Cheeg}(G) \tag{5.1}$$

and our goal in the following will be to show that $\lambda_2 \geqslant \alpha^2$, as then the missing inequality follows. To bring λ_2 into play, we use Lemma 5.19 and choose a minimizer $x \in \mathbb{R}^n$, i.e., we have

$$\lambda_2 = \frac{\sum_{\{i,j\}\in E}(x_i - x_j)^2}{\sum_{i=1}^n x_i^2 d_i} \quad \text{with } x \neq 0 \text{ and } Dx = \sum_{i=1}^n d_i x_i = 0 \tag{5.2}$$

and we now estimate the denominator in the formula for λ_2. First, we observe that we may assume without loss of generality that $x_1 \geqslant x_2 \geqslant \dots \geqslant x_n$ holds. Then we define

$$r := \max\big\{i \in \{1, \dots, n\} \mid 2\operatorname{vol} S_i \leqslant \operatorname{vol} V\big\}$$

and consider the vector $(x_i - x_r)_{i=1,\dots,n}$. By the above, the entries of the latter vector are decreasing; more precisely, for $i < r$ they are greater or equal to zero, for $i = r$ the entry is equal to zero, and for $i > r$ the entries are less or equal to zero. We divide $(x_i - x_r)_{i=1,\dots,n}$ into positive part and negative part:

$$\begin{bmatrix} x_1 - x_r \\ \vdots \\ x_{r-1} - x_r \\ 0 \\ x_{r+1} - x_r \\ \vdots \\ x_n - x_r \end{bmatrix} = \begin{bmatrix} x_1 - x_r \\ \vdots \\ x_{r-1} - x_r \\ 0 \\ 0 \\ \vdots \\ 0 \end{bmatrix} - \begin{bmatrix} 0 \\ \vdots \\ 0 \\ 0 \\ x_r - x_{r+1} \\ \vdots \\ x_r - x_n \end{bmatrix} =: p + n.$$

The entries of p are non-negative and decreasing, those of n are also non-negative but increasing. In particular, $n_i \cdot p_i = 0$ for each $i = 1, \dots, n$. Now we are ready to estimate the denominator of λ_2 in (5.2). Notice that in the first line below the estimate holds, as on the right-hand side the second term is zero by (5.2) while the third is obviously positive:

$$\begin{aligned}
\sum_{i=1}^{n} x_i^2 d_i &\leqslant \sum_{i=1}^{n} x_i^2 d_i - 2x_r \sum_{i=1}^{n} x_i d_i + x_r^2 \sum_{i=1}^{n} d_i \\
&= \sum_{i=1}^{n} (x_i - x_r)^2 d_i = \sum_{i=1}^{n} (p_i - n_i)^2 d_i \\
&= \sum_{i=1}^{n} (p_i^2 - 2p_i n_i + n_i^2) d_i = \sum_{i=1}^{n} (p_i^2 + n_i^2) d_i.
\end{aligned} \tag{5.3}$$

To estimate the numerator in the formula for λ_2, we consider the summands individually. We observe

$$(x_i - x_j)^2 = [(p_i - n_i) - (p_j - n_j)]^2 \underset{(*)}{\overset{\uparrow}{\geqslant}} (p_i - p_j)^2 + (n_i - n_j)^2, \tag{5.4}$$

where the first equation follows from our earlier remarks on the entries of the positive/negative part. The inequality marked with $(*)$ is trivial if $i, j \geqslant r$ or $i, j \leqslant r$ holds, since then $p_i = p_j = 0$ or $n_i = n_j = 0$. Therefore only the case $i \leqslant r \leqslant j$ with $i \neq j$ needs to be considered further. Here we have $p_i, n_j \geqslant 0$ and $p_j = n_i = 0$. It thus follows

$$\begin{aligned}
(x_i - x_j)^2 &= \big((x_i - x_r) - (x_j - x_r)\big)^2 \\
&= (x_i - x_r)^2 - 2(x_i - x_r)(x_j - x_r) + (x_j - x_r)^2 \\
&= p_i^2 - 2p_i(-n_j) + (-n_j)^2 \geqslant p_i^2 + (-n_j)^2 = (p_i - p_j)^2 + (n_i - n_j)^2.
\end{aligned}$$

For the following, we note the two inequalities

$$\begin{aligned}
\forall\, a, c \geqslant 0, b, d > 0 &: \frac{a+c}{b+d} \geqslant \min\Big(\frac{a}{b}, \frac{c}{d}\Big), \\
\forall\, u, v \in \mathbb{R} &: (u+v)^2 \leqslant 2(u^2 + v^2),
\end{aligned} \tag{5.5}$$

whose simple proofs we leave to the reader. The application of our previous estimates to the formula for λ_2 noted in (5.2) leads to an expression in which positive and negative parts appear unmixed:

$$\begin{aligned}
\lambda_2 &= \frac{\sum_{\{i,j\} \in E} (x_i - x_j)^2}{\sum_{i=1}^{n} x_i^2 d_i} \\
&\underset{(5.3)\,(5.4)}{\overset{\uparrow}{\geqslant}} \frac{\sum_{\{i,j\} \in E} (p_i - p_j)^2 + (n_i - n_j)^2}{\sum_{i=1}^{n} (p_i^2 + n_i^2) d_i} \\
&= \frac{\sum_{\{i,j\} \in E} (p_i - p_j)^2 + \sum_{\{i,j\} \in E} (n_i - n_j)^2}{\sum_{i=1}^{n} p_i^2 d_i + \sum_{\{i,j\} \in E} n_i^2 d_i}
\end{aligned}$$

$$
\underset{(5.5)}{\overset{\uparrow}{\geqslant}} \min\Big(\frac{\sum_{\{i,j\}\in E}(p_i-p_j)^2}{\sum_{i=1}^n p_i^2 d_i}, \frac{\sum_{\{i,j\}\in E}(n_i-n_j)^2}{\sum_{\{i,j\}\in E} n_i^2 d_i}\Big)
$$

$$
= \min\Big(\frac{\sum_{\{i,j\}\in E}(p_i-p_j)^2}{\sum_{i=1}^n p_i^2 d_i} \cdot \frac{\sum_{\{i,j\}\in E}(p_i+p_j)^2}{\sum_{\{i,j\}\in E}(p_i+p_j)^2}, \dots\Big)
$$

$$
=: \min\Big(\frac{Z}{N}, \dots\Big).
$$

In the following, we will now only estimate the expression Z/N containing exclusively positive parts. In doing so we will observe that a similar estimate for the expression containing only the negative parts can be obtained analogously. We start with

$$
\begin{aligned}
N &= \sum_{i=1}^n p_i^2 d_i \cdot \sum_{\{i,j\}\in E} (p_i+p_j)^2 \underset{(5.5)}{\overset{\uparrow}{\geqslant}} \sum_{i=1}^n p_i d_i \cdot \sum_{\{i,j\}\in E} 2(p_i^2+p_j^2) \\
&= 2\,\Big(\sum_{i=1}^n p_i^2 d_i\Big) \cdot \Big(\sum_{\{i,j\}\in E} p_i^2 + \sum_{\{i,j\}\in E} p_j^2\Big) = 2\,\Big(\sum_{i=1}^n p_i^2 d_i\Big) \cdot 2\,\Big(\sum_{\{i,j\}\in E} p_i^2\Big) \\
&\underset{(\circ)}{\overset{\uparrow}{\geqslant}} 2\,\Big(\sum_{i=1}^n p_i^2 d_i\Big) \cdot \Big(\sum_{i=1}^n p_i^2 d_i\Big) = 2\,\Big(\sum_{i=1}^n p_i^2 d_i\Big)^2,
\end{aligned}
$$

where $(\circ)$ comes about as follows: Instead of summing over all edges, we sum over all vertices and multiply each time by the degree — since each edge then appears twice, we correct this by dropping the pre-factor 2. Formally, one can argue as follows:

$$
\begin{aligned}
2\,\Big(\sum_{\{i,j\}\in E} p_i^2\Big) &= 2\,\Big(\sum_{\substack{\{i,j\}\in E\\ i<j}} p_i^2\Big) = 2\,\Big(\sum_{\substack{i,j=1\\ i<j}}^n \mathbb{1}_E(i,j)\,p_i^2\Big) = \sum_{i,j=1}^n \mathbb{1}_E(i,j)\,p_i^2 \\
&= \sum_{i=1}^n \Big(p_i^2 \sum_{j=1}^n \mathbb{1}_E(i,j)\Big) = \sum_{i=1}^n p_i^2 \cdot \#\{\{i,j\}\in E\} = \sum_{i=1}^n p_i^2 d_i\,,
\end{aligned}
$$

where we used the abbreviation

$$
\mathbb{1}_E\colon V\times V \to \mathbb{R},\ \mathbb{1}_E(i,j) = \begin{cases} 1, & \text{if } \{i,j\}\in E, \\ 0, & \text{otherwise.} \end{cases} \tag{5.6}
$$

Next, we estimate the numerator. We recommend that the reader sketches the (triangular!) area in $\mathbb{N}\times\mathbb{N}$ over which the sum is taken for the equations marked with (Δ):

$$
Z \quad = \quad \sum_{\{i,j\}\in E} (p_i-p_j)^2 \cdot \sum_{\{i,j\}\in E} (p_i+p_j)^2
$$

$$
\begin{aligned}
&= \left\| \left((p_i - p_j)^2\right)_{\{i,j\}\in E} \right\|^2_{\mathbb{R}^{\#E}} \cdot \left\| \left((p_i + p_j)^2\right)_{\{i,j\}\in E} \right\|^2_{\mathbb{R}^{\#E}} \\
&\underset{\substack{\uparrow \\ \text{CSB}}}{\geqslant} \left\langle \left((p_i - p_j)\right)_{\{i,j\}\in E}, \left((p_i + p_j)\right)_{\{i,j\}\in E} \right\rangle^2_{\mathbb{R}^{\#E}} \\
&\underset{\substack{\uparrow \\ (5.6)}}{=} \Big(\sum_{\substack{i,j=1 \\ i<j}}^{n} \mathbb{1}_E(i,j) \cdot (p_i - p_j)(p_i + p_j) \Big)^2 \\
&\underset{\substack{\uparrow \\ \text{telescopic} \\ \text{sum}}}{=} \Big(\sum_{i=1}^{n-1} \sum_{j=i+1}^{n} \mathbb{1}_E(i,j) \sum_{k=i}^{j-1} p_k^2 - p_{k+1}^2 \Big)^2 \\
&\underset{\substack{\uparrow \\ (\Delta)}}{=} \Big(\sum_{i=1}^{n-1} \sum_{k=i}^{n-1} \sum_{j=k+1}^{n} \mathbb{1}_E(i,j) \cdot (p_k^2 - p_{k+1}^2) \Big)^2 \\
&\underset{\substack{\uparrow \\ (\Delta)}}{=} \Big(\sum_{k=1}^{n-1} \sum_{i=1}^{k} \sum_{j=k+1}^{n} \mathbb{1}_E(i,j) \cdot (p_k^2 - p_{k+1}^2) \Big)^2 \\
&= \Big(\sum_{k=1}^{n-1} (p_k^2 - p_{k+1}^2) \sum_{\substack{i\leqslant k<j \\ \{i,j\}\in E}} 1 \Big)^2 \\
&= \Big(\sum_{k=1}^{n-1} (p_k^2 - p_{k+1}^2) \#\partial S_k \Big)^2 \\
&\underset{\substack{\uparrow \\ \text{Dfn } \alpha \\ (5.1)}}{\geqslant} \Big(\sum_{k=1}^{n-1} (p_k^2 - p_{k+1}^2) \alpha \min(\operatorname{vol} S_k, \operatorname{vol} S_k^{\mathrm{c}}) \Big)^2 \\
&\underset{\substack{\uparrow \\ p_k=0 \text{ for} \\ k\geqslant r \text{ and } 5.20}}{=} \Big(\sum_{k=1}^{r} (p_k^2 - p_{k+1}^2) \alpha \operatorname{vol} S_k \Big)^2 \\
&= \alpha^2 \Big(\sum_{k=1}^{r} p_k^2 \operatorname{vol} S_k - \sum_{k=1}^{r} p_{k+1}^2 \operatorname{vol} S_k \Big)^2 \\
&\underset{\substack{\uparrow \\ \text{index shift} \\ \operatorname{vol} S_0=0 \\ \textit{and } p_r=0}}{=} \alpha^2 \Big(\sum_{k=0}^{r-1} p_{k+1}^2 \operatorname{vol} S_{k+1} - \sum_{k=0}^{r-1} p_{k+1}^2 \operatorname{vol} S_k \Big)^2 \\
&\underset{\substack{\uparrow \\ \text{Lemma} \\ 5.20}}{=} \alpha^2 \Big(\sum_{k=0}^{r-1} p_{k+1}^2 d_{k+1} \Big)^2 \\
&\underset{\substack{\uparrow \\ \text{index shift} \\ \textit{and } p_k=0 \\ \textit{for } k\geqslant r}}{=} \alpha^2 \Big(\sum_{k=1}^{n} p_k^2 d_k \Big)^2 .
\end{aligned}
$$

Combining the estimates for numerator and denominator yields

$$
\frac{Z}{N} \geqslant \frac{\alpha^2 \left(\sum_{k=1}^{n} p_k^2 d_k\right)^2}{2\left(\sum_{i=1}^{n} p_i^2 d_i\right)^2} = \frac{\alpha^2}{2}.
$$

For the term with the negative parts, we obtain exactly the same estimate by analogous reasoning and by using the "dual" statements in Lemma 5.20 and taking into account that $n_k = 0$ holds for $k \leqslant r$. We leave this calculation to the reader. As already explained after (5.1), it follows that

$$\lambda_2 \geqslant \alpha^2/2 \geqslant \mathrm{Cheeg}(G)^2/2$$

holds. □

We note that Cheeger's inequality is named after Jeff Cheeger, who proved a continuous version, i.e., for manifolds instead of graphs, in 1970, which explains the terms "volume", "boundary", etc., as we have used them in this chapter. The proof above is due to Fan Chung from 2007.

Finally, we want to note that Theorem 5.21 implicitly provides an upper bound for the eigenvalue $\lambda_2(\mathcal{L})$.

Corollary 5.22. *Let $G = (V, E)$ be a graph with $\deg(i) > 0$ for all $i \in V$. Then for the second smallest eigenvalue λ_2 of the normalized Laplace matrix of G, the estimate $\lambda_2 \leqslant 8$ holds which means that the Cheeger constant satisfies $\mathrm{Cheeg}(G) \leqslant 4$.* □

References

The concepts of spectral graph theory used in this chapter are all standard and can be found practically everywhere. The proof of Theorem 5.21, including the preceding lemmas, is taken from [Chu07]. The author would like to thank Thomas Sauerwald for many insightful discussions about the content of this chapter.

Exercises

Exercise 5.1. Let G be a d-regular graph with adjacency matrix A and Laplacian L. Show that $\sigma(L) = d - \sigma(A)$ holds. More precisely: show that $Av = \mu v$ holds if and only if $Lv = (d - \mu)v$.

Remark: Sometimes it is easier to compute eigenvalues and eigenvectors of A and then use the above to get the eigenvalues/eigenvectors of the Laplacian.

Exercise 5.2. Determine the Laplacian L, its eigenvalues and a corresponding orthonormal system for the following disconnected graph G:

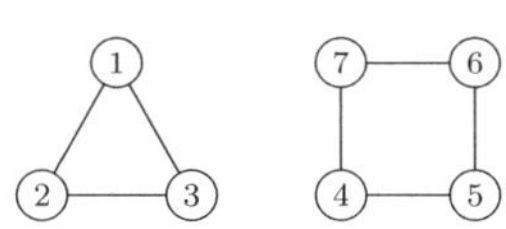

Do the same for the *complete graph* C_8:

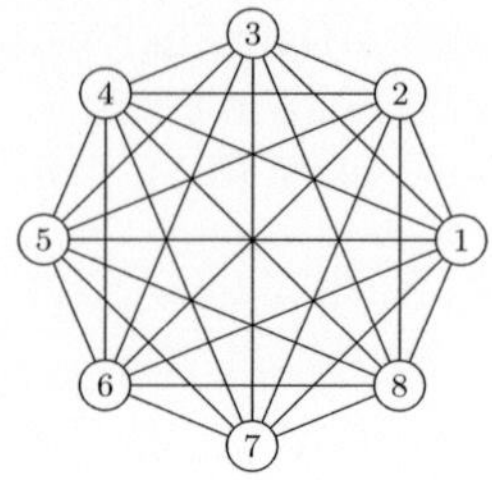

Start with guessing two eigenvalues and compute their eigenspaces. Once this is done, it will become apparent how to continue.

Exercise 5.3. Consider the graph G, given by the following adjacency matrix.

$$A = \begin{bmatrix} 0 & 0 & 1 & 1 & 0 & 0 & 1 & 0 \\ 0 & 0 & 0 & 0 & 1 & 1 & 1 & 0 \\ 1 & 0 & 0 & 1 & 0 & 0 & 0 & 1 \\ 1 & 0 & 1 & 0 & 0 & 0 & 1 & 0 \\ 0 & 1 & 0 & 0 & 0 & 1 & 0 & 1 \\ 0 & 1 & 0 & 0 & 1 & 0 & 0 & 1 \\ 1 & 1 & 0 & 1 & 0 & 0 & 0 & 0 \\ 0 & 0 & 1 & 0 & 1 & 1 & 0 & 0 \end{bmatrix}$$

(i) Draw G by hand: Arrange the vertices as in the complete graph C_8 in Exercise 5.2 and then draw the edges.

(ii) Calculate the Laplacian L and determine the second smallest eigenvalue λ_2 and a corresponding eigenvector v_2 with norm 1.

(iii) Now redraw the graph, representing the nodes as points in $\mathbb{R}^2$, such that node i has the i-th entry of v_2 as its x-coordinate. Choose the y-coordinates such that the picture looks good. Can you now see the clusters?

(iv) Do the same as in (iii) for the two graphs from Exercise 5.2 and for the graph G, which is given by the adjacency matrix available at

`https://mathematicaldatascience.github.io/graph_en.html.`

(v) In this chapter, we propagated the heuristic that, when proceeding as in (iii) and (iv), points which end up close together are more likely to be connected by an edge than points that end up far apart. Go through the examples from above again and watch out for effects that show that one should take this heuristic cautiously.

Hint: In many programming languages there are ready-made functions available to determine the Laplacian based on the adjacency matrix, or calculate eigenvalues along with corresponding eigenvectors.

Exercise 5.4. Let G be a d-regular graph, L the (normal) Laplacian and $\mathcal{L}$ the normalized Laplacian. Show that $\sigma(\mathcal{L}) = \frac{1}{d} \cdot \sigma(L)$ holds. More precisely: show that $Lv = \lambda v$ holds if and only if $\mathcal{L}v = \frac{\lambda}{d}v$.

Exercise 5.5. Calculate the Cheeger constant Cheeg(G) of the graph G:

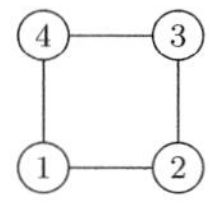

Come up with a conjecture for the Cheeger constant of the *circle* K_n with $n \geqslant 2$ nodes and prove it:

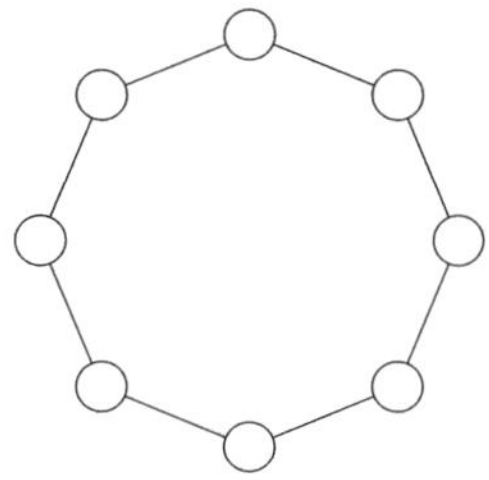

Exercise 5.6. (i) Assume that a graph G has the Laplace matrix L and it holds

$$\det(L - \lambda) = \lambda^3(\lambda - 2)(\lambda - 3)^4(\lambda - 4)^2.$$

How many connected components does G have?

(ii) Determine the Cheeger constant of the following graph:

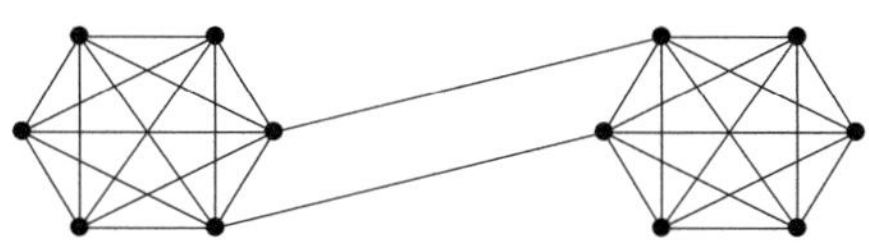

6

Best-Fit Subspaces

In Chapter 2, we fitted an affine linear function $f\colon \mathbb{R}^d \to \mathbb{R}$, or, in other words, the affine linear subspace $\operatorname{gr}(f) \subseteq \mathbb{R}^{d+1}$, to a labeled dataset $\{(x_i, y_i) \in \mathbb{R}^d \times \mathbb{R} \mid i = 1, \dots, n\}$. The least squares method previously used minimizes the sum over the *distances in label direction* between the data points and the affine linear subspace, cf. the pictures before Theorems 2.1 and 2.9.

The theory presented in this chapter is closely related to that of the singular value decomposition, which we will develop in the next Chapter 7. There we will also discuss precisely how best-fit subspaces and SVD are related.

In contrast to affine linear regression, in this chapter we consider an unlabeled dataset $\{x_1, \dots, x_n\} \subseteq \mathbb{R}^d$, and approximate it by a subspace via minimizing the sum of the squares of the Euclidean distances between points and subspace. In the following picture, we seek to minimize the sum of the squares of the lengths of the dotted lines.

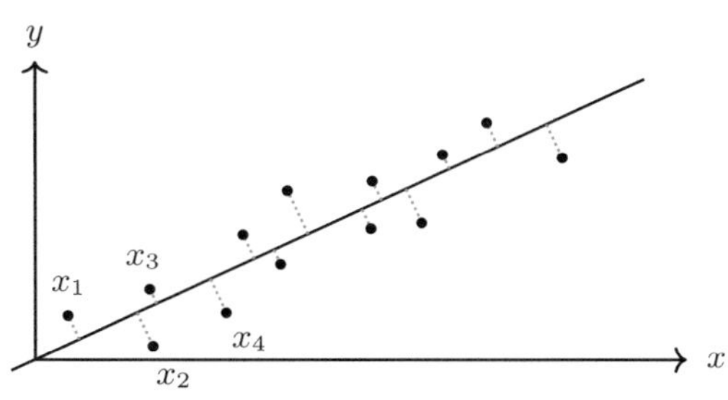

Regarding the restriction to linear subspaces (instead of affine linear ones), we note that if the former does not yield good results, then one may try to center the entire dataset first, cf. Chapter 3.3.

Definition 6.1. Let $D := \{x_1, \dots, x_n\} \subseteq \mathbb{R}^d$ and $1 \leqslant k \leqslant d$. A subspace $V \subseteq \mathbb{R}^d$ is called a *best-fit subspace of dimension* k, or for short, a k*-best fit subspace* for D, if $\dim V = k$ and for every k-dimensional subspace $W \subseteq \mathbb{R}^d$

S. A. Wegner, *Mathematical Introduction to Data Science*,
https://doi.org/10.1007/978-3-662-69426-8_6

the following estimate holds:

$$\sum_{i=1}^{n} \operatorname{dist}(x_i, V)^2 \leqslant \sum_{i=1}^{n} \operatorname{dist}(x_i, W)^2.$$

Here, $\operatorname{dist}(x, V) = \min_{v \in V} \|x - v\|$ is the distance from $x \in \mathbb{R}^d$ to V.

It is easy to see that best-fit subspaces for a given dataset D are in general not unique, see Exercise 6.3.

In the following, the space $\mathbb{R}^d$ is always equipped with the standard scalar product $\langle \cdot, \cdot \rangle$ and the Euclidean norm $\| \cdot \|$ induced by it. We write $x \perp y$ for orthogonal vectors, $x \perp V$, if a vector x is orthogonal to a subspace $V \subseteq \mathbb{R}^d$ and $V^\perp$ for the orthogonal complement of the subspace V. We denote with

$$\pi_V \colon \mathbb{R}^d \to \mathbb{R}^d, \; u \mapsto \pi_V(x)$$

the orthogonal projection and note that we can calculate its values

$$\pi_V(x) = \sum_{j=1}^{k} \langle x, b_j \rangle b_j$$

if $\{b_1, \ldots, b_k\}$ is any orthonormal basis of V. We remind the reader that we know from linear algebra that $x - \pi_V(x) \perp V$ and $\pi_V(x) = \operatorname{argmin}_{v \in V} \|x - v\|$ always hold. If we now consider the projection of a data point $x \in D$ onto a potentially best fitting subspace V, we get the following picture.

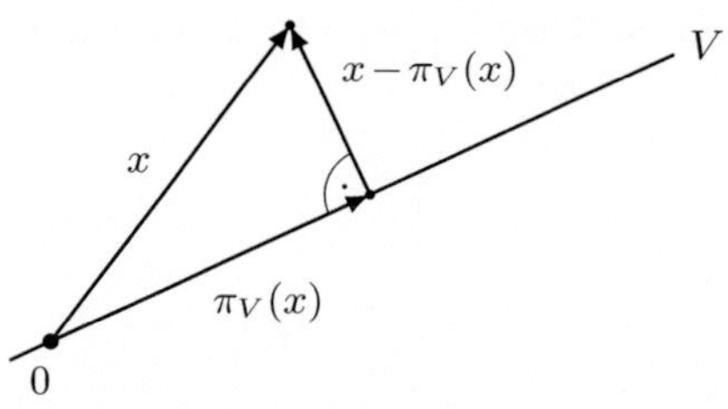

Since we are optimizing $V \subseteq \mathbb{R}^d$ for a fixed dataset, the picture suggests that instead of minimizing the distance squared from x to V, we can also maximize the norm squared of the projection. In fact, the following holds.

Lemma 6.2. *Let a dataset $D = \{x_1, \ldots, x_n\} \in \mathbb{R}^d$ be given with coordinates $x_i = (x_{i1}, \ldots, x_{id})$ for $i = 1, \ldots, n$. We consider the* data matrix

$$X = \begin{bmatrix} x_1 \\ \vdots \\ x_n \end{bmatrix} = \begin{bmatrix} x_{11} & \cdots & x_{1d} \\ \vdots & & \vdots \\ x_{n1} & \cdots & x_{nd} \end{bmatrix} \in \mathbb{R}^{n \times d},$$

whose i-th row contains the coordinates of the point x_i. Let further $1 \leqslant k \leqslant d$ and $V \subseteq \mathbb{R}^d$ be a subspace with orthonormal basis $\{v_1, \ldots, v_k\}$. Then V is *a k-best fitting subspace for D if and only if for every orthonormal system $\{w_1, \ldots, w_k\} \subseteq \mathbb{R}^d$ the inequality*

$$\sum_{j=1}^{n} \|Xv_j\|^2 \geqslant \sum_{j=1}^{n} \|Xw_j\|^2$$

holds.

Proof. Let first $B \subseteq \mathbb{R}^d$ be any subspace with an orthonormal basis $\{b_1, \ldots, b_k\}$ and let $\pi_B \colon \mathbb{R}^d \to \mathbb{R}^d$ be the orthogonal projection onto B. Then it holds

$$\begin{aligned}\sum_{i=1}^{n} \|x_i\|^2 &\underset{\substack{\uparrow\\ \text{Pytha-}\\ \text{goras}}}{=} \sum_{i=1}^{n} \|x_i - \pi_B(x_i)\|^2 + \sum_{i=1}^{n} \|\pi_B(x_i)\|^2 \\ &= \sum_{i=1}^{n} \operatorname{dist}(x_i, B)^2 + \sum_{i=1}^{n}\sum_{j=1}^{k} |\langle x_i, b_j\rangle|^2 \\ &= \sum_{i=1}^{n} \operatorname{dist}(x_i, B)^2 + \sum_{j=1}^{k} \|Xb_j\|^2.\end{aligned}$$

If now $V = \operatorname{span}\{v_1, \ldots, v_k\}$ is as in the lemma and $W \subseteq \mathbb{R}^d$ is any subspace with orthonormal basis $\{w_1, \ldots, w_k\}$, then the above implies the equivalence

$$\sum_{i=1}^{n} \operatorname{dist}(x_i, V)^2 \leqslant \sum_{i=1}^{n} \operatorname{dist}(x_i, W)^2 \iff \sum_{j=1}^{k} \|Xv_j\|^2 \geqslant \sum_{j=1}^{k} \|Xw_j\|^2.$$

By Definition 6.1, V is a k-best fitting subspace if and only if the left inequality holds for all k-dimensional $W \subseteq \mathbb{R}^d$. The latter is therefore equivalent to the right inequality holding for all orthonormal systems $\{w_1, \ldots, w_k\}$. □

If a dataset $D = \{x_1, \ldots, x_n\} \subseteq \mathbb{R}^d$ is given, we can form the associated data matrix X according to Lemma 6.2 and consider the following optimization task:

$$\{v_1, \ldots, v_k\} \in \underset{\substack{\{\tilde{v}_1,\ldots,\tilde{v}_k\} \subseteq \mathbb{R}^d\\ \text{orthonormal}\\ \text{basis}}}{\operatorname{argmax}} \sum_{j=1}^{n} \|X\tilde{v}_j\|^2.$$

If a corresponding maximizer is found, we obtain a k-best fitting subspace by putting $V := \operatorname{span}\{v_1, \ldots, v_k\}$. At first glance, it is not clear whether a maximizer exists. However, if we consider the special case $k = 1$, the optimization Exercise reduces to the maximization of a continuous function over the compact unit sphere, and the existence of

$$v_1 \in \underset{\|v\|=1}{\operatorname{argmax}} \|Xv\|^2 = \underset{\|v\|=1}{\operatorname{argmax}} \|Xv\|$$

is guaranteed. Accordingly, we obtain a 1-best fitting subspace $V_1 = \operatorname{span}\{v_1\}$.

The crucial idea is now, for $k = 2$, not to start again from scratch, but (somewhat boldly!) to speculate that a 2-best fitting subspace V_2 exists *that contains* V_1 *as a subspace.* This leads to the optimization task

$$v_2 \in \operatorname*{argmax}_{\substack{\|v\|=1 \\ v \perp v_1}} \|Xv\|,$$

in which again a continuous function is maximized, this time over the compact subset $\partial \mathrm{B}_1(0) \cap V_1^\perp$. The existence of v_2 is therefore guaranteed, too. What we need to show, however, is that $V_2 := \operatorname{span}\{v_1, v_2\}$ is indeed a 2-best fitting subspace! In the following theorem, we will do this, not only for $k = 2$, but with an induction argument right away for any $1 \leqslant k \leqslant d$.

Theorem 6.3. *Let $D = \{x_1, \dots, x_n\} \in \mathbb{R}^d$ be given and $X = \begin{bmatrix} x_1 \\ \vdots \\ x_n \end{bmatrix} \in \mathbb{R}^{n \times d}$ be the corresponding data matrix. For $1 \leqslant k \leqslant d$ we can then choose successively*

$$v_j \in \operatorname*{argmax}_{\substack{\|v\|=1 \\ v \perp v_1, \dots, v_{j-1}}} \|Xv\|$$

for $j = 1, \dots, n$, where we understand the orthogonality condition in the case $j = 1$ as empty. Then, $V_k := \operatorname{span}\{v_1, \dots, v_k\}$ is a k-best fitting subspace for D.

Proof. Since for each j a continuous function is maximized over a compact set, the existence of the $v_1, \dots, v_k$ follows. To show that V_k is a k-best fitting subspace, we define

$$K := \big\{ k \in \{1, \dots, d\} \mid V_k \text{ is a } k\text{-best fitting subspace} \big\}.$$

Then $1 \in K$ holds according to Lemma 6.2 or according to the explanations before Theorem 6.3. We now claim that the implication

$$\forall\, k \in \{1, \dots, d-1\} \colon k \in K \implies k+1 \in K$$

holds. Let $1 \leqslant k < d$ with $k \in K$. We choose the $v_1, \dots, v_{k+1}$ as above and define V_k and V_{k+1} as the corresponding spans. To show that $k+1 \in K$ holds, we fix an arbitrary $(k+1)$-dimensional subspace $W \subseteq \mathbb{R}^d$ and define

$$U := \operatorname{span}\{\pi_W(v_1), \dots, \pi_W(v_k)\} \subseteq W.$$

Since $\dim U \leqslant k$ holds, we can choose a (suggestively named!) vector $w_{k+1} \in W$ with $w_{k+1} \perp U$ and $\|w_{k+1}\| = 1$. For $i = 1, \dots, k$ we then get

$$\langle w, v_i \rangle = \langle w, \underbrace{v_i - \pi_W(v_i)}_{\perp W} \rangle + \langle w, \underbrace{\pi_W(v_i)}_{\in U} \rangle = 0 + 0 = 0$$

and thus $w_{k+1} \perp v_1, \ldots, v_k$. Combined with $\|w_{k+1}\| = 1$, this implies

$$\|Xw_{k+1}\|^2 \leqslant \max_{\substack{\|v\|=1 \\ v \perp v_1,\ldots,v_k}} \|Xv\|^2 \underset{\substack{\text{Dfn} \\ v_{k+1}}}{\overset{}{\underset{\uparrow}{=}}} \|Xv_{k+1}\|^2.$$

On the other hand, we may extend the singleton $\{w_{k+1}\}$ to an orthonormal basis $\{w_1, \ldots, w_k, w_{k+1}\}$ of W. For the orthonormal system $\{w_1, \ldots, w_k\} \subseteq \mathbb{R}^d$, the inequality

$$\sum_{j=1}^{k} \|Xw_j\|^2 \leqslant \sum_{j=1}^{k} \|Xv_j\|^2,$$

holds according to Lemma 6.2, because $V_k = \operatorname{span}\{v_1, \ldots, v_k\}$ is a k-best fitting subspace by assumption. Adding up the last two inequalities yields

$$\sum_{j=1}^{k+1} \|Xw_j\|^2 \leqslant \sum_{j=1}^{k+1} \|Xv_j\|^2,$$

from which it follows by another application of Lemma 6.2 that V_{k+1} is a $(k+1)$-best fitting subspace. □

The algorithm for calculating V_k via Theorem 6.3 is a so-called *greedy algorithm*: Notice, that by choosing the v_j via the recursion formula, it could have occurred

$$\min_{\substack{V \subseteq \mathbb{R}^d \\ \dim V = 2}} \sum_{i=1}^{n} \operatorname{dist}(x_i, V)^2 < \min_{\substack{v_1 \in V \subseteq \mathbb{R}^d \\ \dim V = 2}} \sum_{i=1}^{n} \operatorname{dist}(x_i, V)^2.$$

If this were the case, then the "greedy" choice of the first basis vector v_1, which was chosen to get a best-fit subspace of dimension 1, would have made it impossible to find a minimizer among all 2-dimensional subspaces in the second step. Theorem 6.3 shows however that this does *not* happen and thus the greedy strategy is successful in this case.

If a dataset shall be approximated via a best-fit subspace, the question arises of which dimension k one should choose for the approximating space. We start with the following observation.

Lemma 6.4. *Let $D = \{x_1, \ldots, x_n\} \subseteq \mathbb{R}^d$ be a dataset, $r := \dim \operatorname{span} D$ and let V_r be as in Theorem 6.3. Then $V_r = \operatorname{span} D$.*

Proof. Firstly, it is clear that $V := \operatorname{span}\{x_1, \ldots, x_n\}$ is an r-best fitting subspace, since $\operatorname{dist}(x_i, V) = 0$ for all i. As V_r is r-best fitting too, we have

$$\sum_{i=1}^{n} \operatorname{dist}(x_i, V_r)^2 = \sum_{i=1}^{n} \operatorname{dist}(x_i, V)^2 = 0$$

by definition. Since $V_r \subseteq \mathbb{R}^d$ is closed, this is only possible if $x_i \in V_r$ for all i. For dimensional reasons, it then follows that $V_r = V$. □

The above means that the interesting choices for the dimension are $k < r$: For $k = r$, we just get the span of the dataset, and if we choose $k > r$, the $v_{r+1}, \ldots, v_k$ chosen according to the recursion formula in Theorem 6.3 are merely an arbitrary extension of $v_1, \ldots, v_k$ to an orthonormal system.

If a dataset D is given, it may however be difficult to determine the dimension of $\operatorname{span} D$, which makes the above only conditionally useful. The following result shows how to recognize, during the recursion process of Theorem 6.3, when we have arrived at v_r.

Proposition 6.5. *Let* $D = \{x_1, \ldots, x_n\} \in \mathbb{R}^d$ *be given and let* $X = \begin{bmatrix} x_1 \\ \vdots \\ x_n \end{bmatrix} \in \mathbb{R}^{n \times d}$ *be the associated data matrix. For* $1 \leqslant k \leqslant d$ *let*

$$\sigma_k := \|Xv_k\| = \max_{\substack{\|v\|=1 \\ v \perp v_1, \ldots, v_{k-1}}} \|Xv\|,$$

where we again understand for $k = 1$ *the orthogonality condition as empty. Then* $\sigma_1 \geqslant \sigma_2 \geqslant \cdots \geqslant \sigma_r > 0$ *and* $\sigma_{r+1} = \cdots = \sigma_d = 0$ *hold with* $r = \dim \operatorname{span} D$.

Proof. That the σ_k's are decreasing and non-negative is clear by construction. For $k \geqslant r+1$, we have

$$\sigma_k^2 = \max_{\substack{\|v\|=1 \\ v \perp v_1, \ldots, v_{k-1}}} \|Xv\|^2 = \min_{\substack{\|v\|=1 \\ v \perp V_{k-1}}} \sum_{i=1}^{n} \operatorname{dist}(x_i, V_{k-1})^2 = 0$$

according to Lemma 6.4 and since $V_{k-1} \supseteq V_r$. Thus, $\sigma_{r+1}, \ldots, \sigma_d = 0$. We now show by contradiction, that $\sigma_1, \ldots, \sigma_r > 0$ holds: Suppose there is a $1 \leqslant k \leqslant r$ with $\sigma_k = \|Xv_k\| = 0$. Because of $\sigma_1 \geqslant \sigma_2 \geqslant \cdots \geqslant \sigma_d \geqslant 0$ it follows that the $\sigma_{k+1}, \ldots, \sigma_d$ are also zero. The latter then means

$$\|Xv_k\| = \|Xv_{k+1}\| = \cdots = \|Xv_d\| = 0.$$

With this we have $v_k, \ldots, v_d \in \ker X$ and consequently $\dim \ker X \geqslant d - (k-1)$. Furthermore, $\operatorname{rk} X = \dim \operatorname{span}\{x_1, \ldots, x_n\} = r$, as row rank=column rank. It follows

$$\dim \ker X + \dim \operatorname{ran} X \geqslant (d - k + 1) + r = d + 1 + (r - k) \underset{k \leqslant r}{\overset{\uparrow}{\geqslant}} d + 1$$

in contradiction to the rank theorem. □

The σ_j in Proposition 6.5 are important characteristics of the matrix X, which we will examine more closely in the following chapter on *singular value* decomposition. In particular, in Corollary 7.19, we will express the approximation quality of V_k, i.e., the sum $\operatorname{dist}(x_1, V_k)^2 + \cdots + \operatorname{dist}(x_n, V_k)^2$ via the numbers $\sigma_{k+1}, \ldots, \sigma_r$.

References

This chapter mainly follows [BHK20, Chapter 3], but has been modified numerous times during several courses taught by the author. In particular, in the present text, the best-fit subspaces are introduced independently of the singular value decomposition, which will only be treated in the following Chapter 7. The numerical example from Exercise 6.1 is taken from [Ham].

Exercises

Exercise 6.1. Calculate for the dataset $D = \{(3,2,2),(2,3,-2)\} \subseteq \mathbb{R}^3$ the vectors v_1, v_2, v_3 using the recursion formula in Theorem 6.3. Create a plot of the 1- and 2-best fitting subspaces with a suitable software.

Hint: Use Lagrange multipliers and check your computations via a CAS.

Exercise 6.2. Determine, for example by using suitable packages in Python, a 1-best fitting subspace for the data from Exercise 2.3 and plot it. Compare with the regression line from Exercise 2.3.

Exercise 6.3. Show that n-many data points in $\mathbb{R}^d$ in general do not determine a k-best fitting subspace uniquely. This is true even if the rank of the data matrix $X \in \mathbb{R}^{n\times d}$ is strictly greater than k.

Note: An example where more than one best-fit subspaces "obviously" exist can be found quickly. However, for the formal proof, it is recommended to use Method 1 or Method $1\frac{1}{2}$ from the following Chapter 7.

7 Singular Value Decomposition

If $A \in \mathbb{R}^{n\times n}$ is a symmetric matrix with spectrum $\sigma(A) = \{\lambda_1, \dots, \lambda_n\}$, then the orthogonal diagonalization theorem from linear algebra implies that there exists an orthogonal matrix $B \in \mathbb{R}^{n\times n}$ such that the matrix equation

$$B^{\mathsf{T}}AB = \operatorname{diag}(\lambda_1, \dots, \lambda_n)$$

holds. In this chapter, we are going to generalize the above to arbitrary non-square matrices $A \in \mathbb{R}^{n\times d}$. For this, we first need to generalize the concept of eigenvalues; note in particular that the usual eigenvalue equation $Av = \lambda v$ for $A \in \mathbb{R}^{n\times d}$ is not well-defined if $n \neq d$.

Definition 7.1. Let $A \in \mathbb{R}^{n\times d}$. A number $\sigma > 0$ is called a *singular value* of A, if there exist $v \in \mathbb{R}^d\backslash\{0\}$ and $u \in \mathbb{R}^n\backslash\{0\}$ such that $Av = \sigma u$ and $A^{\mathsf{T}}u = \sigma v$ both hold. We call v and u the *singular vectors* associated with σ; more precisely, v is called a *right* and u a *left singular vector*. We denote the set of all singular values by $\mathrm{s}(A)$.

Singular values and singular vectors of a matrix $A \in \mathbb{R}^{n\times d}$ are closely related to the eigenvalues and eigenvectors of the matrices $A^{\mathsf{T}}A \in \mathbb{R}^{d\times d}$ and $AA^{\mathsf{T}} \in \mathbb{R}^{n\times n}$. For later use, we note the following properties.

Lemma 7.2. *Let $A \in \mathbb{R}^{n\times d}$. Then the following hold:*

(i) $A^{\mathsf{T}}A$ and AA^{T} are symmetric and positive semidefinite.

(ii) We have $\sigma(A^{\mathsf{T}}A)\backslash\{0\} = \sigma(AA^{\mathsf{T}})\backslash\{0\}$.

(iii) For eigenvalues $\lambda \neq 0$, we have $\dim\ker(A^{\mathsf{T}}A - \lambda) = \dim\ker(AA^{\mathsf{T}} - \lambda)$.

(iv) We have $\operatorname{rk} A = \operatorname{rk} A^{\mathsf{T}}A = \operatorname{rk} AA^{\mathsf{T}} = \operatorname{rk} A^{\mathsf{T}}$.

(v) Counted with (geometric = algebraic) multiplicity, $A^{\mathsf{T}}A$ and AA^{T} have each precisely $\operatorname{rk}(A)$-many non-zero eigenvalues.

Proof. (i) The symmetry is clear; to see that $A^{\mathsf{T}}A$ and AA^{T} are positive semi-

S. A. Wegner, *Mathematical Introduction to Data Science*,
https://doi.org/10.1007/978-3-662-69426-8_7

definite, we compute for $x \in \mathbb{R}^d$

$$\langle x, A^{\mathsf{T}}Ax\rangle = \langle Ax, Ax\rangle = \|Ax\|^2 \geqslant 0$$

and for $y \in \mathbb{R}^n$

$$\langle y, AA^{\mathsf{T}}y\rangle = \langle A^{\mathsf{T}}y, A^{\mathsf{T}}y\rangle = \|A^{\mathsf{T}}y\|^2 \geqslant 0.$$

(ii) Let $\lambda \in \sigma(A^{\mathsf{T}}A)\backslash\{0\}$. Then there exists $v \in \mathbb{R}^d\backslash\{0\}$ with $A^{\mathsf{T}}Av = \lambda v$. Multiplication with A from the left yields $AA^{\mathsf{T}}Av = \lambda Av$. We put $u := Av \in \mathbb{R}^n$ and obtain $AA^{\mathsf{T}}u = \lambda u$ where

$$\|u\|^2 = \|Av\|^2\langle Av, Av\rangle = \langle v, A^{\mathsf{T}}Av\rangle = \langle v, \lambda v\rangle = \lambda\|v\|^2 \neq 0,$$

thus $u \neq 0$, and therefore $\lambda \in \sigma(AA^{\mathsf{T}})\backslash\{0\}$. The other inclusion can be shown analogously.

(iii) Let $v_1, \ldots, v_k \in \ker(A^{\mathsf{T}}A-\lambda)$ be linearly independent. We put $u_j := Av_j$ and note $u_j \in \ker(AA^{\mathsf{T}} - \lambda)$ for $j = 1, \ldots, k$ due to (ii). Now let $\alpha_1 u_1 + \cdots + \alpha_k u_k = 0$. Then it follows

$$0 = \sum_{j=1}^{k} \alpha_j A v_j = \sum_{j=1}^{k} \alpha_j \lambda v_j = \lambda \sum_{j=1}^{k} \alpha_j v_j \overset{\lambda\neq 0}{\Longrightarrow} \sum_{j=1}^{k} \alpha_j v_j = 0$$

and thus $\alpha_1 = \cdots = \alpha_k = 0$, which shows that the $u_1, \ldots, u_k$ are linearly independent. Analogously, one can see that linearly independent $u_1, \ldots, u_k \in \ker(AA^{\mathsf{T}} - \lambda)$ lead to k-many linearly independent vectors in $\ker(A^{\mathsf{T}}A - \lambda)$ via $v_j := A^{\mathsf{T}}u_j$.

(iv) We show that $\ker A = \ker A^{\mathsf{T}}A$ holds; from whence it follows $\operatorname{rk} A = \operatorname{rk} A^{\mathsf{T}}A$ according to the rank theorem. The inclusion "$\subseteq$" is clear. For "$\supseteq$" let $A^{\mathsf{T}}Ax = 0$. We compute

$$0 = \langle x, A^{\mathsf{T}}Ax\rangle = \langle Ax, Ax\rangle = \|Ax\|^2$$

and get $Ax = 0$. The middle of the claimed equations is clear and the last one can be proved analogously to the first.

(v) According to the orthogonal diagonalization theorem, there is an orthogonal matrix $B \in \mathbb{R}^{d\times d}$ with

$$B^{\mathsf{T}}(A^{\mathsf{T}}A)B = \operatorname{diag}(\lambda_1, \ldots, \lambda_n),$$

where $\lambda_1, \ldots, \lambda_n$ are the eigenvalues of $A^{\mathsf{T}}A$ with multiplicities. The right-hand side of the equation thus has as rank the number of non-zero eigenvalues, if we count these with multiplicities. The rank of the left-hand side in the matrix equation is equal to $\operatorname{rk} A^{\mathsf{T}}A$. □

Now we establish the connection between singular values/vectors of A and eigenvalues and eigenvalues/vectors of $AA^\intercal$ and $A^\intercal A$.

Lemma 7.3. *Let $A \in \mathbb{R}^{n\times d}$. It holds:*

(i) If u and v are singular vectors to the singular value $\sigma \in \mathrm{s}(A)$, then v is an eigenvector of $A^\intercal A$ and u is an eigenvector of $AA^\intercal$, each with corresponding eigenvalue σ^2.

Conversely, we have:

(ii) If v is an eigenvector of $A^\intercal A$ to the positive eigenvalue $\lambda > 0$, then v and $u := \frac{1}{\sqrt{\lambda}} Av$ are singular vectors to the singular value $\sqrt{\lambda}$.

(iii) If u is an eigenvector of $AA^\intercal$ for the positive eigenvalue $\lambda > 0$, then u and $v := \frac{1}{\sqrt{\lambda}} A^\intercal u$ are singular vectors with respect to the singular value $\sqrt{\lambda}$.

Proof. (i) We compute

$$AA^\intercal v = A^\intercal \sigma u = \sigma\sigma v = \sigma^2 v \quad \text{and} \quad AA^\intercal u = A\sigma v = \sigma\sigma u = \sigma^2 u.$$

(ii) Here we have $Av = \sqrt{\lambda} u$ by definition and thus

$$A^\intercal u = A^\intercal\big(\frac{1}{\sqrt{\lambda}} Av\big) = \frac{1}{\sqrt{\lambda}} A^\intercal Av = \frac{\lambda}{\sqrt{\lambda}} v = \sqrt{\lambda} v.$$

(iii) Again, $A^\intercal u = \sqrt{\lambda} u$ holds by definition and we compute

$$Av = A\big(\frac{1}{\sqrt{\lambda}} A^\intercal u\big) = \frac{1}{\sqrt{\lambda}} AA^\intercal u = \frac{\lambda}{\sqrt{\lambda}} u = \sqrt{\lambda} u,$$

which concludes the proof. □

With Lemma 7.3, it follows that for $A \in \mathbb{R}^{n\times d}$ there can be at most $p := \min(n, d)$-many pairwise different singular values. As with eigenvalues, we want to count singular values "with multiplicities". That this is indeed possible follows from Lemma 7.3, in combination with Lemma 7.2. According to these we have for $\sigma \in \mathrm{s}(A)$:

$$\begin{aligned} \mathrm{m}(\sigma) &:= \dim\left\{ v \in \mathbb{R}^d \;\middle|\; \begin{array}{c} v = 0 \text{ or } v \text{ is a right singular vector of } A \\ \text{corresponding to the singular value } \sigma \end{array} \right\} \\ &= \dim\left\{ u \in \mathbb{R}^n \;\middle|\; \begin{array}{c} u = 0 \text{ or } u \text{ is a left singular vector of } A \\ \text{corresponding to the singular value } \sigma \end{array} \right\} \\ &= \text{multiplicity of the eigenvalue } \sigma^2 \text{ of } A^\intercal A, \\ &\quad\ \text{or, equivalently, of } AA^\intercal. \end{aligned}$$

We call the above number the *multiplicity of the singular value* $\sigma > 0$. If we count with multiplicities, there are precisely $r := \mathrm{rk}(A)$-many singular values according to Lemma 7.2.

Lemma 7.4. *Let $A \in \mathbb{R}^{n\times d}$ and let $\{v_1, \dots, v_r\}$ be an orthonormal system of eigenvectors of $A^\intercal A$ corresponding to the eigenvalues $\sigma_1^2 \geqslant \dots \geqslant \sigma_r^2 > 0$. Then $\{u_1, \dots, u_r\}$ with $u_i = \frac{1}{\sigma_i} A v_i$ is an orthonormal system. The same applies if an orthonormal system $\{u_1, \dots, u_r\}$ is given, for $\{v_1, \dots, v_r\}$ with $v_i = \frac{1}{\sigma_i} A^\intercal u_i$.*

Proof. It holds

$$\langle u_i, u_j \rangle = \langle \frac{1}{\sigma_i} A v_i, \frac{1}{\sigma_j} A v_j \rangle = \frac{1}{\sigma_i \sigma_j} \langle v_i, A^\intercal A v_j \rangle = \frac{\sigma_j}{\sigma_i} \langle v_i, v_j \rangle,$$

i.e., for $i \neq j$ we have $\langle u_i, u_j \rangle = 0$ and for $i = j$ we get $\|u_j\|^2 = 1$, thus $\|u_j\| = 1$. The second part can be handled with a similar calculation. □

If we extend the orthonormal systems from Lemma 7.4 to orthonormal bases $\mathcal{V} = \{v_1, \dots, v_d\}$ and $\mathcal{U} = \{u_1, \dots, u_n\}$, it follows that the matrix of the linear map $\mathbb{R}^d \to \mathbb{R}^n$, $x \mapsto Ax$, is precisely the "diagonal matrix" $\mathrm{diag}(\sigma_1, \dots, \sigma_r) \in \mathbb{R}^{n\times d}$, suitably filled with zeros, if we equip $\mathbb{R}^d$ with the basis $\mathcal{V}$ and $\mathbb{R}^n$ with the basis $\mathcal{U}$. If we organize the basis vectors in matrices, we obtain a corresponding factorization of A.

Definition 7.5. Let $A \in \mathbb{R}^{n\times d}$ and $r = \mathrm{rk}\, A$. A factorization

$$A = U\Sigma V^\intercal$$

with orthogonal matrices $V \in \mathbb{R}^{d\times d}$, $U \in \mathbb{R}^{n\times n}$ and a diagonal matrix $\Sigma = \mathrm{diag}(\sigma_1, \dots, \sigma_r) \in \mathbb{R}^{n\times d}$ with $\sigma_1 \geqslant \dots \geqslant \sigma_r > 0$ is called a *singular value decomposition* or *SVD* of A.

Remark 7.6. (i) Since U and V are orthogonal, we can rearrange the equation $A = U\Sigma V^\intercal$ in Definition 7.5 to

$$U^\intercal A V = \mathrm{diag}(\sigma_1, \dots, \sigma_r) \in \mathbb{R}^{n\times d}$$

and obtain the analogue of the orthogonal diagonalization theorem announced at the beginning of the chapter.

(ii) If we put $\sigma_{r+1} := \dots := \sigma_p := 0$ for $p = \min(n, d)$, then for the diagonal matrix in the singular value decomposition it holds

$$\Sigma \in \left\{ \begin{bmatrix} \sigma_1 & & \\ & \ddots & \\ & & \sigma_p \end{bmatrix}, \begin{bmatrix} \sigma_1 & & & 0 & \cdots & 0 \\ & \ddots & & \vdots & & \vdots \\ & & \sigma_p & 0 & \cdots & 0 \end{bmatrix}, \begin{bmatrix} \sigma_1 & & \\ & \ddots & \\ & & \sigma_p \\ 0 & \cdots & 0 \\ \vdots & & \vdots \\ 0 & \cdots & 0 \end{bmatrix} \right\}$$

depending on whether $n = d$, $n < d$ or $n > d$. Note however, that the $\sigma_{r+1}, \ldots, \sigma_p$ are not singular values according to Definition 7.1.

We note the following simple, but very helpful, representation of matrix-matrix-matrix products of the form $U\Sigma V^{\mathsf{T}}$.

Lemma 7.7. *Let $U = (u_{ij}) \in \mathbb{R}^{n\times n}$ and $V = (v_{ij}) \in \mathbb{R}^{d\times d}$ be orthogonal matrices with columns $u_1, \ldots, u_n \in \mathbb{R}^n$ and $v_1, \ldots, v_d \in \mathbb{R}^d$. Let $\Sigma = \operatorname{diag}(\sigma_1, \ldots, \sigma_p) \in \mathbb{R}^{n\times d}$ be a matrix with $\sigma_i \in \mathbb{R}$ and $p = \min(n, d)$. Then we have*

$$U\Sigma V^{\mathsf{T}} = \sum_{i=1}^{p} u_i \sigma_i v_i^{\mathsf{T}} \quad \text{and} \quad (U\Sigma V^{\mathsf{T}})_{ij} = \sum_{\ell=1}^{p} u_{i\ell}\sigma_\ell v_{j\ell}.$$

If $1 \leqslant k < p$ and $\sigma_{k+1} = \cdots = \sigma_p = 0$ holds, then the above equation holds with k instead of p on the right-hand side.

Proof. We calculate

$$\begin{aligned}
\sum_{i=1}^{p} u_i\sigma_i v_i^{\mathsf{T}} &= \sum_{i=1}^{p} \begin{bmatrix} u_{1i} \\ \vdots \\ u_{ni} \end{bmatrix} \begin{bmatrix} \sigma_i v_{1i} & \cdots & \sigma_i v_{di} \end{bmatrix} \\
&= \sum_{i=1}^{p} \begin{bmatrix} u_{1i}\sigma_i v_{1i} & \cdots & u_{1i}\sigma_i v_{di} \\ \vdots & & \vdots \\ u_{ni}\sigma_i v_{1i} & \cdots & u_{ni}\sigma_i v_{di} \end{bmatrix} \\
&= \begin{bmatrix} \sum_{i=1}^{p} u_{1i}\sigma_i v_{1i} & \cdots & \sum_{i=1}^{p} u_{1i}\sigma_i v_{di} \\ \vdots & & \vdots \\ \sum_{i=1}^{p} u_{ni}\sigma_i v_{1i} & \cdots & \sum_{i=1}^{p} u_{ni}\sigma_i v_{di} \end{bmatrix} \\
&= \begin{bmatrix} u_{11} & \cdots & u_{1p} \\ \vdots & & \vdots \\ u_{n1} & \cdots & u_{np} \end{bmatrix} \cdot \begin{bmatrix} \sigma_1 v_{11} & \cdots & \sigma_1 v_{d1} \\ \vdots & & \vdots \\ \sigma_p v_{1p} & \cdots & \sigma_p v_{dp} \end{bmatrix} \\
&= \begin{bmatrix} u_{11} & \cdots & u_{1p} \\ \vdots & & \vdots \\ u_{n1} & \cdots & u_{np} \end{bmatrix} \cdot \begin{bmatrix} \sigma_1 & & \\ & \ddots & \\ & & \sigma_p \end{bmatrix} \cdot \begin{bmatrix} v_{11} & \cdots & v_{d1} \\ \vdots & & \vdots \\ v_{1p} & \cdots & v_{dp} \end{bmatrix} = U\Sigma V^{\mathsf{T}},
\end{aligned}$$

where the last equation is clear in the case $n = p$. In the other two cases, compare Remark 7.6(ii), one can see that removing rows from V^{T} or columns from U has no influence on the entries of $U\Sigma V^{\mathsf{T}}$. □

We are now ready for our first theorem on singular value decomposition.

Theorem 7.8. (Method 1 for determining an SVD) *Let $A \in \mathbb{R}^{n\times d}$ and $r = \operatorname{rk} A$. The following steps lead to a singular value decomposition of A:*

1. *Form the matrix $A^{\mathsf{T}}A$, determine its strictly positive eigenvalues and number them in descending order $\lambda_1 \geqslant \cdots \geqslant \lambda_r > 0$ (counting with multiplicities, there are precisely r-many; in particular, one does not need to know r a priori!). Determine a corresponding orthonormal system of eigenvectors $v_1, \ldots, v_r$.*

2. *Put* $\sigma_i = \sqrt{\lambda_i}$ *for* $i = 1, \ldots, r$.
3. *Put* $u_i = \frac{1}{\sigma_i} A v_i$ *for* $i = 1, \ldots, r$; *obtain an orthonormal set* $\{u_1, \ldots, u_r\}$.
4. *Complete to orthonormal bases* $\mathcal{V} = \{v_1, \ldots, v_d\}$ *and* $\mathcal{U} = \{u_1, \ldots, u_n\}$.
5. *Put* $V = [v_1 \ \cdots \ v_d]$, $U = [u_1 \ \cdots \ u_n]$ *and* $\Sigma = \operatorname{diag}(\sigma_1, \ldots, \sigma_r) \in \mathbb{R}^{n \times d}$.

If $AA^{\intercal}$ *has a smaller format than* $A^{\intercal}A$, *it may be easier to start with this matrix and proceed analogously. We refer to this as* Method $1\frac{1}{2}$.

Proof. It follows from Lemmas 7.2, 7.3 and 7.4 that steps 1–5 lead to two orthogonal matrices U and V, whose columns are singular vectors to the singular values σ_i. It remains to be shown that $A = U\Sigma V^{\intercal}$ holds. According to Lemma 7.7, it is sufficient to check that the two linear maps

$$A\colon \mathbb{R}^d \to \mathbb{R}^n \quad \text{and} \quad B := \sum_{i=1}^{r} u_i \sigma_i v_i^{\intercal} : \mathbb{R}^d \to \mathbb{R}^n$$

coincide on the basis $\mathcal{V}$. For $1 \leqslant j \leqslant r$ we have

$$Bv_j = \Big(\sum_{i=1}^{r} u_i \sigma_i v_i^{\intercal}\Big) v_j = \sum_{i=1}^{r} u_i \sigma_i v_i^{\intercal} v_j = \sum_{i=1}^{r} u_i \sigma_i \langle v_i, v_j \rangle = \sigma_j u_j = Av_j.$$

For $r < j \leqslant d$ we have $Bv_j = 0$ and we only need to check that $Av_j = 0$ holds. We know $\sigma_1, \ldots, \sigma_r \neq 0$ and as $A^{\intercal}u_i = \sigma_i v_i$, the $v_1, \ldots, v_r$ must lie in the range of $A^{\intercal}$. Since $\operatorname{ran} A^{\intercal}$ has dimension r, the vectors $v_{r+1}, \ldots, v_d$ must be orthogonal to $\operatorname{ran} A^{\intercal}$. That is, for any $u \in \mathbb{R}^d$ and $r < j \leqslant d$ we have

$$\langle Av_j, u \rangle = \langle v_j, A^{\intercal}u \rangle = 0.$$

But this is only possible if Av_j is zero. □

Due to the multiple choices that we have to make in the recipe established in Theorem 7.8, the matrices U and V will in general not be uniquely determined by A. However, the matrix Σ is, due to our convention of always sorting the singular values $\sigma_1 \geqslant \sigma_2 \geqslant \cdots \geqslant \sigma_r > 0$ in descending order.

Proposition 7.9. *Let* $A = U\Sigma V^{\intercal}$ *with orthogonal matrices* $U \in \mathbb{R}^{n \times n}$, $V = \mathbb{R}^{d \times d}$ *and a diagonal matrix* $\Sigma = \operatorname{diag}(\sigma_1, \ldots, \sigma_p) \in \mathbb{R}^{n \times d}$, *where* $p = \min(n, d)$. *Let* $r = \operatorname{rk} A$. *Then precisely* r*-many of the* σ_i *are zero and the rest of them are the singular values of* A *and appear in the diagonal matrix according to their multiplicity.*

Proof. Since U and V have full rank, it follows that $\operatorname{rk}\Sigma = \operatorname{rk} A = r$ and thus the first statement holds. For $j \in \{1, \ldots, p\}$ with $\sigma_j \neq 0$ we get from Lemma 7.7

$$Av_j = U\Sigma V^{\intercal} v_j = \sum_{i=1}^{r} u_i \sigma_i v_i^{\intercal} v_j = \sigma_j u_j$$

and

$$A^{\intercal}u_j = (U\Sigma V^{\intercal})^{\intercal}u_j = \sum_{i=1}^{r}(u_i\sigma_i v_i^{\intercal})^{\intercal}u_j = \sum_{i=1}^{r} v_i\sigma_i u_i^{\intercal}u_j = \sigma_j v_j.$$

Consequently, σ_j is a singular value with the j-th column v_j of V and the j-th column u_j of U as associated singular vectors. □

Next, we establish the connection between singular value decompositions and best-fit subspaces, which was already announced at the end of Chapter 6. For this, we need the Courant-Fischer formula, but in a slightly different version than in Theorem 5.13.

Theorem 7.10. (Courant-Fischer formula for eigenvalues sorted in descending order) *Let $M \in \mathbb{R}^{n\times n}$ be a symmetric matrix with eigenvalues $\lambda_1 \geqslant \lambda_2 \geqslant \cdots \geqslant \lambda_n$ and an orthonormal basis of corresponding eigenvectors $\{v_1, \dots, v_n\}$. Then for $k = 1, \dots, n$ we have*

$$\lambda_k = \max_{\substack{x\neq 0 \\ \langle x,v_i\rangle=0 \\ \text{for } i=1,\dots,k-1}} \frac{\langle x,Mx\rangle}{\langle x,x\rangle} \quad \text{and} \quad v_k \in \operatorname*{argmax}_{\substack{x\neq 0 \\ \langle x,v_i\rangle=0 \\ \text{for } i=1,\dots,k-1}} \frac{\langle x,Mx\rangle}{\langle x,x\rangle},$$

where we understand the orthogonality condition to be empty for $k = 1$.

Proof. Let M and $\lambda_1 \geqslant \lambda_2 \geqslant \cdots \geqslant \lambda_n$ be as in the theorem. The matrix $-M$ then has the eigenvalues $-\lambda_1 \leqslant -\lambda_2 \leqslant \cdots \leqslant -\lambda_n$ and the $v_1, \dots, v_n$ form (without changing the sign!) an orthonormal basis of eigenvectors for $-M$. According to Theorem 5.13 we get for $k = 1, \dots, n$

$$-\lambda_k = \min_{\substack{x\neq 0 \\ \langle x,v_i\rangle=0 \\ \text{for } i=1,\dots,k-1}} \frac{\langle x,-Mx\rangle}{\langle x,x\rangle} = -\max_{\substack{x\neq 0 \\ \langle x,v_i\rangle=0 \\ \text{for } i=1,\dots,k-1}} \frac{\langle x,Mx\rangle}{\langle x,x\rangle},$$

as well as

$$v_k \in \operatorname*{argmin}_{\substack{x\neq 0 \\ \langle x,v_i\rangle=0 \\ \text{for } i=1,\dots,k-1}} \frac{\langle x,-Mx\rangle}{\langle x,x\rangle} = \operatorname*{argmin}_{\substack{x\neq 0 \\ \langle x,v_i\rangle=0 \\ \text{for } i=1,\dots,k-1}} \frac{\langle x,Mx\rangle}{\langle x,x\rangle},$$

which concludes the proof. □

Firstly, we will now show that the iterative method to calculate the k-best fitting subspace presented in Chapter 6 provides a singular value decomposition.

Theorem 7.11. (Method 2 for determining an SVD) *Let $A \in \mathbb{R}^{n\times d}$ and $r = \operatorname{rk} A$. The following steps lead to a singular value decomposition of A:*

1. *For $k \geqslant 1$ determine iteratively*

$$\sigma_k = \max_{\substack{\|v\|=1\\ v\perp v_1,\dots,v_{k-1}}} \|Av\|, \quad v_k \in \operatorname*{argmax}_{\substack{\|v\|=1\\ v\perp v_1,\dots,v_{k-1}}} \|Av\|, \quad u_k := \frac{1}{\sigma_k} Av_k,$$

as long as $\sigma_k \neq 0$ holds (this is precisely the case for $k = 1,\dots,r$; so we do not need to know r a priori!).

2. *Complete to orthonormal bases $\mathcal{V} = \{v_1,\dots,v_d\}$ and $\mathcal{U} = \{u_1,\dots,u_n\}$.*
3. *Put $V = [v_1 \ \cdots \ v_d]$, $U = [u_1 \ \cdots \ u_d]$ and $\Sigma = \operatorname{diag}(\sigma_1,\dots,\sigma_r) \in \mathbb{R}^{n\times d}$.*

If we start with $A^\intercal$ instead of A and proceed analogously, we also get a singular value decomposition. We refer to this as Method $2\frac{1}{2}$.

Proof. Let $\lambda_1 \geqslant \dots \geqslant \lambda_r > 0$ be the strictly positive eigenvalues of $A^\intercal A$ as in Theorem 7.8 and let $\{v_1,\dots,v_r\}$ be an orthonormal system of corresponding eigenvectors. The Courant-Fischer formula in Theorem 7.10 (with substitution $v := x/\|x\|$) then implies

$$\sigma_k = \lambda_k^{1/2} = \max_{\substack{\|v\|=1\\ v\perp v_1,\dots,v_{k-1}}} (\langle v, A^\intercal Av\rangle)^{1/2} = \max_{\substack{\|v\|=1\\ v\perp v_1,\dots,v_{k-1}}} (\langle Av, Av\rangle)^{1/2} = \max_{\substack{\|v\|=1\\ v\perp v_1,\dots,v_{k-1}}} \|Av\|.$$

By construction, the $v_1,\dots,v_r$ form an orthonormal system. What now needs to be shown is that v_i is an eigenvector with eigenvalue λ_i for $i = 1,\dots,r$. Since $A^\intercal A$ is symmetric, we can choose an orthogonal matrix $B \in \mathbb{R}^{d\times d}$ with $B^\intercal A^\intercal AB = \operatorname{diag}(\lambda_1,\dots,\lambda_r,0,\dots,0)$. We consider the isometric bijection

$$B^\intercal\colon \mathbb{R}^d \to \mathbb{R}^d,\ v \mapsto B^\intercal v$$

and note that by virtue of this, $v \in \ker(A^\intercal A - \lambda_i)$ holds if and only if $B^\intercal v \in \ker(\operatorname{diag}(\lambda_1,\dots,\lambda_r,0,\dots,0) - \lambda_i)$. The latter means that $(B^\intercal v)_j = 0$ when j satisfies $\lambda_j = \lambda_i$. Next, we calculate

$$\begin{aligned}\|Av\|^2 &= \|ABB^\intercal v\|^2 = \|B^\intercal ABB^\intercal v\|^2\\ &= \|\operatorname{diag}(\lambda_1,\dots,\lambda_r,0,\dots,0)B^\intercal v\|^2 = \sum_{j=1}^r \lambda_j^2 (B^\intercal v)_j^2.\end{aligned}$$

Let i_1 be the multiplicity of λ_1 and $1 \leqslant i \leqslant i_1$. We claim

$$\operatorname*{argmax}_{\|x\|=1} \sum_{j=1}^r \lambda_j^2 x_j^2 = \{x \in \partial \mathrm{B}_1(0) \mid x_j = 0 \text{ for } j > i_1\} =: M_1.$$

Indeed, if we plug any $x \in M_1$ into the function to be maximized on the left-hand side, we get the value λ_1^2. If $x = (x_1,\dots,x_d)$ is given with $\|x\| = 1$ and there exists $j_0 > i_1$ with $x_{j_0} \neq 0$, then we have

$$\sum_{j=1}^{r} \lambda_j^2 x_j^2 \leqslant \sum_{j=1}^{r} \lambda_1^2 x_j^2 \leqslant \lambda_1^2 \|x\|^2 = \lambda_1^2,$$

where in the case that $j_0 \leqslant r$, the first estimate is strict because of the j_0-th summand. If $j_0 > r$, then the sum over the x_i^2 is strictly less than $\|x\|^2$ and therefore the second estimate is strict. It follows that M_1 is precisly the specified set of maximizers. According to the above, we get $B^\mathsf{T} v_i \in M_1$ and it follows $v_i \in \ker(A^\mathsf{T} A - \lambda_1)$ for $1 \leqslant i \leqslant i_1$. Furthermore, we note

$$M_1 = \operatorname{span}\{B^\mathsf{T} v_1, \ldots, B^\mathsf{T} v_{i_1}\} \cap \partial \mathrm{B}_1(0).$$

Let i_2 be the multiplicity of λ_{i_1+1} and from now on $i_1 < i \leqslant i_1 + i_2$. We claim

$$\underset{\substack{\|x\|=1 \\ x \perp B^\mathsf{T} v_1, \ldots, B^\mathsf{T} v_{i_1}}}{\operatorname{argmax}} \sum_{j=1}^{r} \lambda_j^2 x_j^2 = \{x \in \partial \mathrm{B}_1(0) \mid x_j = 0 \text{ for } j \leqslant i_1 \text{ and for } j > i_2\} =: M_2.$$

Let $x \in M_2$. Then $\|x\| = 1$ and $x \perp B^\mathsf{T} v_1, \ldots, B^\mathsf{T} v_{i_1}$. Subbing x into the sum on the left-hand side yields $\lambda_{i_1+1}^2$. Let x be given with $\|x\| = 1$ and $x \perp B^\mathsf{T} v_1, \ldots, B^\mathsf{T} v_{i_1}$, i.e., $x \perp M_1$ and thus $x_j = 0$ for $j \leqslant i_1$. If there is a $j_0 > i_2$ with $x_{j_0} \neq 0$, then we see

$$\sum_{j=1}^{r} \lambda_j^2 x_j^2 = \sum_{j=i_1+1}^{r} \lambda_j^2 x_j^2 \leqslant \sum_{j=i_1+1}^{r} \lambda_{i_1+1}^2 x_j^2 \leqslant \lambda_{i_1+1}^2 \|x\|^2 = \lambda_{i_1+1}^2,$$

where again at least one of the inequalities is strict. With this, we have the equation for M_2 and it follows that $v_{i_1+1}, \ldots, v_{i_1+i_2}$ are eigenvectors to λ_{i_1+1}. We repeat the same arguments for all further eigenvalues. □

The condition according to which the $v_1, \ldots, v_r$ are chosen in Theorem 7.11 coincides with that in Theorem 6.3. The calculation of the k-best fitting subspaces for $k = 1, \ldots, r$ with respect to a dataset $\{a_1, \ldots, a_n\} \subseteq \mathbb{R}^d$ thus "incidentally" provides a singular value decomposition of the matrix A, whose rows are the a_i. The next theorem shows that the converse is also true. Note that the matrix V is not uniquely determined by A.

Theorem 7.12. *Let $D = \{a_1, \ldots, a_n\} \subseteq \mathbb{R}^d$ be a dataset consisting of points $a_i = (a_{i1}, \ldots, a_{id})$ for $i = 1, \ldots, n$. We consider the* data matrix

$$A = \begin{bmatrix} a_{11} & \cdots & a_{1d} \\ \vdots & & \vdots \\ a_{n1} & \cdots & a_{nd} \end{bmatrix} \in \mathbb{R}^{n \times d},$$

whose i-th row contains the coordinates of the point a_i. Let $A = U\Sigma V^\mathsf{T}$ be any singular value decomposition of A. Then the first k-many columns $v_1, \ldots, v_k$

of V span a k-best fitting subspace for D.

Proof. Let $r = \operatorname{rk} A$ and consider firstly $k \leqslant r$. Then, according to Proposition 7.9, the columns $v_1, \ldots, v_k$ of V are right singular vectors with singular values $\sigma_1, \ldots, \sigma_k > 0$. According to Lemma 7.3, v_i is therefore an eigenvector of $A^\intercal A$ corresponding to the eigenvalue σ_i for $i = 1, \ldots, k$. As in the proof of the previous Theorem 7.11, it follows

$$v_i \in \underset{\substack{\|v\|=1 \\ v \perp v_1, \ldots, v_{i-1}}}{\operatorname{argmax}} \|Av\|$$

for $i = 1, \ldots, k$. According to Theorem 6.3, the $v_1, \ldots, v_k$ therefore span a k-best fitting subspace for D. If now $k > r$ holds, Lemma 6.4 implies

$$\operatorname{span}\{a_1, \ldots, a_n\} \subseteq \operatorname{span}\{v_1, \ldots, v_r\} \subseteq \operatorname{span}\{v_1, \ldots, v_k\}$$

and the latter space is obviously k-best fitting. □

Remark 7.13. Instead of proving the existence of a singular value decomposition by applying the classic orthogonal diagonalization theorem, one may also use the algorithm of Theorem 7.11 ("algorithmic proof"): For this, let U, V and Σ be defined as in Theorem 7.11. We know by construction that $V^\intercal V = \mathrm{id}_{\mathbb{R}^d}$ and $U^\intercal U = \mathrm{id}_{\mathbb{R}^n}$. What needs to be shown is $U^\intercal A V = \operatorname{diag}(\sigma_1, \ldots, \sigma_r)$. In order to do this, we compute the matrix product, where we denote the rows of $U^\intercal$ with $u_i^\intercal$:

$$U^\intercal A V = \begin{bmatrix} u_1^\intercal \\ \vdots \\ u_n^\intercal \end{bmatrix} A \left[v_1 \cdots v_n\right] = \begin{bmatrix} u_1^\intercal \\ \vdots \\ u_n^\intercal \end{bmatrix} \left[Av_1 \cdots Av_n\right] = \begin{bmatrix} u_1^\intercal A v_1 & \cdots & u_1^\intercal A v_d \\ \vdots & & \vdots \\ u_n^\intercal A v_1 & \cdots & u_n^\intercal A v_d \end{bmatrix}.$$

For $1 \leqslant i \leqslant r$ we obtain $u_i^\intercal A v_i = u_i^\intercal \sigma_i u_i = \sigma_i \langle u_i, u_i \rangle = \sigma_i$, where we have used the definition $u_i = \frac{1}{\sigma_i} A v_i$. For $1 \leqslant i < k \leqslant r$ it holds $u_k^\intercal A v_i = u_k^\intercal \sigma_k u_i = \sigma_k \langle u_k, u_i \rangle = 0$. For $k > r$ we have $\|A v_k\| = 0$ and thus all corresponding entries in the product matrix are zero:

$$U^\intercal A V = \begin{bmatrix} \sigma_1 & \boxed{u_1^\intercal A v_2 \; u_1^\intercal A v_2 \; \cdots \; u_1^\intercal A v_d} \\ 0 & \sigma_2 \; \boxed{u_2^\intercal A v_2 \; \cdots \; u_2^\intercal A v_d} \\ \vdots & \vdots \\ 0 & 0 \; \cdots\cdots \; \boxed{u_n^\intercal A v_d} \end{bmatrix}.$$

We now show iteratively that each framed row is zero, too. For this purpose, let us abbreviate the latter by $w_1^\intercal$, $w_2^\intercal$, $\ldots$, $w_n^\intercal$ from top to bottom. Furthermore, we denote the submatrix that is located below $w_i^\intercal$ as B_i. For $i = 1$ this yields

$$U^\intercal A V = \begin{bmatrix} \sigma_1 & w_1^\intercal \\ 0 & B_1 \end{bmatrix}$$

and we claim that $w_1 = 0$. To prove the latter, we compute

$$\left\|U^\mathsf{T}AV\left[\begin{smallmatrix}\sigma_1\\ w_1\end{smallmatrix}\right]\right\|^2 = \left\|\left[\begin{smallmatrix}\sigma_1 & w_1^\mathsf{T}\\ 0 & B_1\end{smallmatrix}\right]\left[\begin{smallmatrix}\sigma_1\\ w_1\end{smallmatrix}\right]\right\|^2 = \left\|\left[\begin{smallmatrix}\sigma_1^2+\langle w_1,w_1\rangle\\ B_1w_1\end{smallmatrix}\right]\right\|^2 \geqslant (\sigma_1^2 + \|w_1\|^2)^2$$

and

$$\left\|\left[\begin{smallmatrix}\sigma_1\\ w_1\end{smallmatrix}\right]\right\|^2 = \sigma_1^2 + \|w_1\|^2.$$

Since U^T and V are isometries, and with $v := \left[\begin{smallmatrix}\sigma_1\\ w_1\end{smallmatrix}\right] / \left\|\left[\begin{smallmatrix}\sigma_1\\ w_1\end{smallmatrix}\right]\right\|$, we get

$$\begin{aligned}\sigma_1 &= \max_{\|x\|=1}\|Ax\| = \|A\|_{\mathrm{op}} = \|U^\mathsf{T}\|_{\mathrm{op}}\|A\|_{\mathrm{op}}\|V\|_{\mathrm{op}}\\ &\geqslant \|U^\mathsf{T}AV\|_{\mathrm{op}} = \max_{\|x\|=1}\|U^\mathsf{T}AVx\| \geqslant \frac{\left\|U^\mathsf{T}AV\left[\begin{smallmatrix}\sigma_1\\ w_1\end{smallmatrix}\right]\right\|}{\left\|\left[\begin{smallmatrix}\sigma_1\\ w_1\end{smallmatrix}\right]\right\|}\\ &\geqslant \frac{\sigma_1^2 + \|w_1\|^2}{\sqrt{\sigma_1^2 + \|w_1\|^2}} = \sqrt{\sigma_1^2 + \|w_1\|^2} \geqslant \sigma_1,\end{aligned}$$

which shows that all inequalities were indeed equalities. But this is only possible if $w_1 = 0$. If $w_1, \dots, w_k = 0$ are already zero, then $w_{k+1} = 0$ is shown by repeating the last argument, but with

$$v = \begin{bmatrix}0\\ \vdots\\ 0\\ w_{k+1}\\ B_{k+1}\end{bmatrix} \Bigg/ \left\|\begin{bmatrix}0\\ \vdots\\ 0\\ w_{k+1}\\ B_{k+1}\end{bmatrix}\right\|.$$

Before we move on to further applications of the singular value decomposition (besides the calculation of best-fit subspaces), we want to show that the singular value decomposition indeed generalizes the orthogonal diagonalization theorem.

Proposition 7.14. *Let $A \in \mathbb{R}^{n\times n}$ be a symmetric matrix, let $r = \operatorname{rk} A$ and let $\sigma_1 \geqslant \cdots \geqslant \sigma_r > 0$ be its singular values with right singular vectors $v_1, \dots, v_r$, which we extend to an orthonormal basis $\{v_1, \dots, v_n\}$. Then $\sigma_1, \dots, \sigma_r$ are the strictly positive eigenvalues of A and zero is an eigenvalue with multiplicity $n - r$. Furthermore, $V^\mathsf{T}AV = \operatorname{diag}(\sigma_1, \dots, \sigma_r, 0, \dots, 0)$ with $V = [v_1 \ \cdots \ v_n]$.*

Proof. Let λ be a non-zero eigenvalue. Then $Av = \lambda v$ with $v \neq 0$. Since A is symmetric, it follows that $A^\mathsf{T}v = \lambda v$, i.e., λ is a singular value with v as right as well as left singular vector. We obtain $A = V\Sigma V^\mathsf{T}$ if we extend the orthonormal systems $\mathcal{U} = \mathcal{V} = \{v_1, \dots, v_n\}$ in the same way to orthonormal bases consisting of eigenvectors corresponding to eigenvalues $\sigma_{r+1}, \dots, \sigma_n$. □

Multiplication with V^T from the left in Proposition 7.14 thus yields exactly an orthogonal diagonalization $V^\mathsf{T}AV = \operatorname{diag}(\sigma_1, \dots, \sigma_r, 0, \dots, 0)$ as stated at the very beginning of the chapter. The reader is advised however to be cautious:

If A is *square but not symmetric*, then the σ_i in general do *not* coincide with the eigenvalues of A!

7.1 Dimensionality Reduction

When calculating the singular value decomposition in concrete applications, e.g., rating matrices A, see Example 7.22, or image matrices A, see Exercise 7.4, one notices that the singular values $\sigma_1 \geqslant \sigma_2 \geqslant \cdots \geqslant \sigma_r > 0$ often *decrease very quickly*, i.e., already for relatively small k we have

$$\sigma_{k+1} \approx \cdots \approx \sigma_r \approx 0.$$

The central idea is now to replace these small singular values by exact zero in the singular value decomposition and in this way to define an approximation

$$\breve{A} := U\breve{\Sigma}V^{\mathsf{T}} \quad \text{with} \quad \breve{\Sigma} := \operatorname{diag}(\sigma_1, \dots, \sigma_k) \in \mathbb{R}^{n\times d} \tag{7.1}$$

of the original matrix A. To compare A with $\breve{A}$, we use the following norm on the space $\mathbb{R}^{n\times d}$.

Definition 7.15. Let $A \in \mathbb{R}^{n\times d}$ be given with $A = (a_{ij})$. Then the *Frobenius norm* of A is defined by

$$\|A\|_{\mathrm{F}} := \Big(\sum_{i=1}^{n}\sum_{j=1}^{d} |a_{ij}|^2\Big)^{1/2}.$$

One can see that the Frobenius norm is just the Euclidean norm of the vector $(a_{11}, a_{12}, \dots, a_{nd}) \in \mathbb{R}^{n\cdot d}$, in which we wrote all matrix entries one after the other. Therefore, it is clear that $\|\cdot\|_{\mathrm{F}}$ is indeed a norm on $\mathbb{R}^{n\times d}$. We note the following important relationship between Frobenius norm and singular values.

Proposition 7.16. *Let $A \in \mathbb{R}^{n\times d}$ be given with rank r and singular values $\sigma_1 \geqslant \cdots \geqslant \sigma_r > 0$. Then it holds*

$$\|A\|_{\mathrm{F}} = \Big(\sum_{j=1}^{r} \sigma_j^2\Big)^{1/2} = \left\|\begin{bmatrix}\sigma_1\\ \vdots\\ \sigma_r\end{bmatrix}\right\|,$$

where on the right-hand side $\|\cdot\|$ stands for the Euclidean norm on $\mathbb{R}^r$.

Proof. In the following, we denote by trace(B), for a square matrix B, the sum of the entries on the main diagonal of B. It is then easy to check that

$$\operatorname{trace}(AA^{\mathsf{T}}) = \sum_{i=1}^{n}\sum_{j=1}^{d} a_{ij}a_{ij} = \sum_{j=1}^{d}\sum_{i=1}^{n} |a_{ij}|^2 = \|A\|_{\mathrm{F}}^2$$

holds. By calculating the main diagonal of matrix-matrix products XY and YX for $X \in \mathbb{R}^{m\times \ell}$ and $Y \in \mathbb{R}^{\ell\times m}$ we see $\mathrm{tr}(XY) = \mathrm{tr}(YX)$. Now we compute

$$\begin{aligned}\|A\|_{\mathrm{F}}^2 &= \mathrm{trace}(AA^{\mathsf{T}}) = \mathrm{trace}(U\Sigma V^{\mathsf{T}}(U\Sigma V^{\mathsf{T}})^{\mathsf{T}}) \\ &= \mathrm{trace}(U\Sigma V^{\mathsf{T}} V \Sigma^{\mathsf{T}} U^{\mathsf{T}}) = \mathrm{trace}(U\Sigma\Sigma^{\mathsf{T}} U^{\mathsf{T}}) \\ &= \mathrm{trace}(U^{\mathsf{T}} U \Sigma\Sigma^{\mathsf{T}}) = \mathrm{trace}(\Sigma\Sigma^{\mathsf{T}}) = \sigma_1^2 + \cdots + \sigma_r^2\end{aligned}$$

and are done. □

Next, we want to quantify the approximation quality that we achieve by zeroing the singular values $\sigma_{k+1}, \ldots, \sigma_r$, i.e., if we replace the data matrix A with the approximation $\breve{A}$ defined in (7.1).

Theorem 7.17. (zeroing singular values) *Let $U\Sigma V^{\mathsf{T}}$ be a singular value decomposition of $A = (a_{ij}) \in \mathbb{R}^{n\times d}$ with singular values $\sigma_1 \geqslant \cdots \geqslant \sigma_r > 0$. For fixed $1 \leqslant k \leqslant r$, let $\breve{A} = (\breve{a}_{ij})$ be defined as in (7.1). We denote by v_j the j-th column of V and by a_i and $\breve{a}_i$ the i-th row of A and $\breve{A}$, respectively. Both are understood as (column) vectors in $\mathbb{R}^d$. Then the following holds:*

(i) $\|A - \breve{A}\|_{\mathrm{F}} = \sqrt{\sigma_{k+1}^2 + \cdots + \sigma_r^2}$.

(ii) $\forall\, i = 1, \ldots, n\colon \breve{a}_i = \pi_{\mathrm{span}\{v_1,\ldots,v_k\}}(a_i)$.

Proof. (i) Since $A - \breve{A} = U(\Sigma - \breve{\Sigma})V^{\mathsf{T}} = U \operatorname{diag}(0, \ldots, 0, \sigma_{k+1}, \ldots, \sigma_r)V^{\mathsf{T}}$ holds, we get from Proposition 7.9 that the $\sigma_{k+1} \geqslant \cdots \geqslant \sigma_r$ are precisely the singular values of $A - \breve{A}$. Employing Proposition 7.16, we get the formula in (i).

(ii) Lemma 7.7 provides a formula for the entries of the i-th row of A, namely

$$\begin{aligned}a_i = (a_{i1}, \ldots, a_{id}) &= \Big(\sum_{\ell=1}^{r} u_{i\ell}\sigma_\ell v_{1\ell}, \ldots, \sum_{\ell=1}^{r} u_{i\ell}\sigma_\ell v_{d\ell}\Big) \\ &= \sum_{\ell=1}^{r} u_{i\ell}\sigma_\ell (v_{1\ell}, \ldots, v_{d\ell}) = \sum_{\ell=1}^{r} u_{i\ell}\sigma_\ell v_\ell,\end{aligned}$$

where $v_\ell = (v_{1\ell}, \ldots, v_{d\ell})$. The same calculation applies to $\breve{A}$, except that the summation stops at $k \leqslant r$. Using both leads to

$$\begin{aligned}\pi_{\mathrm{span}\{v_1,\ldots,v_k\}}(a_i) &= \sum_{\mu=1}^{k} \langle a_i, v_\mu\rangle v_\mu = \sum_{\mu=1}^{k} \Big\langle \sum_{\ell=1}^{r} u_{i\ell}\sigma_\ell v_\ell, v_\mu\Big\rangle v_\mu \\ &= \sum_{\mu=1}^{k}\sum_{\ell=1}^{r} u_{i\ell}\sigma_\ell \langle v_\ell, v_\mu\rangle v_\mu = \sum_{\ell=1}^{k} u_{i\ell}\sigma_\ell v_\ell = \breve{a}_i\end{aligned}$$

as claimed. □

If we expand the matrix-matrix-matrix product $U\check{\Sigma}V^\intercal$ in Theorem 7.17, compare Lemma 7.7, we get

$$\check{A} = U\check{\Sigma}V^\intercal = \begin{bmatrix} u_{11} \cdots u_{1k} \\ \vdots \quad\quad \vdots \\ u_{n1} \cdots u_{nk} \end{bmatrix} \begin{bmatrix} \sigma_1 & & \\ & \ddots & \\ & & \sigma_k \end{bmatrix} \begin{bmatrix} v_{11} \cdots v_{d1} \\ \vdots \quad\quad \vdots \\ v_{1k} \cdots v_{dk} \end{bmatrix},$$

which simplifies the calculation of $\check{A}$ in examples. Furthermore, we see now that $\check{A}$ can be reconstructed from $(k\cdot n + k + k\cdot d)$-many real numbers. We will use this in Exercise 7.4 for image compression via singular value decomposition.

According to Theorem 7.17(i), zeroing singular values approximates the original matrix in the Frobenius norm and we can control the Frobenius distance by the number of singular values that we set to zero. Due to $|a_{ij} - \check{a}_{ij}| \leqslant \|A-\check{A}\|_{\mathrm{F}}$ we obtain in particular a *pointwise* approximation of the matrix $A = (a_{ij})$ by $\check{A} = (\check{a}_{ij})$. Moreover, we can control the rank of the matrix $\check{A}$ by choosing k, thus obtaining an *approximation of A by matrices of lower rank* — which turns out to be optimal in the following sense.

Corollary 7.18. (best approximation of lower rank) *Under the assumptions of Theorem 7.17 we have*

$$\check{A} \in \operatorname*{argmin}_{\substack{B\in\mathbb{R}^{n\times d} \\ \operatorname{rk} B = k}} \|A - B\|_{\mathrm{F}}.$$

Proof. Assume there exists $B \in \mathbb{R}^{n\times d}$ with $\|A - B\|_{\mathrm{F}} < \|A - \check{A}\|_{\mathrm{F}}$. Put

$$V_k := \operatorname{span}\{v_1, \ldots, v_k\},\ W := \operatorname{span}\{b_1, \ldots, b_n\} \text{ and } \check{V} := \operatorname{span}\{\check{a}_1, \ldots, \check{a}_n\},$$

where $b_i \in \mathbb{R}^d$ is the vector that has as entries those of the i-th row of B. According to Theorem 7.12, V_k is a k-best fitting subspace for $\{a_1, \ldots, a_n\}$. We claim that $\check{V} = V_k$. By Theorem 7.17(ii), we have $\check{a}_i \in V_k$ for $i = 1, \ldots, n$ and thus $\check{V} \subseteq V_k$. The equality then follows from $\dim \check{V} = \operatorname{rk} \check{A} = k = \dim V_k$. Consequently, we obtain for each $i = 1, \ldots, n$ the estimates

$$\|a_i - b_i\| \underset{\substack{\uparrow \\ b_i\in W}}{\geqslant} \|a_i - \pi_W(a_i)\| \underset{\substack{\uparrow \\ V_k\ k\text{-best} \\ \text{fitting for} \\ a_1,\ldots,a_n}}{\geqslant} \|a_i - \pi_{V_k}(a_i)\| \underset{\substack{\uparrow \\ V_k=\check{V}}}{=} \|a_i - \pi_{\check{V}}(a_i)\| \underset{\substack{\uparrow \\ \text{Thm} \\ 7.17\text{(ii)}}}{=} \|a_i - \check{a}_i\|.$$

Squaring and summing up yields

$$\|A - \check{A}\|_{\mathrm{F}}^2 = \sum_{i=1}^{n} \|a_i - \check{a}_i\|^2 \leqslant \sum_{i=1}^{n} \|a_i - b_i\|^2 = \|A - B\|_{\mathrm{F}}^2$$

which contradicts our assumption. □

Next, we will prove a quantitative version of Proposition 6.5: Assuming that all singular values $\sigma_1 \geqslant \cdots \geqslant \sigma_r > 0$ are known to us, the following Corollary 7.19 will allow us to choose k such that the approximation error $\|A - \breve{A}\|_{\mathrm{F}}$ stays below a specified threshold.

Corollary 7.19. (finetuning of best-fit subspaces) *Under the assumptions of Theorem 7.17 we have*

$$\sum_{j=k+1}^{r} \sigma_j^2 = \sum_{i=1}^{n} \operatorname{dist}(a_i, \operatorname{span}\{v_1, \dots, v_k\})^2 = \min_{\substack{V \subseteq \mathbb{R}^d \\ \dim V = k}} \sum_{i=1}^{n} \operatorname{dist}(a_i, V)^2.$$

Proof. Let $V_k := \operatorname{span}\{v_1, \dots, v_k\}$. By Theorem 7.12, V_k is a k-best fitting subspace for $\{a_1, \dots, a_n\}$. The first equation in the corollary then follows from

$$\sum_{i=1}^{n} \operatorname{dist}(x_i, V_k)^2 = \sum_{i=1}^{n} \|a_i - \pi_{V_k}(a_i)\|^2 \underset{\substack{\uparrow \\ \text{Thm} \\ 7.17\text{(ii)}}}{=} \|A - \breve{A}\|_{\mathrm{F}}^2 \underset{\substack{\uparrow \\ \text{Thm} \\ 7.17\text{(i)}}}{=} \sum_{j=k+1}^{r} \sigma_j^2.$$

The second equation holds because V_k is a k-best fitting subspace. □

Let a dataset in $\mathbb{R}^d$ be given and assume that all data points are close to a subspace W of $\mathbb{R}^d$ of lower dimension. Then one might try to reduce the "dimensionality" of the dataset by orthogonally projecting the data points onto W. If we think, e.g., about the k-NN algorithm discussed in Chapter 3 it is highly desirable that this projection does not change pairwise distances between data points too much. The following picture suggests that a best-fit subspace with not too small dimension could be a good candidate for W.

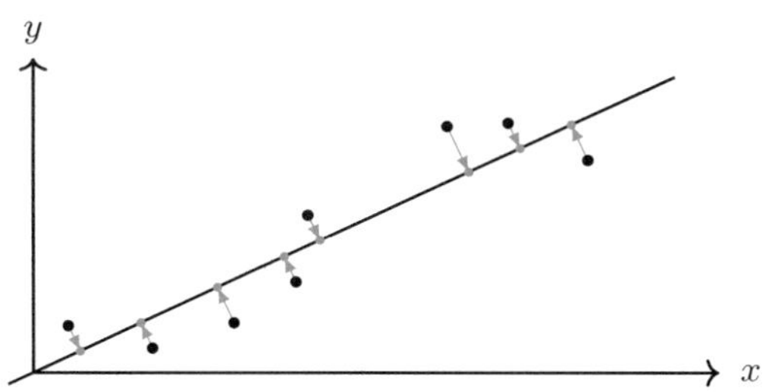

We will now give this intuition a try. In order to reduce not only the dimension, but also the actual number of coordinates, we will not use the projection π_{V_k} itself, but the mapping

$$T_{V_k} \colon \mathbb{R}^d \to \mathbb{R}^k, \; x \mapsto \begin{bmatrix} v_{11} & \cdots & v_{d1} \\ \vdots & & \vdots \\ v_{1k} & \cdots & v_{dk} \end{bmatrix} x. \tag{7.2}$$

Since $T_{V_k} x = (\langle v_1, x \rangle, \dots, \langle v_k, x \rangle)$ holds, $T_{V_k} x$ is the coordinate vector of the orthogonal projection of x onto $V_k = \operatorname{span}\{v_1, \dots, v_k\}$ with respect to the basis

$\mathcal{V}_k = \{v_1, \ldots, v_k\}$.

Corollary 7.20. (dimensionality reduction via SVD) *Under the assumptions of Theorem 7.19, let* T_{V_k} *be defined as in* (7.2). *Then we have*

$$\forall\, i,j \in \{1,\ldots,n\}\colon \big|\|T_{V_k}a_i - T_{V_k}a_j\| - \|a_i - a_j\|\big| \leqslant 2\Big(\sum_{\ell=k+1}^{r} \sigma_\ell\Big)^{1/2}.$$

Proof. We denote by $\mathcal{V}_k$ the orthonormal basis $\{v_1, \ldots, v_k\}$ of the space $V_k = \operatorname{span}\{v_1, \ldots, v_k\}$. According to Theorem 7.17(ii), we then have

$$(\check{a}_i)_{\mathcal{V}_k} = (\pi_{V_k}(a_i))_{\mathcal{V}_k} = (\langle v_1, a_i\rangle, \ldots, \langle v_k, a_i\rangle) = T_{V_k}a_i$$

for $1 \leqslant i \leqslant n$. For i and j as in the corollary, we get

$$\|T_{V_k}a_i - T_{V_k}a_j\| = \|\check{a}_i - \check{a}_j\| \leqslant \|\check{a}_i - a_i\| + \|a_i - a_j\| + \|a_j - \check{a}_j\|$$

and

$$\begin{aligned}\|a_i - a_j\| &\leqslant \|a_i - \check{a}_i\| + \|\check{a}_i - \check{a}_j\| + \|\check{a}_j - a_j\| \\ &= \|a_i - \check{a}_i\| + \|T_{V_k}a_i - T_{V_k}a_j\| + \|\check{a}_j - a_j\|.\end{aligned}$$

Subtracting $\|a_i - a_j\|$ from the first inequality and subtracting $\|T_{V_k}a_i - T_{V_k}a_j\|$ from the second inequality yields

$$\begin{aligned}\big|\|T_{V_k}a_i - T_{V_k}a_j\| - \|a_i - a_j\|\big| &\leqslant \|a_i - \check{a}_i\| + \|a_j - \check{a}_j\| \\ &= \Big(\sum_{\ell=1}^{d} |a_{j\ell} - \check{a}_{j\ell}|^2\Big)^{1/2} + \Big(\sum_{\ell=1}^{d} |a_{i\ell} - \check{a}_{i\ell}|^2\Big)^{1/2} \\ &\leqslant 2\Big(\sum_{j=1}^{n}\sum_{\ell=1}^{d} |a_{j\ell} - \check{a}_{j\ell}|^2\Big)^{1/2} = 2\,\|A - \check{A}\|_{\mathrm{F}},\end{aligned}$$

which, with Theorem 7.17(i), concludes the proof. □

Unlike in Corollary 7.18, V_k is, in the sense that pairwise distances get minimally corrupted under projection to V_k, not optimal among the k-dimensional subspaces. More formally, it holds

$$V_k \notin \operatorname*{argmin}_{\substack{V \subseteq \mathbb{R}^d \\ \dim V = k}} \sum_{i,j=1}^{n} \big|\|\pi_V(a_i) - \pi_V(a_j)\| - \|a_i - a_j\|\big|$$

and we will demonstrate this with the following example.

Example 7.21. For $d = n = 2$ we consider the matrix $A = \left[\begin{smallmatrix} 2 & 1 \\ 1 & 2 \end{smallmatrix}\right]$ with singular value decomposition

$$A = \begin{bmatrix} \sqrt{2}/2 & -\sqrt{2}/2 \\ \sqrt{2}/2 & \sqrt{2}/2 \end{bmatrix} \begin{bmatrix} 3 & 0 \\ 0 & 1 \end{bmatrix} \begin{bmatrix} \sqrt{2}/2 & \sqrt{2}/2 \\ -\sqrt{2}/2 & \sqrt{2}/2 \end{bmatrix}.$$

For the dataset $\{(2,1),(1,2)\}$, the space $V_1 = \operatorname{span}\{(\sqrt{2}/2, \sqrt{2}/2)\}$ is thus 1-best fitting. The following picture shows, however, that when projecting onto the subspace V, the distances between the two points remain unchanged, while when projecting onto V_1, the distance after projection is zero.

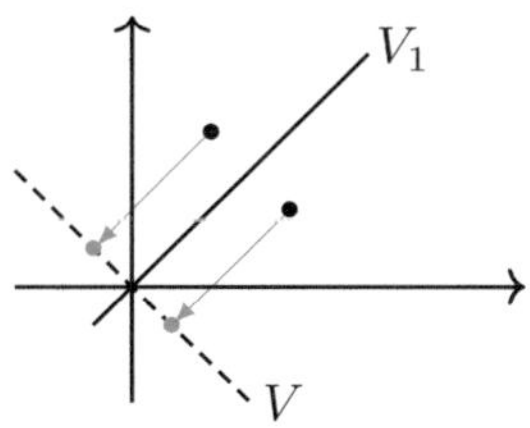

The next example illustrates on the one hand a classic application of lower rank approximation and shows how the latter can be used as a data compression method. On the other hand, we will use the example to explain both *principal component analysis (PCA)* as well as SVD-based *collaborative filtering.*

Example 7.22. We consider the table of movie ratings (from [LRU12])

	Alien	Casablanca	Star Wars	Titanic	The Matrix
Abbie	0	2	0	2	1
Bailey	1	0	1	0	1
Catherine	5	0	5	0	5
Darlene	0	4	0	4	2
Elena	3	0	3	0	3
Fatima	0	5	0	5	0
Gladys	4	0	4	0	4

respectively the associated data matrix

$$A := \begin{bmatrix} 0 & 2 & 0 & 2 & 1 \\ 1 & 0 & 1 & 0 & 1 \\ 5 & 0 & 5 & 0 & 5 \\ 0 & 4 & 0 & 4 & 2 \\ 3 & 0 & 3 & 0 & 3 \\ 0 & 5 & 0 & 5 & 0 \\ 4 & 0 & 4 & 0 & 4 \end{bmatrix}$$

and determine a singular value decomposition using Python:

$$A = \underbrace{\begin{bmatrix} 0.07 & 0.29 & 0.32 & 0.51 & 0.66 & 0.18 & -0.23 \\ 0.13 & -0.02 & -0.01 & -0.79 & 0.59 & -0.02 & -0.06 \\ 0.68 & -0.11 & -0.05 & -0.05 & -0.24 & 0.56 & -0.35 \\ 0.15 & 0.59 & 0.65 & -0.25 & -0.33 & -0.09 & 0.11 \\ 0.41 & -0.07 & -0.03 & 0.10 & -0.02 & -0.78 & -0.43 \\ 0.07 & 0.73 & -0.67 & 0.00 & -0.00 & 0.00 & 0.00 \\ 0.55 & 0.09 & -0.04 & 0.17 & 0.17 & -0.11 & 0.78 \end{bmatrix}}_{U} \underbrace{\begin{bmatrix} 12.4 & 0.0 & 0.0 & 0.0 & 0.0 \\ 0.0 & 9.5 & 0.0 & 0.0 & 0.0 \\ 0.0 & 0.0 & 1.3 & 0.0 & 0.0 \\ 0.0 & 0.0 & 0.0 & 0.0 & 0.0 \\ 0.0 & 0.0 & 0.0 & 0.0 & 0.0 \\ 0.0 & 0.0 & 0.0 & 0.0 & 0.0 \\ 0.0 & 0.0 & 0.0 & 0.0 & 0.0 \end{bmatrix}}_{\Sigma} \underbrace{\begin{bmatrix} 0.56 & 0.09 & 0.56 & 0.09 & 0.59 \\ -0.12 & 0.69 & -0.12 & 0.69 & 0.02 \\ -0.40 & -0.09 & -0.40 & -0.09 & 0.80 \\ 0.41 & 0.09 & 0.40 & 0.09 & -0.80 \\ 0.51 & 0.48 & -0.51 & -0.48 & -0.00 \\ 0.48 & -0.51 & -0.48 & 0.51 & -0.00 \end{bmatrix}}_{V^\intercal}$$

$$= \begin{bmatrix} 0.07 & 0.29 & 0.32 \\ 0.13 & -0.02 & -0.01 \\ 0.68 & -0.11 & -0.05 \\ 0.15 & 0.59 & 0.65 \\ 0.41 & -0.07 & -0.03 \\ 0.07 & 0.73 & -0.67 \\ 0.55 & 0.09 & -0.04 \end{bmatrix} \begin{bmatrix} 12.4 & & \\ & 9.5 & \\ & & 1.3 \end{bmatrix} \begin{bmatrix} 0.56 & 0.09 & 0.56 & 0.09 & 0.59 \\ -0.12 & 0.69 & -0.12 & 0.69 & 0.02 \\ -0.40 & -0.09 & -0.40 & -0.09 & 0.80 \end{bmatrix}.$$

Notice that we wrote above "$A = \cdots$" even though the entries of A were truncated after two decimal places; the symbol "$\approx$" shall for now be reserved to indicate approximations that we achieve by zeroing singular varlues. Indeed, if we set σ_3 to zero we get the best approximation of rank 2

$$\breve{A} = \begin{bmatrix} 0.15 & 1.97 & 0.15 & 1.97 & 0.56 \\ 0.92 & 0.01 & 0.92 & 0.01 & 1.94 \\ 4.84 & 0.03 & 4.84 & 0.03 & 4.95 \\ 0.36 & 4.03 & 0.36 & 4.03 & 1.20 \\ 2.92 & 0.00 & 2.92 & 0.00 & 2.98 \\ -0.34 & 4.86 & -0.34 & 4.86 & 0.65 \\ 3.71 & 1.20 & 3.71 & 1.20 & 4.04 \end{bmatrix} = \begin{bmatrix} 0.07 & 0.29 \\ 0.13 & -0.02 \\ 0.68 & -0.11 \\ 0.15 & 0.59 \\ 0.41 & -0.07 \\ 0.07 & 0.73 \\ 0.55 & 0.09 \end{bmatrix} \begin{bmatrix} 12.4 & \\ & 9.5 \end{bmatrix} \begin{bmatrix} 0.56 & 0.09 & 0.56 & 0.09 & 0.59 \\ -0.12 & 0.69 & -0.12 & 0.69 & 0.02 \end{bmatrix}$$

with $\|A - \breve{A}\|_F = 1.3$. To store the matrix $\breve{A}$ based on the singular value decomposition, we would need to store $14 + 2 + 10 = 26$ real numbers as opposed to $5 \cdot 7 = 35$ for the original matrix A. At first glance, this does not seem very impressive, but of course the ratio will improve *rapidly* when considering larger data matrices.

Before we next delve into principal component analysis, we note that in this chapter the term "dimensionality" has not yet been formally defined. In the sense of Corollaries 7.18 and 7.20, this can easily be rectified by designating the dimension of span D or, equivalently, the rank of the data matrix $A = (a_{ij})$ as the *dimensionality* of D or, equivalently, of A.

7.2 Principal Component Analysis

As announced, we will stick with Example 7.22 and now explain the method of *principal component analysis (PCA)*. In order to do so, we first consider the rating of the movie Titanic by reviewer Bailey, i.e., the penultimate entry in the second row of the matrix:

$$\begin{array}{c} \phantom{\text{Bailey} \rightarrow} \qquad\qquad \text{Titanic} \\ \phantom{\text{Bailey} \rightarrow} \qquad\qquad \downarrow \\ \text{Bailey} \rightarrow \begin{bmatrix} 0 & 2 & 0 & 2 & 1 \\ 1 & 0 & 1 & \mathbf{0} & 1 \\ 5 & 0 & 5 & 0 & 5 \\ 0 & 4 & 0 & 4 & 2 \\ 3 & 0 & 3 & 0 & 3 \\ 0 & 5 & 0 & 5 & 0 \\ 4 & 0 & 4 & 0 & 4 \end{bmatrix}. \end{array}$$

We can retrieve this entry from the matrix by multiplying from the left with a row vector that has a one in the second entry and zeros elsewhere, and from the right with a column vector that has a one in the penultimate entry and zeros elsewhere. The entries of these row or column vectors can be understood as Cartesian coordinates of Bailey in the *space of reviewers* and of Titanic in the *space of movies*, respectively. Formally, we define

$$R := \mathbb{R}^{\{\text{Abbie, Bailey}, \dots\}} \quad \text{and} \quad M := \mathbb{R}^{\{\text{Alien, Casablanca}, \dots\}}$$

as vector spaces of mappings. If we use $\mathcal{R} := \{\mathbb{1}_{\text{Abbie}}, \mathbb{1}_{\text{Bailey}}, \dots\}$ and $\mathcal{M} := \{\mathbb{1}_{\text{Alien}}, \mathbb{1}_{\text{Casablanca}}, \dots\}$ as bases of R and M respectively, then we can identify $R \cong \mathbb{R}^5$ and $M \cong \mathbb{R}^7$ via isomorphisms such that the unit vectors in $\mathbb{R}^5$ and $\mathbb{R}^7$ correspond to the reviewers and movies.

We now call

$$\text{rat}\colon R \times M \to \mathbb{R}, \ (\text{Reviewer}, \text{Movie}) \mapsto (\text{Reviewer})_{\mathcal{R}}^{\mathsf{T}} \cdot A \cdot (\text{Movie})_{\mathcal{M}}$$

the *rating map*. In the case of Bailey and Titanic we get, employing the singular value decomposition of A given in Example 7.22:

$$\begin{aligned}
\text{rat}(\text{Bailey}, \text{Titanic}) &= \begin{bmatrix} 0 & 1 & 0 & \cdots & 0 \end{bmatrix} \begin{bmatrix} 0 & 2 & 0 & 2 & 1 \\ 1 & 0 & 1 & 0 & 1 \\ 5 & 0 & 5 & 0 & 5 \\ 0 & 4 & 0 & 4 & 2 \\ 3 & 0 & 3 & 0 & 3 \\ 0 & 5 & 0 & 5 & 0 \\ 4 & 0 & 4 & 0 & 4 \end{bmatrix} \begin{bmatrix} 0 \\ \vdots \\ 0 \\ 1 \\ 0 \end{bmatrix} \\
&= \begin{bmatrix} 0 & 1 & 0 & \cdots & 0 \end{bmatrix} \begin{bmatrix} 0.07 & 0.29 & \cdots & -0.23 \\ 0.13 & -0.02 & & -0.06 \\ 0.68 & -0.11 & & -0.35 \\ 0.15 & 0.59 & & 0.11 \\ 0.41 & -0.07 & & -0.43 \\ 0.07 & 0.73 & & 0.00 \\ 0.55 & 0.09 & \cdots & 0.78 \end{bmatrix} \begin{bmatrix} 12.4 & & & \\ & 9.5 & & \\ & & 1.3 & \\ & & & \ddots \end{bmatrix} \begin{bmatrix} 0.56 & 0.09 & 0.56 & 0.09 & 0.59 \\ -0.12 & 0.69 & -0.12 & 0.69 & 0.02 \\ \vdots & & & & \vdots \\ 0.48 & -0.51 & -0.48 & 0.51 & 0.00 \end{bmatrix} \begin{bmatrix} 0 \\ \vdots \\ 0 \\ 1 \\ 0 \end{bmatrix} \\
&= \begin{bmatrix} 0.13 & -0.02 & \cdots & -0.06 \end{bmatrix} \begin{bmatrix} 12.4 & & & \\ & 9.5 & & \\ & & 1.3 & \\ & & & \ddots \end{bmatrix} \begin{bmatrix} 0.09 \\ 0.69 \\ \vdots \\ 0.00 \end{bmatrix}
\end{aligned}$$

Bailey in Cartesian coordinates (arrow to $\begin{bmatrix} 0 & 1 & 0 & \cdots & 0 \end{bmatrix}$); Titanic in Cartesian coordinates (arrow to $\begin{bmatrix} 0 & \cdots & 0 & 1 & 0 \end{bmatrix}^{\mathsf{T}}$); Bailey in $\mathcal{U}$-coordinates (arrow to $\begin{bmatrix} 0.13 & -0.02 & \cdots & -0.06 \end{bmatrix}$); Titanic in $\mathcal{V}$-coordinates (arrow to $\begin{bmatrix} 0.09 & 0.69 & \cdots & 0.00 \end{bmatrix}^{\mathsf{T}}$)

In the last line, we may again consider the entries of the row vector on the left and the column vector on the right as coordinates of Bailey and Titanic — but now with respect to *new bases* $\mathcal{U} = \{u_1, \dots, u_5\}$ of R and $\mathcal{V} = \{v_1, \dots, v_7\}$ of M! Their entries are obtained by

$$\begin{aligned}
u_1 &= \ \ 0.07 \cdot \text{Abbie} + 0.13 \cdot \text{Bailey} + \cdots + 0.55 \cdot \text{Gladys}, \\
u_2 &= \ \ 0.29 \cdot \text{Abbie} - 0.02 \cdot \text{Bailey} + \cdots + 0.09 \cdot \text{Gladys}, \\
&\ \, \vdots \\
u_5 &= -0.23 \cdot \text{Abbie} - 0.06 \cdot \text{Bailey} + \cdots + 0.78 \cdot \text{Gladys}
\end{aligned}$$

and by

$$\begin{aligned} v_1 &= 0.56 \cdot \text{Alien} + 0.09 \cdot \text{Casablanca} + \cdots + 0.59 \cdot \text{The Matrix}, \\ v_2 &= -0.12 \cdot \text{Alien} + 0.69 \cdot \text{Casablanca} + \cdots + 0.02 \cdot \text{The Matrix}, \\ &\ \vdots \\ v_7 &= 0.48 \cdot \text{Alien} - 0.51 \cdot \text{Casablanca} + \cdots + 0.00 \cdot \text{The Matrix} \end{aligned}$$

from the old basis vectors. If we denote with $\mathbb{R}^5_{\mathcal{R}}$ the space of coordinate vectors with respect to the basis $\mathcal{R}$ of the space R etc., we obtain the following base change diagram:

$$\begin{array}{ccccc} (\text{Movie})_{\mathcal{M}} & \mathbb{R}^5_{\mathcal{M}} & \xrightarrow{\ A\ } & \mathbb{R}^7_{\mathcal{R}} & (\text{Reviewer})_{\mathcal{R}} \\ \big\downarrow & V^{\mathsf{T}} \big\downarrow \sim & & \sim \big\downarrow U^{\mathsf{T}} & \big\downarrow \\ (\text{Movie})_{\mathcal{V}} & \mathbb{R}^5_{\mathcal{V}} & \xrightarrow[\ \Sigma\ ]{} & \mathbb{R}^7_{\mathcal{U}} & (\text{Reviewer})_{\mathcal{U}}. \end{array}$$

At the beginning of the chapter, we denoted by $\mathcal{U} = \{u_1, \ldots, u_n\}$ and $\mathcal{V} = \{v_1, \ldots, v_n\}$ those bases of $\mathbb{R}^n$ and $\mathbb{R}^d$ whose elements u_j and v_j were the columns of the matrices U and V, respectively. Here, of course, we get exactly the same if we identify the abstract spaces R and M with $\mathbb{R}^5_{\mathcal{V}}$ and $\mathbb{R}^7_{\mathcal{U}}$.

The new basis vectors $u_1, \ldots, u_7$ and $v_1, \ldots, v_5$ are, by definition, elements of the space R of reviewers and the space M of movies. Since they do not appear in our original table, we refer to them as *artificial reviewers* and *artificial movies*. Due to the basis property, we can reconstruct the original reviewers and movies from the artificial ones, e.g.,

$$\begin{aligned} \text{Abbie} &= \begin{bmatrix} u_1 & u_2 & \cdots & u_7 \end{bmatrix} \cdot (\text{first row of } U)^{\mathsf{T}} \\ &= 0.07 \cdot u_1 + 0.29 \cdot u_2 + \cdots - 0.23 \cdot u_7, \end{aligned}$$

$$\begin{aligned} \text{Alien} &= \begin{bmatrix} v_1 & v_2 & \cdots & v_5 \end{bmatrix} \cdot (\text{first row of } V)^{\mathsf{T}} \\ &= \begin{bmatrix} v_1 & v_2 & \cdots & v_5 \end{bmatrix} \cdot \text{first column of } V^{\mathsf{T}} \\ &= 0.56 \cdot v_1 - 0.12 \cdot v_2 + \cdots + 0.48 \cdot v_5. \end{aligned}$$

The crucial point is that we have chosen the bases $\mathcal{U}$ and $\mathcal{V}$ in such a way that for all $r \in R$ and all $m \in M$, the equation

$$\text{rat}(r, m) = (b)^{\mathsf{T}}_{\mathcal{U}} \cdot \Sigma \cdot (m)_{\mathcal{V}}$$

holds. Since Σ has only three non-zero diagonal elements, every possible rating is a sum of three numbers, in which the singular values can be interpreted as (rapidly) decreasing weights. Due to Theorem 7.17(i), we also know that the ratings of the original table are only minimally distorted if we zero $\sigma_3 = 1.3$. For Bailey's rating of Titanic, we have, e.g.,

$$\begin{aligned}\text{rat}(\text{Bailey}, \text{Titanic}) &= \begin{bmatrix} 0.13 & \text{-}0.02 & \cdots & \text{-}0.06 \end{bmatrix} \begin{bmatrix} 12.4 & & & \\ & 9.5 & & \\ & & 1.3 & \\ & & & \ddots \end{bmatrix} \begin{bmatrix} 0.09 \\ 0.69 \\ \vdots \\ 0.00 \end{bmatrix} \\ &= 0.13 \cdot 12.4 \cdot 0.09 - 0.02 \cdot 9.5 \cdot 0.69 + 0.01 \cdot 1.3 \cdot 0.09 \\ &\approx 0.13 \cdot 12.4 \cdot 0.09 - 0.02 \cdot 9.5 \cdot 0.69.\end{aligned}$$

It follows that every rating is almost completely given by the first two entries of the $\mathcal{U}$- and $\mathcal{V}$-coordinates of reviewer and movie, resepctively. These so-called *principal components* can be easily read off the rank-2 approximation of A given in Example 7.22 and in our case correspond to two *hidden concepts*:

Abbie, Bailey, Catherine, Darlene, Elena, Fatima, Gladys (rows); Alien, Casablanca, Star Wars, Titanic, The Matrix (columns)

$$\check{A} = \begin{bmatrix} 0.07 & 0.29 \\ 0.13 & \text{-}0.02 \\ 0.68 & \text{-}0.11 \\ 0.15 & 0.59 \\ 0.41 & \text{-}0.07 \\ 0.07 & 0.73 \\ 0.55 & 0.09 \end{bmatrix} \begin{bmatrix} 12.4 & \\ & 9.5 \end{bmatrix} \begin{bmatrix} 0.56 & 0.09 & 0.56 & 0.09 & 0.59 \\ \text{-}0.12 & 0.69 & \text{-}0.12 & 0.69 & 0.02 \end{bmatrix}$$

Looking at the principal components of the movies, it is not far-fetched to identify the hidden concepts as "SF" and "Romance": Alien, Star Wars and The Matrix have large first and small second entries, while it is the other way around for Casablanca and Titanic. Due to the singular values $\sigma_1 = 12.4$ and $\sigma_2 = 9.5$, "SF" has a greater influence on the ratings than "Romance". The principal components of the reviewers can be used to determine how strongly they react to the two concepts. For example, Bailey is not very receptive to the first concept and not at all to the second. For Catherine's rating, on the other hand, the first concept plays a major role, and for Fatima's rating, the second one.

To formalize the above, we first note that although we need all artificial reviewers and all movies to reconstruct the original ones individually, the first two are sufficient if we want to reconstruct only the ratings. We define the *concept spaces of reviewers and movies*

$$U_2 := \text{span}\{u_1, u_2\} \subseteq R \quad \text{and} \quad V_2 := \text{span}\{v_1, v_2\} \subseteq M,$$

and equip them with the orthonormal bases $\check{\mathcal{U}} = \{u_1, u_2\}$ and $\check{\mathcal{V}} = \{u_1, u_2\}$. Furthermore, we denote with

$$\check{U} = [u_1 \ u_2], \ \check{V} = [v_1 \ v_2] \ \text{and} \ \check{\Sigma} = \begin{bmatrix} 12.4 & \\ & 9.5 \end{bmatrix}$$

the truncated matrices. Due to $A \approx \breve{A} = \breve{U}\breve{\Sigma}\breve{V}^{\mathsf{T}}$ we obtain the "almost" commutative diagram

$$\begin{array}{ccccccc}
(\text{Movie})_{\mathcal{M}} & & \mathbb{R}^5_{\mathcal{M}} & \xrightarrow{\ A\ } & \mathbb{R}^7_{\mathcal{R}} & & (\text{Reviewer})_{\mathcal{R}} \\
\big\downarrow & & \breve{V}^{\mathsf{T}}\big\downarrow & \begin{smallmatrix}\text{commutes}\\ \text{"almost"}\end{smallmatrix} & \big\downarrow\breve{U}^{\mathsf{T}} & & \big\downarrow \\
(\pi_{V_2}(\text{Movie}))_{\breve{\mathcal{V}}} & & \mathbb{R}^2_{\breve{\mathcal{V}}} & \xrightarrow[\ \breve{\Sigma}\ ]{} & \mathbb{R}^2_{\breve{\mathcal{U}}} & & (\pi_{U_2}(\text{Reviewer}))_{\breve{\mathcal{U}}},
\end{array}$$

in which the entries of the vectors $(\pi_{U_2}(\text{Reviewer}))_{\breve{\mathcal{U}}}$ and $(\pi_{V_2}(\text{Movie}))_{\breve{\mathcal{V}}}$ are precisely the principal components. If we omit π_{V_2} and π_{U_2} to simplify our notation, the expressions $(\text{Reviewer})_{\breve{\mathcal{U}}}$ and $(\text{Movie})_{\breve{\mathcal{V}}}$ suggest that they stand for the coordinates of Reviewer and Movie in concept space. This is *not* correct, for example:

$$\text{Bailey} \not\approx 0.13 \cdot u_1 - 0.02 \cdot u_2 \quad \text{and} \quad \text{Titanic} \not\approx 0.09 \cdot v_1 + 0.69 \cdot v_2.$$

What *is* correct is that

$$\text{rat}(\text{Bailey}, \text{Titanic}) \approx \begin{bmatrix} 0.13 & \text{-}0.02 \end{bmatrix} \begin{bmatrix} 12.4 & \\ & 9.5 \end{bmatrix} \begin{bmatrix} 0.09 \\ 0.69 \end{bmatrix}$$

holds. To think of $(\text{Bailey})_{\breve{\mathcal{U}}}$ and $(\text{Titanic})_{\breve{\mathcal{V}}}$ as representing the actual Bailey and Titanic in the corresponding concept spaces thus nevertheless makes sense, because $\left[\begin{smallmatrix} 0.13 \\ \text{-}0.02 \end{smallmatrix}\right]$ and $\left[\begin{smallmatrix} 0.09 \\ \text{-}0.69 \end{smallmatrix}\right]$ apparently contain all of the information about Bailey and Titanic that is necessary to reconstruct ratings of Bailey for the original movies, or ratings of original reviewers for Titanic.

The artificial reviewers u_1, u_2 can be seen as *prototypes of reviewers* that are receptive to one and only one of the concepts. Similarly, the artificial movies v_1, v_2 can be thought of as *prototypes of movies* that are based on precisely one of the concepts. In our example, v_1 is a *pure SF movie* and u_1 is a *pure SF fan*, while v_2 is a *pure romance movie* and u_2 is a *pure romance fan*. In fact

$$\begin{aligned}
\text{rat}(u_1, \text{The Matrix}) &\approx (u_1)^{\mathsf{T}}_{\breve{\mathcal{U}}} \begin{bmatrix} 12.4 & \\ & 9.5 \end{bmatrix} (\text{The Matrix})_{\breve{\mathcal{V}}} \\
&= \begin{bmatrix} 1 & 0 \end{bmatrix} \begin{bmatrix} 12.4 & \\ & 9.5 \end{bmatrix} \begin{bmatrix} 0.59 \\ 0.02 \end{bmatrix} \\
&= 7.19
\end{aligned}$$

even exceeds the rating scale! This shows that in applications it may be necessary to pre- and postprocess the data first, see the remarks at the very end of the chapter.

Some authors call the u_i *typical reviewers* and the v_i *typical movies*. Here, "typical" is not to be understood in the sense of being average, but in the sense of a movie type explained above.

Finally, we note that in our low-dimensional example, the hidden concepts can also be read off directly from the rating matrix if we only rearrange it in a suitable way:

	Alien	The Matrix	Star Wars	Titanic	Casablanca
Bailey	**1**	**1**	**1**	0	0
Elena	**3**	**3**	**3**	0	0
Gladys	**4**	**4**	**4**	0	0
Catherine	**5**	**5**	**5**	0	0
Darlene	0	2	0	**4**	**4**
Fatima	0	0	0	**5**	**5**
Abbie	0	1	0	**2**	**2**

The crucial point is however, that the SVD-based method *automatically* finds the principal components and concept spaces — even, if we have no a priori knowledge about what hidden concepts might be behind a data matrix! In fact, it may even be the case that we still have no good interpretation after carrying out the principal component analysis, for example as we are dealing with ratings of movies which are completely unknown to us.

7.3 Collaborative Filtering

In Chapter 3, we have already discussed an instance of collaborative filtering, and in Remark 3.13(ii) we explained in particular that the notion stems from the fact that predictions for ratings by one reviewer are based not only on his own past ratings but on ratings by other reviewers as well. In Example 3.12, we used the cosine similarity of the rating vectors to compare reviewers. In the following, we will instead use (a linear extension of) the truncated diagonal matrix $\check{\Sigma}$.

We start with a scenario in which, in addition to the rating matrix from Example 7.22, we are given another reviewer of whom we only know the ratings for the movies Alien and Casablanca:

	Alien	Casablanca	Star Wars	Titanic	The Matrix
Hannah	4	1	n/a	n/a	?

To predict Hannah's rating for The Matrix, we first use the given two ratings to obtain Hannah's coordinates in the basis $\mathcal{U}_2$. Indeed, the former lead to the linear system of equations

$$\operatorname{rat}(\text{Hannah}, \text{Alien}) = [x\ y]\begin{bmatrix} 12.4 & \\ & 9.5 \end{bmatrix}\left[\begin{smallmatrix} 0.56 \\ -0.12 \end{smallmatrix}\right] = 6.94x - 1.14y \stackrel{!}{=} 4$$

$$\operatorname{rat}(\text{Hannah}, \text{Casablanca}) = [x\ y]\begin{bmatrix} 12.4 & \\ & 9.5 \end{bmatrix}\left[\begin{smallmatrix} 0.09 \\ 0.69 \end{smallmatrix}\right] = 1.12x + 6.55y \stackrel{!}{=} 1.$$

The system has the solution $x = 0.58$ and $y = 0.04$ and it follows $(\text{Hannah})_{\mathcal{U}_2} = \left[\begin{smallmatrix} 0.58 \\ 0.04 \end{smallmatrix}\right]$. Now, having Hannah's coordinates, we are able to compute

$$\text{rat}(\text{Hannah}, \text{The Matrix}) = [0.58\ 0.04]\begin{bmatrix} 12.4 & \\ & 9.5 \end{bmatrix}\left[\begin{smallmatrix} 0.59 \\ 0.02 \end{smallmatrix}\right] = 4.24,$$

and predict that Hannah will rate The Matrix with 4.

SVD-based collaborative filtering is well suited for so-called *out-of-sample predictions* as we have just seen. Note in particular that it is not necessary for a new reviewer to have a movie taste similar to that an existing reviewer, unlike, for example, when predictions are made using cosine similarity. Above, a similar movie taste to a given reviewer was evident, but happened only because our data matrix was very small and had only two hidden concepts involved.

We have already noted that the method, as we explained it above, does not automatically provide predictions in the interval $[0, 5]$. To correct such effects, one needs to adjust the input data and/or the SVD-based predictions, e.g., by normalizing beforehand and capping afterwards. In addition, it might happen, that basis vectors appear, which are more difficult to interpret than what was the case in our artificial example.

Complementing the above, we will now consider *in-sample predictions*, i.e., all reviewers and all movies are given at the beginning, but in any row and any column of our rating matrix entries are missing:

$$A = \begin{bmatrix} 1 & 1 & \square & 0 & 5 & \square \\ 2 & \square & 3 & \square & 4 & \square \\ \square & \square & 5 & 0 & 0 & \square \\ \square & 2 & 2 & \square & \square & 5 \end{bmatrix}.$$

Analogously to Example 3.11, one could provisionally fill the empty cells with row or column mean, calculate the singular value decomposition and then replace the provisionally filled in entries by SVD-predicted ratings. Alternatively we can, instead of using the factorization $A = U\Sigma V^\intercal$, use $A = PQ^\intercal$, where $P \in \mathbb{R}^{n\times k}$ and $Q \in \mathbb{R}^{d\times k}$ have orthogonal columns. In the latter factorization, the singular values no longer show up, but knowing them explicitly is not necessary for calculating predictions! We get then

$$(P, Q) \in \underset{\substack{(P,Q)\text{ have} \\ \text{orthogonal} \\ \text{columns}}}{\text{argmin}} \|A - PQ^\intercal\|_{\text{F}}^2 = \underset{\substack{(P,Q)\text{ have} \\ \text{orthogonal} \\ \text{columns}}}{\text{argmin}} \sum_{i,j} (a_{ij} - \langle p_i, q_i \rangle)^2,$$

where p_i denotes the i-th row of P and q_j the j-th row of Q. If now some of the a_{ij}'s are missing, we may simply omit these summands and optimize over the rest. With this approach, Simon Funk achieved a top position in the Netflix Challenge 2006.

References

The technique of singular value decomposition is of considerable theoretical and practical importance in many applications [Kal96]. Our approach at the beginning of this chapter follows [SSBD14, Appendix C.4]. The algorithmic proof for the existence of the SVD in Remark 7.13 follows [DH08, Theorem 5.14]. The daring teacher of a basic linear algebra course may start with the latter and then, only as a special case, present the orthogonal diagonalization theorem. The applications in Subchapters 7.3 –7.1 follow [LRU12]. We also refer to [Agg16] for a very detailed discussion of collaborative filtering. More information about the Netflix Challenge can be found in [Gow14]; [Fun06] contains an original blog post by Simon Funk.

Exercises

Exercise 7.1. Find examples for matrices $A \in \mathbb{R}^{n\times d}$ with

(i) $\sigma(A^\intercal A) \neq \sigma(AA^\intercal)$,

(ii) $\dim\ker(A^\intercal A - 0) \neq \dim\ker(AA^\intercal - 0)$.

Exercise 7.2. Let $A \in \mathbb{R}^{n\times d}$ and $\sigma > 0$. Show that $v \in \mathbb{R}^d$ can be an eigenvector of $A^\intercal A$ and $u \in \mathbb{R}^n$ can be an eigenvector of $AA^\intercal$ each with eigenvalue σ^2, without v and u being (a corresponding pair of) singular vectors with singular value σ.

Exercise 7.3. Find the 1-best fitting subspaces for the points given below using Method 1 or $1\frac{1}{2}$:

(i) $\begin{bmatrix} 4 \\ 0 \end{bmatrix}, \begin{bmatrix} 3 \\ -5 \end{bmatrix} \in \mathbb{R}^2$, (ii) $\begin{bmatrix} 2 \\ 2 \end{bmatrix}, \begin{bmatrix} -1 \\ 1 \end{bmatrix} \in \mathbb{R}^2$, (iii) $\begin{bmatrix} 3 \\ 2 \end{bmatrix}, \begin{bmatrix} 2 \\ 6 \end{bmatrix} \in \mathbb{R}^2$,

(iv) $\begin{bmatrix} 3 \\ 2 \\ 2 \end{bmatrix}, \begin{bmatrix} 2 \\ 3 \\ -2 \end{bmatrix} \in \mathbb{R}^3$, (v) $\begin{bmatrix} 1 \\ 1 \\ 0 \\ 0 \end{bmatrix}, \begin{bmatrix} 0 \\ 0 \\ 1 \\ 1 \end{bmatrix} \in \mathbb{R}^4$.

Draw pictures for (i)–(iv).

Exercise 7.4. The following grayscale image

is given by the matrix $A \in \mathbb{R}^{320\times 240}$ available at `https://mathematicaldatascience.github.io/MFO_en.html`. Compute a singular value decomposition for A. Then consider for $k = 1, 5, 10, 50$ the matrices $\check{A}$ according to (7.1).

(i) Plot the grayscale image given by each $\breve{A}$.

(ii) Compute each $\left\|\begin{bmatrix}\sigma_1\\ \vdots\\ \sigma_k\end{bmatrix}\right\| / \left\|\begin{bmatrix}\sigma_1\\ \vdots\\ \sigma_r\end{bmatrix}\right\|$ and interpret this number.

(iii) Compute each time how many real numbers need to be stored to reconstruct the image corresponding to $\breve{A}$ from these.

Exercise 7.5. Assume that in addition to the information in Example 7.22, we know that the movie Notting Hill was rated 4 by Abbie and 1 by Bailey. What rating by Catherine and what rating by Fatima does the method from Chapter 7.3 predict?

Exercise 7.6. Consider the following table of book ratings.

	Hunger Games	Jane Eyre	Twilight	Animal Farm	Da Vinci Code
Abdullah	1	0	1	0	1
Bryan	0	2	0	2	1
Calvin	5	1	5	0	5
David	1	4	1	4	3
Elias	3	1	3	0	3
Finn	0	5	0	5	0
Gustavo	4	1	4	0	4

(i) Suppose Holden rates Hunger Games with 2 and Jane Eyre with 5. Which of the other books would you recommend to Holden?

(ii) What could be the hidden concepts behind the rating table?

Note: Use without proof that the rating matrix A has the following SVD:

$$A = \begin{bmatrix} 0.12 & 0.07 & 0.01 & -0.54 & 0.57 & 0.57 & -0.14\\ 0.12 & -0.27 & 0.32 & 0.22 & 0.00 & 0.00 & -0.86\\ 0.62 & 0.28 & -0.09 & -0.40 & -0.57 & 0.00 & -0.14\\ 0.36 & -0.48 & 0.66 & -0.08 & 0.00 & 0.00 & 0.43\\ 0.38 & 0.14 & -0.11 & 0.68 & 0.00 & 0.57 & 0.14\\ 0.19 & -0.72 & -0.65 & -0.06 & 0.00 & 0.00 & 0.00\\ 0.50 & 0.21 & -0.10 & 0.13 & 0.57 & -0.57 & 0.00 \end{bmatrix} \begin{bmatrix} 13.2 & 0.0 & 0.0 & 0.0 & 0.0\\ 0.0 & 8.9 & 0.0 & 0.0 & 0.0\\ 0.0 & 0.0 & 1.3 & 0.0 & 0.0\\ 0.0 & 0.0 & 0.0 & 0.3 & 0.0\\ 0.0 & 0.0 & 0.0 & 0.0 & 0.0\\ 0.0 & 0.0 & 0.0 & 0.0 & 0.0\\ 0.0 & 0.0 & 0.0 & 0.0 & 0.0 \end{bmatrix} \begin{bmatrix} 0.51 & 0.31 & 0.51 & 0.20 & 0.57\\ 0.25 & -0.61 & 0.25 & -0.68 & 0.11\\ -0.39 & -0.20 & -0.39 & 0.02 & 0.80\\ -0.11 & 0.69 & -0.11 & -0.69 & 0.07\\ -0.70 & 0.00 & 0.70 & 0.00 & 0.00 \end{bmatrix}.$$

8 Curse and Blessing of High Dimensionality

Datasets often consist not only of a huge number of data points, but in many cases each data point also has a huge number of features. We then deal with sets

$$D = \{(x^{(i)}, y^{(i)}) \mid i = 1, \ldots, n\} \subseteq \mathbb{R}^d \times \mathbb{R}$$

where n and d are both very large. In Example 3.17 we saw that the vectorizations of relatively short texts already lead to elements of a space with a dimension greater than or equal to 100. If we reduce the vectorization to English words in their basic form, replacing, e.g., "drove", "driven" and "drives" each by the infinitive "drive" before vectorizing, we still potentially arrive at a vector space with dimension $d \approx 170\,000$. If we now take as texts, e.g., all articles of the English Wikipedia, we obtain $n \approx 6.8$ million data points. Indeed, in other examples the dimension d may even be much larger than the cardinality n of the data set: Consider, e.g., all information about the student body of a university. If one includes not only biographical data and text contributions on social media but also every photo and every video a student ever shared, it is clear that the dimension quickly reaches into the billions, while the number of data points remains in the (ten-)thousands.

In the previous chapters, we have seen some more or less advanced mathematical methods to study sets D as above, as well as datasets with multi-dimensional labels $y \in \mathbb{R}^m$. From a pure math perspective, it seems at first glance irrelevant how large n and d are in the end, since the linear algebra, analysis or probability theory used does not depend on the actual sizes of n and d. On second glance, we see however that the size of n and d is indeed very relevant, since calculating nearest neighbors, cosine similarities, singular value decompositions or the minimizer of a cost function becomes more expensive the larger n and d are. In view of a large d, this is often referred to as the *curse of high dimensionality*. For this reason, techniques for dimensionality reduction

S. A. Wegner, *Mathematical Introduction to Data Science*,
https://doi.org/10.1007/978-3-662-69426-8_8

are desirable, and techniques that are not relying on a mathematical method, that itself may cause difficulties in high dimensions, are of particular interest. After some preparations, we will introduce such a method in Chapter 11.

In addition to the issue of concrete computations, spaces of high dimension have some initially very unintuitive geometric properties, which can best be explained by considering the distribution of randomly selected points. Although they take some getting used to, these high-dimensional effects lead to methods that are not available in moderately large dimensions. This is referred to as the *blessing of high dimensionality*. In this first of several chapters on the matter, we will make ourselves familiar with some of the unintuitive effects. General results will be presented in later chapters, and we will then also return to their role in dealing with high-dimensional data.

In the following, we denote by $\|\cdot\|$ the Euclidean norm, by $\mathcal{B}^d$ the σ-algebra of Borel sets, and by λ^d the Lebesgue measure on $\mathbb{R}^d$.

Definition 8.1. Let $(\Omega, \Sigma, \mathrm{P})$ be a probability space, let $B \in \mathcal{B}^d$ be a Borel set, and let $X\colon \Omega \to \mathbb{R}^d$ be a random vector.

(i) We say that X is *uniformly distributed on* B if

$$\mathrm{P}[X \in A] = \frac{\lambda^d(A \cap B)}{\lambda^d(B)}$$

holds for every Borel set $A \in \mathcal{B}^d$. We write $X \sim \mathcal{U}(B)$ in this case, and note that we will later particularly consider the *hypercube* $\mathrm{H}_d := [-1,1]^d$ and the closed *unit ball* $\mathrm{B}_{\mathbb{R}^d} := \overline{\mathrm{B}}_1(0) \subseteq \mathbb{R}^d$ as examples for B.

(ii) We say that X is *(spherically) Gaussian distributed with mean* $\mu \in \mathbb{R}^d$ *and variance* $\sigma > 0$ if

$$\mathrm{P}[X \in A] = \frac{1}{(2\pi\sigma^2)^{d/2}} \int_A \mathrm{e}^{-\frac{\|x-\mu\|^2}{2\sigma^2}} \,\mathrm{d}\lambda^d(x)$$

holds for every Borel set $A \in \mathcal{B}^d$. We write $X \sim \mathcal{N}(0, \sigma^2, \mathbb{R}^d)$, or, if there is no risk of confusion, just $X \sim \mathcal{N}(0, \sigma^2)$. A simple calculation, see Appendix A.2, shows that $X = (X_1, \dots, X_d) \sim \mathcal{N}(0, \sigma^2, \mathbb{R}^d)$ holds if and only if $X_i \sim \mathcal{N}(0, \sigma^2, \mathbb{R})$ for all $i = 1, \dots, d$ and the X_i are independent. If $\mu = 0$ and $\sigma = 1$, we speak of *standard normally distributed* random vectors.

We now begin with the d-dimensional hypercube $\mathrm{H}_d = [-1,1]^d$, which is obviously a compact and (absolutely) convex subset of $\mathbb{R}^d$. If we consider its "corners" $v = (\pm 1, \pm 1, \dots, \pm 1) \in \mathrm{H}_d$, we see that

$$\|v - 0\| = \sqrt{(\pm 1)^2 + (\pm 1)^2 + \cdots + (\pm 1)^2} = \sqrt{d} \xrightarrow{d \to \infty} \infty$$

holds, and thus their distances to the center $0 = (0, 0, \dots, 0)$ of the hypercube become larger with increasing dimension. Moreover, with increasing d, there are always more, namely 2^d-many, corners for which the above applies. If we denote by $\mathrm{e}_i = (0, \ldots, 0, 1, 0, \ldots, 0)$ the i-th unit vector, then for $i \neq j$ we have

$$\| \pm \mathrm{e}_i - (\pm \mathrm{e}_j)\| = \|(0, \ldots, 0, \pm 1, 0, \ldots, 0, \pm 1, 0, \ldots, 0)\| = \sqrt{2}$$

and further

$$\| \mathrm{e}_i - (- \mathrm{e}_i)\| = \|(0, \ldots, 0, 2, 0, \ldots, 0)\| = 2.$$

If we read the $\pm \mathrm{e}_i$ as the centers of two different "sides" of the hypercube, then the above means that the distances of any different side centers are constant, i.e., independent of the dimension. Both calculations suggest in combination that the corners "protrude" from the hypercube and resemble increasingly longer spikes if we increase the dimension — while the side centers remain close to the origin:

If distances do not behave homogeneously with increasing d, it is to be expected that the volume in the high-dimensional hypercube is distributed inhomogeneously as well. To see this, we consider for $0 < \varepsilon < 1$ the subset

$$S_{\varepsilon,d} = \mathrm{H}_d \setminus (1 - \varepsilon)\, \mathrm{H}_d$$

of H_d, which we can think of as an ε-thick outer shell, and which is illustrated in the following three pictures for dimensions 1, 2 and 3:

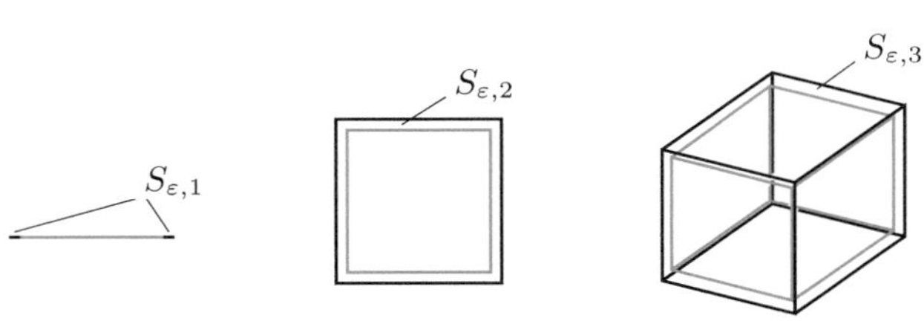

Since $(1 - \varepsilon)\, \mathrm{H}_d = [-1 + \varepsilon, 1 - \varepsilon]^d$ is a cuboid, we are able calculate the ratio of the volume of $S_{\varepsilon,d}$ to the volume of the entire hypercube as follows

$$\begin{aligned}\frac{\lambda^d(S_{\varepsilon,d})}{\lambda^d(\mathrm{H}_d)} &= \frac{\lambda^d(\mathrm{H}^d)-\lambda^d((1-\varepsilon)\,\mathrm{H}_d)}{\lambda^d(\mathrm{H}_d)} \\ &= \frac{2^d-((2(1-\varepsilon))^d}{2^d} = 1-(1-\varepsilon)^d \xrightarrow{d\to\infty} 1,\end{aligned}$$

where $0 < \varepsilon < 1$ is arbitrary. The latter means that the vast majority of the hypercube's volume is located near its surface.

The volume of the entire hypercube $\mathrm{H}_d = [-1,1]^d$, as we have considered it so far, tends towards infinity. If we instead consider a hypercube with side length one, e.g., $[0,1]^d$, we have $\lambda^d([0,1]^d) = 1$ for every dimension d. The volume in an ε-thick shell $\lambda^d([0,1]^d \backslash [\varepsilon, 1-\varepsilon]^d) = 1-(1-2\varepsilon)^d$ converges to one, as long as $0 < \varepsilon < 1/2$. So here too, we have the effect that for large d almost the entire volume of the hypercube is located near its surface.

On the other hand, not much volume of $[-1,1]^d$ or $[0,1]^d$ can be located near the "corners": For a corner $v = (v_1,\ldots,v_d) \in \mathrm{H}_d$ with $v_i \in \{1,-1\}$ we consider the set

$$\begin{aligned}K_v := \big\{x \in \mathbb{R}^d \mid x_i \in [1-\varepsilon, 1],\ &\text{if } v_i = 1 \\ \text{and } x_i \in [-1, -1+\varepsilon],\ &\text{if } v_i = -1\big\},\end{aligned}$$

i.e., a cuboid with side length $0 < \varepsilon < 1/2$ located in the v-th corner of H_d, as illustrated in the following picture for $d = 2$:

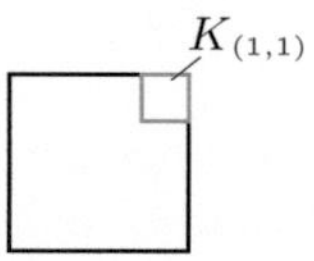

It follows that $\lambda^d(K_v) = \varepsilon^d$, and since there are 2^d corners, the volume near the corners tends to zero

$$\sum_{v\in V} \lambda^d(K_v) = 2^d \cdot \varepsilon^d \xrightarrow{d\to\infty} 0,$$

where we denote by V the set of corners of H_d. To summarize, the volume of H_d is located in a thin layer just below the surface of the hypercube, and at the same time accumulates around the midpoints of the faces.

The following pictures illustrate the three very different properties of the hypercube that we have discussed above. We advise the reader to take these pictures cautiously and even with a bit of suspicion: In addition to the fact that only one specific aspect is illustrated in each picture and the remaining properties are ignored, these are 2-dimensional drawings of a high-dimensional object:

The hypercube is compact and convex.

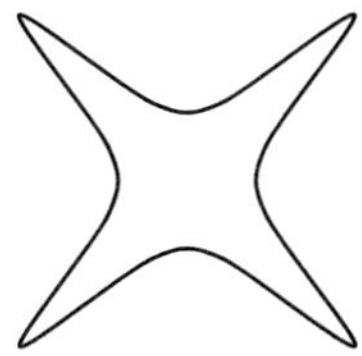

Distances between corners grow with the dimension, while distances between faces remain constant.

The volume concentrates at the surface and in the middle of the faces.

If we interpret our above findings about the volume distribution in a probabilistic way, it follows that points selected uniformly at random from H_d are likely to be close to H_d's surface and close to the center of a face. So if we have, for example, a dataset whose features are uniformly distributed in H_d, the above suggests that the pairwise distances of points, as well as their angles, will all be very close to each other: The far right picture suggests $\measuredangle(x, y) \approx 90°$ for two randomly selected $x, y \in \mathrm{H}_d$ and in Exercise 8.1 we will experimentally corroborate that $\|x - y\| \approx \sqrt{\frac{3}{2}d}$ is to be expected. Both results together mean that we will run into difficulties when using predictors based on distances or, for example, on cosine similarity.

Next, we consider points that are randomly selected from the entire space $\mathbb{R}^d$ according to a Gaussian distribution. For simplicity, we assume that the distribution was mean zero and variance one. As explained in Definition 8.1(ii), we can formalize this through a random vector $X : \Omega \to \mathbb{R}^d$, whose coordinates are independent standard normally distributed random variables. The following picture shows the distribution of the norms of 50 000 points $x^{(i)} \in \mathbb{R}^{100}$, which were randomly selected in this way.

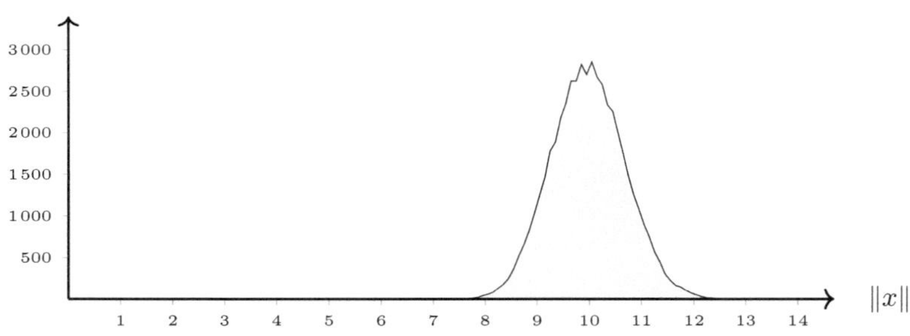

The simulation illustrates that on average $\|x^{(i)}\| \approx 10$ and that indeed almost all of the random points are located near the surface of a sphere with radius

10 in $\mathbb{R}^{100}$. If we repeat this experiment several times and vary the dimension d, the results suggest that $\mathrm{E}(\|X\|) \approx \sqrt{d}$ for $X \sim \mathcal{N}(0,1,\mathbb{R}^d)$. In addition, the sample variance of the norms seems to be bounded by a constant independent of the dimension, leading to the conjecture that $\mathrm{V}(X)$ is bounded.

d	1	10	100	1 000	10 000	100 000	1 000 000
$\frac{1}{100}\sum_{i=1}^{100}\|x^{(i)}\|$	0.73	3.05	10.10	31.61	100.03	316.21	1000.03
$\sqrt{d}$	1.00	3.16	10.00	31.62	100.00	316.22	1000.00
Variance	0.33	0.48	0.54	0.45	0.52	0.36	0.44

We leave it to the reader as Exercise 8.2 to carry out the above experiments, while we now focus on expectation and variance of $X \sim \mathcal{N}(0,1,\mathbb{R}^d)$ in an abstract mathematical sense. Since the coordinate functions $X_i \sim \mathcal{N}(0,1)$ are independent and normally distributed, we obtain

$$\begin{aligned} \mathrm{E}(\|X\|^2) &= \mathrm{E}\big(X_1^2 + \cdots + X_d^2\big) = \mathrm{E}(X_1^2) + \cdots + \mathrm{E}(X_d^2) \\ &= \mathrm{E}\big((X_1 - \mathrm{E}(X_1))^2\big) + \cdots + \mathrm{E}\big((X_d - \mathrm{E}(X_d))^2\big) \\ &= \mathrm{V}(X_1) + \cdots + \mathrm{V}(X_d) = d \end{aligned} \tag{8.1}$$

without difficulty, where we first consider the norm squared, as it is not a priori clear how the root function interacts with the expected value. Analogously, for the variance we get

$$\begin{aligned} \mathrm{V}(\|X\|^2) &= \mathrm{V}(X_1^2) + \cdots + \mathrm{V}(V_d^2) \\ &= d \cdot \mathrm{V}(X_1^2) = d \cdot \big(\mathrm{E}(X_1^4) - \mathrm{E}(X_1^2)^2\big) \\ &= d \cdot \big(\tfrac{1}{\sqrt{2\pi}} \int_{\mathbb{R}} x^4 \exp(-x^2/2)\,\mathrm{d}x - \mathrm{V}(X_1)\big) \\ &= (3-1)d = 2d. \end{aligned} \tag{8.2}$$

After this preliminary work, we are ready to prove the following more precise result, which will confirm the picture drawn by the experiments: For a d-dimensional random vector X we have $\mathrm{E}(X) \approx \sqrt{d}$ asymptotically for $d \to \infty$ and the variance is bounded by a constant independent of d.

Theorem 8.2. *Let* $X \sim \mathcal{N}(0,1,\mathbb{R}^d)$. *Then*

(i) $\forall\, d \in \mathbb{N}\colon |\mathrm{E}(\|X\| - \sqrt{d})| \leqslant 1/\sqrt{d}$,

(ii) $\forall\, d \in \mathbb{N}\colon\ \mathrm{V}(\|X\|) \leqslant 2$.

Proof. (i) We start with the equation

$$\|X\| - \sqrt{d} = \frac{\|X\|^2 - d}{2\sqrt{d}} - \frac{(\|X\|^2 - d)^2}{2\sqrt{d}\,(\|X\| + \sqrt{d})^2} =: S_d - R_d,$$

which follows from $\|X\|^2 - d = (\|X\| - \sqrt{d})(\|X\| + \sqrt{d})$. Then we use (8.2) and $\|X\| \geqslant 0$, to estimate:

$$0 \leqslant \mathrm{E}(R_d) \leqslant \frac{\mathrm{E}((\|X\|^2 - d)^2)}{2d^{3/2}} = \frac{\mathrm{V}(\|X\|^2)}{2d^{3/2}} = \frac{2d}{2d^{3/2}} = \frac{1}{\sqrt{d}}.$$

It follows that $\mathrm{E}(R_d) \to 0$ for $d \to \infty$. Since $\mathrm{E}(\|X\|^2) = d$, we get $\mathrm{E}(S_n) = 0$ and consequently

$$|\,\mathrm{E}(\|X\| - \sqrt{d})| = |\,\mathrm{E}(S_d - R_d)| = |-\mathrm{E}(R_d)| \leqslant \frac{1}{\sqrt{d}},$$

as claimed.

(ii) For the variance we calculate

$$\begin{aligned}
\mathrm{V}(\|X\|) &= \mathrm{V}(\|X\| - \sqrt{d}) = \mathrm{E}\big((\|X\| - \sqrt{d})^2\big) - \big(\mathrm{E}(\|X\| - \sqrt{d})\big)^2 \\
&\leqslant \mathrm{E}\big((\|X\| - \sqrt{d})^2\big) = \mathrm{E}\big(\|X\|^2 - 2\|X\|\sqrt{d} + d\big) \\
&= \mathrm{E}(\|X\|^2) - 2\sqrt{d}\,\mathrm{E}(\|X\|) + d \\
&= 2d - 2\sqrt{d}\,\mathrm{E}\big(\|X\| - \sqrt{d} + \sqrt{d}\big) \\
&= 2\sqrt{d}\,\mathrm{E}(R_d) \\
&\leqslant 2,
\end{aligned}$$

which proves (ii). □

Next, we want to examine the distance between two independent standard normally distributed random vectors $X, Y \sim \mathcal{N}(0, 1, \mathbb{R}^d)$. For the squared distance, we again obtain

$$\begin{aligned}
\mathrm{E}(\|X - Y\|^2) &= \sum_{i=1}^{d} \mathrm{E}\big((X_i - Y_i)^2\big) \\
&= \sum_{i=1}^{d} \mathrm{E}(X_i^2) + 2\,\mathrm{E}(X_i)\,\mathrm{E}(Y_i) + \mathrm{E}(Y_i^2) \qquad (8.3)\\
&= \sum_{i=1}^{d} (1 + 2\cdot 0\cdot 0 + 1) = 2d
\end{aligned}$$

by direct calculation, which suggests $\mathrm{E}(\|X - Y\|) \approx \sqrt{2d}$. Experiments confirm this, see Exercise 8.2, and indeed similar arguments as those in the proof of Theorem 8.2 show the following.

Theorem 8.3. *Let $X, Y \sim \mathcal{N}(0, 1, \mathbb{R}^d)$. Then*

(i) $\forall\, d \in \mathbb{N}\colon\ \mathrm{E}(\|X - Y\| - \sqrt{2d}) \leqslant 1/\sqrt{2d}$,

(ii) $\forall\, d \in \mathbb{N}$: $\mathrm{V}(\|X - Y\|) \leqslant 3$. □

We leave it to the reader, as Exercise 8.3, to work out the details of the proof.

To determine the expected value of the angle $\measuredangle(X, Y)$, we apply the already known equations $\mathrm{E}(\|X\|^2) = \mathrm{E}(\|Y\|^2) = d$ and $\mathrm{E}(\|X - Y\|^2) = 2d$ and obtain the following result.

Theorem 8.4. *Let* $X, Y \sim \mathcal{N}(0, 1, \mathbb{R}^d)$ *and let* $\xi \in \mathbb{R}^d$ *be constant. Then*

(i) $\mathrm{E}(\langle X, Y\rangle) = 0$ *and* $\mathrm{V}(\langle X, Y\rangle) = d$,

(ii) $\mathrm{E}(\langle X, \xi\rangle) = 0$ *and* $\mathrm{V}(\langle X, \xi\rangle) = \|\xi\|^2$.

Proof. (i) We use the law of cosines

$$\|X - Y\|^2 = \|X\|^2 + \|Y\|^2 - 2\|X\|\|Y\| \cos(\theta),$$

where θ is the angle between X and Y as in the following picture.

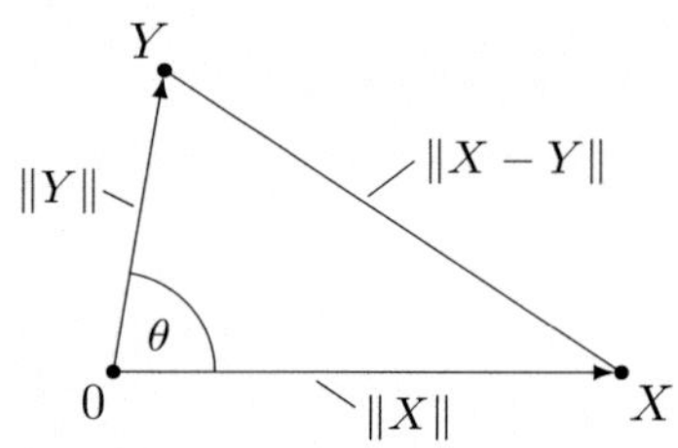

Then we use $\cos\theta = \langle X/\|X\|, Y/\|Y\|\rangle$ and obtain

$$\langle X, Y\rangle = \tfrac{1}{2}\big(\|X\|^2 + \|Y\|^2 - \|X - Y\|^2\big).$$

The expected value of this scalar product can be calculated by exploiting our previous work in (8.1) and (8.3) as

$$\mathrm{E}(\langle X, Y\rangle) = \tfrac{1}{2}\big(\mathrm{E}(\|X\|^2) + \mathrm{E}(\|Y\|^2) - \mathrm{E}(\|X - Y\|^2)\big) = \tfrac{1}{2}(d + d - 2d) = 0.$$

For the variance, we obtain with (8.2):

$$\begin{aligned}\mathrm{V}(\langle X, Y\rangle) &= \mathrm{V}(X_1Y_1 + \cdots + X_dY_d) = \mathrm{V}(X_1Y_1) + \cdots + \mathrm{V}(X_dY_d)\\ &= d \cdot \mathrm{V}(X_1Y_1) = d \cdot \big(\mathrm{E}(X_1^2)\,\mathrm{E}(Y_1^2) - \mathrm{E}(X_1)^2\,\mathrm{E}(Y_1)^2\big)\\ &= d \cdot \big(\mathrm{E}(X_1^2 - \mathrm{E}(X_1))\,\mathrm{E}(Y_1^2 - \mathrm{E}(Y_1))\big) = d \cdot \mathrm{V}(X_1)\,\mathrm{V}(Y_1) = d.\end{aligned}$$

(ii) With $\xi = (\xi_1, \ldots, \xi_d)$, one can immediately see

$$\mathrm{E}(\langle X, \xi\rangle) = \sum_{i=1}^{d} \mathrm{E}(X_i\xi_i) = \sum_{i=1}^{d} \xi_i\,\mathrm{E}(X_i) = 0$$

and

$$\mathrm{V}(\langle X,\xi\rangle)=\sum_{i=1}^{d}\mathrm{V}(\xi_iX_i)=\sum_{i=1}^{d}\xi_i^2\,\mathrm{V}(X_i)=\|\xi\|^2,$$

as claimed. □

According to Theorem 8.4, it is to be expected that two standard normally distributed randomly chosen points $x,y\in\mathbb{R}^d$ will be almost orthogonal. Unlike to all previous results, however, the sample variances are unbounded and therefore one cannot conclude that indeed for most of the points x, y in a sample $\langle x,y\rangle\approx 0$ holds. If we normalize the vectors before forming the scalar product, simulations, compare Exercise 8.2(iii)–(iv), suggest however that indeed

$$\mathrm{E}\big(\big\langle\tfrac{X}{\|X\|},\tfrac{Y}{\|Y\|}\big\rangle\big)\approx 0 \;\text{ and }\; \mathrm{V}\big(\big\langle\tfrac{X}{\|X\|},\tfrac{Y}{\|Y\|}\big\rangle\big)\xrightarrow{d\to 0}0$$

hold. Note that a direct calculation of expectation and variance is non-trivial, especially since X and $\|X\|$ are not independent.

So far, we have worked with expectation and variance and argued that a small and bounded variance implies that for the elements of a sample the quantities of interest to us (norm, distance, angle) are close to the expected value. Experiments have confirmed this. In the next Chapter 10, we will provide explicit estimates for the probabilities

$$\mathrm{P}\big[\big|\|X\|-\sqrt{d}\,\big|\geqslant\varepsilon\big] \quad\text{and}\quad \mathrm{P}\big[|\langle X,Y\rangle|\geqslant\varepsilon\big]$$

assuming $X,Y\sim\mathcal{N}(0,1,\mathbb{R}^d)$. In the case of a uniform distribution, we will consider the closed unit ball instead of the hypercube and for $X,Y\sim\mathcal{U}(\overline{\mathrm{B}}_1(0))$ we will prove explicit estimates of the above type, which will quantify the results of this chapter. We already now point out that in these estimates one has to carefully consider whether, and if so in what way, $\varepsilon\searrow 0$ and $d\to\infty$ may or must be assumed. In Chapters 9–10, we will show the following asymptotics:

Distribution	$\|X\|$	$\|X-Y\|$	$\langle X,Y\rangle$
$X,Y\sim\mathcal{N}(0,1,\mathbb{R}^d)$	$\approx\sqrt{d}$	$\approx\sqrt{2d}$	≈ 0
$X,Y\sim\mathcal{U}(\overline{\mathrm{B}}_1(0))$	≈ 1	$\approx\sqrt{2}$	≈ 0

Above, for example, $\|X\|\approx\sqrt{d}$ means that the probability for $\big|\|X\|-\sqrt{d}\big|$ being greater than a threshold can be estimated by an expression that is small when the dimension is large enough. Ideally, the threshold is a constant $\varepsilon>0$, which we can arbitrarily specify, and the bound on the probability depends only on d; in fact, however, we will have to live with the fact that often both expressions depend on ε and d. We will then avoid limits and prove theorems that guarantee that for "almost all" data points of a set D, e.g., more than the $(1-1/N)$-th part, a property applies, or that for every point of D a property holds "with high probability", e.g., via showing that $\mathrm{P}[\cdot]$ is greater than $1-1/N$. Here, N can depend on $n = \#D$ and d. This approach is referred to as *non-asymptotic analysis*.

References

This chapter is mainly based on [BHK20] and [Köp13]. The term *blessing of high dimensionality* is taken from [Don04] and the proof of Theorem 8.2 follows [Pin20]. The image of the high-dimensional hypercube on page 117 is in fact an echidnahedron, which can be plotted with Mathematica [Ras14]. In [Bul06, Dub21] similar representations (which are not echidnahedra) are given. In [Köp13, Figure 4] a 2-dimensional representation of a high-dimensional hypercube is shown with vertices that spike out. A pre-version of Chapters 8–12 is available in the form of lecture notes on arXiv, see [Weg21].

Exercises

Exercise 8.1. Write a piece of code, for example in Python, that generates points $x^{(1)}, \dots, x^{(n)}$ in the hypercube H^d by selecting their coordinates uniformly at random from $[-1,1]$. Then verify that for suitably large d

(i) $\frac{1}{n}\sum_{i=1}^{n} \|x^{(i)}\|_\infty \approx 1,$ (ii) $\frac{1}{n}\sum_{i=1}^{n} \|x^{(i)}\|_2, \approx \sqrt{\frac{d}{3}},$

(iii) $\frac{1}{n(n-1)}\sum_{i\neq j} \|x^{(i)} - x^{(j)}\|_2 \approx \sqrt{\frac{2}{3}d},$ (iv) $\frac{1}{n(n-1)}\sum_{i\neq j} \measuredangle(x^{(i)}, x^{(j)}) \approx \frac{\pi}{2}$

hold. Here, $\|\cdot\|_\infty$ denotes the maximum norm, $\|\cdot\|_2$ the Euclidean norm and the angle $\theta = \measuredangle(x,y)$ is given by $\cos\theta = \langle x/\|x\|, y/\|y\|\rangle$. Also plot the distribution of the Euclidean norms. Does the variance increase with d?

Note: Test your code first, e.g., with $n = 10$ and $d = 5$ and increase both parameters later. In Python you may use functions like `average`, `maximum`, `acos` etc. For the plot, you need to divide the interval $\left[\min_i \|x^{(i)}\|, \max_i \|x^{(i)}\|\right]$ into equal subintervals. Their number is an additional parameter, for which you have to find a suitable value by trial and error.

Exercise 8.2. Write a piece of code, e.g., in Python, that generates random points $x^{(1)}, \dots, x^{(n)} \in \mathbb{R}^d$, whose coordinates are standard normally distributed.

(i) For $d = 100$ and $n = 50\,000$, plot the distribution of their norms. The result should look like the graph on page 119.

(ii) Verify that for sufficiently large d

$$\frac{1}{n}\sum_{j=1}^{n} \|x^{(j)}\| \approx \sqrt{d} \quad \text{and} \quad \frac{1}{n(n-1)}\sum_{k\neq j} \|x^{(j)} - x^{(k)}\| \approx \sqrt{2d}$$

hold and that the sample variances are each bounded. Compare your results on the norms with the table on page 120. For the pairwise distances, you will get something very similar.

(iii) Verify that for sufficiently large d

$$\tfrac{1}{n(n-1)}\sum_{i\neq j}\langle x^{(i)}, x^{(j)}\rangle \approx 0$$

holds, but that the sample variance tends to infinity; compare Theorem 8.4, or Exercise 8.3 for a quantitative and non-experimental version.

(iv) Verify that for sufficiently large d

$$\tfrac{1}{n(n-1)}\sum_{i\neq j}\Big\langle \tfrac{x^{(i)}}{\|x^{(i)}\|}, \tfrac{x^{(j)}}{\|x^{(j)}\|}\Big\rangle \approx 0$$

holds and that now the sample variance is not only bounded, but for $d \to \infty$ even converges to zero.

Exercise 8.3. Let $X, Y \sim \mathcal{N}(0, 1, \mathbb{R}^d)$. Show:

(i) $\forall\, d \in \mathbb{N}\colon\ \mathrm{E}(\|X - Y\| - \sqrt{2d}) \leqslant 1/\sqrt{2d}$,

(ii) $\forall\, d \in \mathbb{N}\colon\ \mathrm{V}(\|X - Y\|) \leqslant 3$.

Note: Firstly, calculate $\mathrm{V}((X_i - Y_i)^2) = 3$ by proving $X_i - Y_i \sim \mathcal{N}(0, 2, \mathbb{R})$ and applying a suitable formula for the fourth moment. Then infer $\mathrm{V}(\|X - Y\|^2) \leqslant 3d$. Finally, adapt the arguments with which we calculated or estimated $\mathrm{E}(\|X\| - \sqrt{d})$ and $\mathrm{V}(\|X\|)$ in the proof of Theorem 8.2.

Exercise 8.4. If you enjoy horror movies, watch Andrzej Sekuła's *Hypercube* from 2002.

9 Concentration of Measure

As in the previous Chapter 8, in the following $\|\cdot\|$ denotes the Euclidean norm, $\mathcal{B}^d$ the σ-algebra of Borel sets, and λ^d the Lebesgue measure on $\mathbb{R}^d$. We further denote with

$$\mathrm{B}_r(x_0) = \{x \in \mathbb{R}^d \mid \|x - x_0\| < r\} \quad \text{and} \quad \overline{\mathrm{B}}_r(x_0) = \{x \in \mathbb{R}^d \mid \|x - x_0\| \leqslant r\}$$

the open, respectively closed, ball with center x_0 and radius $r > 0$ in $\mathbb{R}^d$. We abbreviate the closed unit ball with $\mathrm{B}_{\mathbb{R}^d} := \overline{\mathrm{B}}_1(0)$ and remind the reader that $r\,\mathrm{B}_{\mathbb{R}^d} = \overline{\mathrm{B}}_r(x_0)$ holds.

We now begin with the following observation. Let $B \subseteq \mathbb{R}^d$ be a measurable set, such that $\mu \cdot B \subseteq B$ holds for $0 < \mu < 1$. The reader may think for example of a star-shaped domain with center zero. Then we may interpret $B\backslash(1-\varepsilon)B \subseteq B$ as the part of B which is close to the "surface" of B.

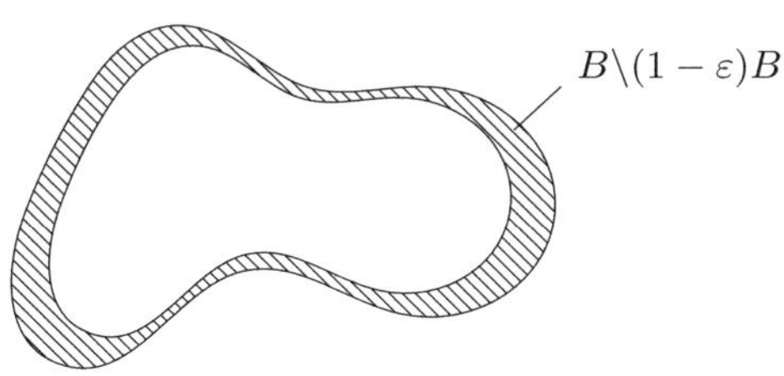

Analogous to the hypercube in Chapter 8, we can calculate the ratio

$$\begin{aligned}\frac{\lambda^d(B\backslash(1-\varepsilon)B)}{\lambda^d(B)} &= \frac{\lambda^d(B)-\lambda^d((1-\varepsilon)B)}{\lambda^d(B)}\\ &= \frac{\lambda^d(B)-(1-\varepsilon)^d\lambda^d(B)}{\lambda^d(B)}\\ &= 1-(1-\varepsilon)^d\end{aligned}$$

and see that the latter is close to one for arbitrarily small $0 < \varepsilon < 1$ if only

S. A. Wegner, *Mathematical Introduction to Data Science*,
https://doi.org/10.1007/978-3-662-69426-8_9

d is large enough, see also Remark 9.3. This generalizes the case of the hypercube and shows that almost the entire measure of B is located just below its "surface", while the "interior" contributes little to the measure. With the well-known estimate $e^t \leqslant \frac{1}{1-t}$, the above provides the following *surface concentration theorem* in the special case of the unit ball $B = B_{\mathbb{R}^d}$.

Theorem 9.1. (Surface concentration theorem) *Let $d \in \mathbb{N}$ and $0 < \varepsilon \leqslant 1$. Then we have*

$$\lambda^d(\overline{B}_1(0) \backslash B_{1-\varepsilon}(0)) \geqslant \big(1 - e^{-\varepsilon d}\big) \cdot \lambda^d(\overline{B}_1(0)),$$

that is, at least the $(1 - e^{-\varepsilon d})$-th part of the volume of the d-dimensional unit ball is ε-close to its surface. □

To make this effect of volume concentration more comprehensible let us consider a ball in $\mathbb{R}^3$: If we partition the ball into spherical shells that all have the same thickness but correspondingly different radii, then shells with a small radius will contribute less to the volume than shells with a large radius. This argument applies indeed for any dimension $d \geqslant 2$ and the more dimensions are available, the stronger the effect of the radius will be.

To formulate the following *waist concentration theorem*, we consider the first coordinate of $\mathbb{R}^d$ as "north" and then look at a 2ε-wide slice W_ε cut out of the unit sphere around its "equator":

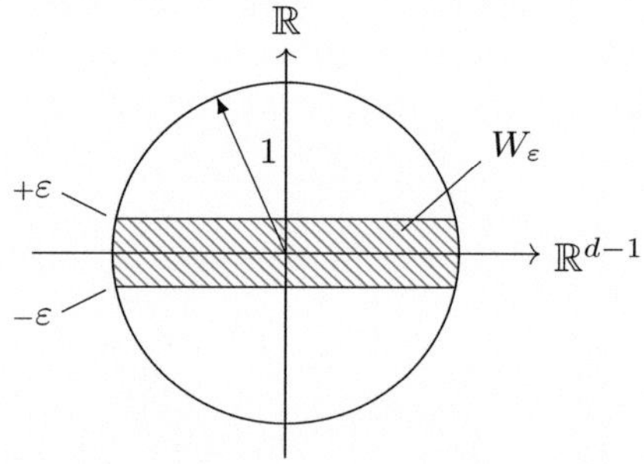

Theorem 9.2 will show that most of the volume of the unit ball is located within this slice, or in other words, at its waist. Just like in the case of surface concentration, this is only surprising at first glance: if we go from the origin in "north" direction, the ball becomes thinner in all of the remaining $(d-1)$-many directions.

Theorem 9.2. (Waist concentration theorem) *Let $d \in \mathbb{N}_{\geqslant 3}$ and $\varepsilon > 0$. Then*

$$\lambda^d(W_\varepsilon) \geqslant \big(1 - \tfrac{2}{\varepsilon\sqrt{d-1}}\, e^{-\frac{\varepsilon^2(d-1)}{2}}\big) \cdot \lambda^d(B_{\mathbb{R}^d})$$

holds, where $W_\varepsilon = \{(x_1, \ldots, x_d) \in B_{\mathbb{R}^d} \mid |x_1| \leqslant \varepsilon\}$.

Proof. We first note that the statement is trivial for $\varepsilon \geqslant 1$ and therefore we assume $\varepsilon < 1$. For $\delta \geqslant 0$ we define $H_\delta = \{(x_1, \ldots, x_d) \in \overline{\mathrm{B}}_1(0) \mid x_1 \geqslant \delta\}$, see the following picture, and use that obviously $\frac{1}{2} \cdot \lambda^d(W_\varepsilon) = \lambda^d(H_0 \backslash H_\varepsilon)$ holds.

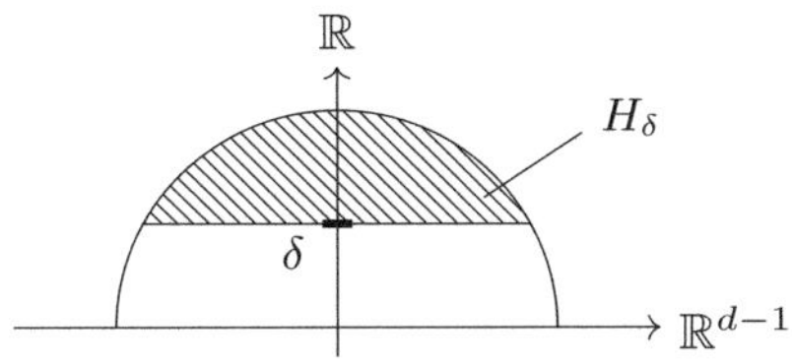

We will now firstly calculate the volume of H_δ. For $x_1 \in [\delta, 1]$ we cut out a slice at height x_1 from H_δ and get

$$
\begin{aligned}
H_{\delta, x_1} &= \big\{(x_2, \ldots, x_d) \in \mathbb{R}^{d-1} \mid (x_1, x_2, \ldots, x_d) \in H_\delta\big\} \\
&= \big\{(x_2, \ldots, x_d) \in \mathbb{R}^{d-1} \mid \|(x_1, x_2, \ldots, x_d)\| \leqslant 1\big\} \\
&= \big\{y \in \mathbb{R}^{d-1} \mid x_1^2 + \|y\|^2 \leqslant 1\big\} \\
&= \overline{\mathrm{B}}_{(1-x_1^2)^{1/2}}(0),
\end{aligned}
$$

where the ball in the last line is taken in $\mathbb{R}^{d-1}$.

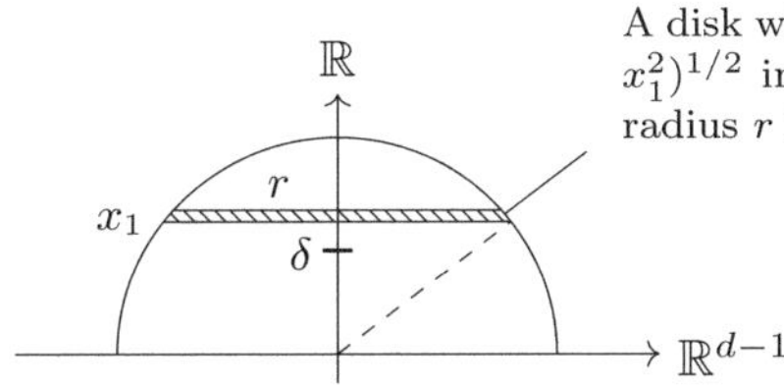

A disk with radius $r = (1-x_1^2)^{1/2}$ in $\mathbb{R}^d$ is a ball with radius r in $\mathbb{R}^{d-1}$.

Now we apply the Cavalieri's Principle and get

$$
\begin{aligned}
\lambda^d(H_\delta) &= \int_\delta^1 \lambda^{d-1}(H_{\delta, x_1}) \, \mathrm{d}\lambda(x_1) \\
&= \int_\delta^1 \lambda^{d-1}(\overline{\mathrm{B}}_{(1-x_1^2)^{1/2}}(0)) \, \mathrm{d}\lambda(x_1) \\
&= \int_\delta^1 (1 - x_1^2)^{\frac{d-1}{2}} \lambda^{d-1}(\overline{\mathrm{B}}_1(0)) \, \mathrm{d}\lambda(x_1) \\
&= \lambda^{d-1}(\mathrm{B}_{\mathbb{R}^{d-1}}) \cdot \int_\delta^1 (1 - x_1^2)^{\frac{d-1}{2}} \, \mathrm{d}\lambda(x_1).
\end{aligned} \tag{9.1}
$$

Since we want to estimate $\lambda^d(W_\varepsilon) = 2 \cdot (\lambda^d(H_0 \backslash H_\varepsilon)) = 2 \cdot (\lambda^d(H_0) - \lambda^d(H_\varepsilon))$ from below, we need an estimate of the integral in the last line of (9.1) from below for $\delta = 0$, and an estimate from above for $\delta = \varepsilon$.

① We start with $\delta = 0$. Reducing the upper limit of the integral from 1 to $\frac{1}{\sqrt{d-1}}$ in the last line of (9.1) will make the integral's value only smaller. By integrating only over $0 \leqslant x_1 \leqslant \frac{1}{\sqrt{d-1}}$, we bound the integrand from below as follows:

$$x_1^2 \leqslant \tfrac{1}{d-1} \implies 1 - \tfrac{1}{d-1} \leqslant 1 - x_1^2 \implies \left(1 - \tfrac{1}{d-1}\right)^{\frac{d-1}{2}} \leqslant \left(1 - x_1^2\right)^{\frac{d-1}{2}}.$$

For the sake of readability, we now rename the integration variable to x. Changing the integration limit and applying the lower bound for the integrand established above, we then get

$$\begin{aligned} \int_0^1 (1-x^2)^{\frac{d-1}{2}}\, \mathrm{d}\lambda(x) &\geqslant \int_0^{\frac{1}{\sqrt{d-1}}} (1-x^2)^{\frac{d-1}{2}}\, \mathrm{d}\lambda(x) \\ &\geqslant \int_0^{\frac{1}{\sqrt{d-1}}} \left(1 - \tfrac{1}{d-1}\right)^{\frac{d-1}{2}} \mathrm{d}\lambda(x) \\ &= \left(1 - \tfrac{1}{d-1}\right)^{\frac{d-1}{2}} \cdot \frac{1}{\sqrt{d-1}} \\ &\underset{\substack{\uparrow \\ \text{Bernoulli's} \\ \text{inequality}}}{\geqslant} \left(1 - \frac{d-1}{2} \cdot \frac{1}{d-1}\right) \cdot \frac{1}{\sqrt{d-1}} = \frac{1}{2\sqrt{d-1}}. \end{aligned}$$

Notice that Bernoulli's inequality applies, since $\frac{d-1}{2} \geqslant 1$ holds due to our assumption $d \geqslant 3$ in the statement of the theorem.

② Next, we estimate the integral for $\delta = \varepsilon > 0$ from above. From the elementary inequality $\mathrm{e}^x \leqslant \frac{1}{1-x}$ it follows that $1 - x^2 \leqslant \mathrm{e}^{-x^2}$ and thus we get

$$\begin{aligned} \int_\varepsilon^1 (1-x^2)^{\frac{d-1}{2}}\, \mathrm{d}\lambda(x) &\leqslant \int_\varepsilon^1 \left(\mathrm{e}^{-x^2}\right)^{\frac{d-1}{2}} \tfrac{x}{\varepsilon}\, \mathrm{d}x \underset{\substack{\uparrow \\ x/\varepsilon \geqslant 1}}{\leqslant} \int_\varepsilon^1 \left(\mathrm{e}^{-x^2}\right)^{\frac{d-1}{2}} \tfrac{x}{\varepsilon}\, \mathrm{d}\lambda(x) \\ &= \tfrac{1}{\varepsilon} \cdot \int_\varepsilon^1 x\, \mathrm{e}^{\frac{d-1}{2}}\, x^2\, \mathrm{d}x = \tfrac{1}{\varepsilon} \cdot \frac{\mathrm{e}^{-\frac{d-1}{2}} - \mathrm{e}^{-\frac{\varepsilon^2(d-1)}{2}}}{-(d-1)} \\ &\leqslant \frac{1}{\varepsilon(d-1)}\, \mathrm{e}^{-\frac{\varepsilon^2(d-1)}{2}}. \end{aligned}$$

Subbing the results of ① and ② into (9.1) yields the two estimates

$$\lambda^d(H_0) \geqslant \lambda^d(\mathrm{B}_{\mathbb{R}^{d-1}}) \frac{1}{2\sqrt{d-1}} \quad \text{and} \quad \lambda^d(H_\varepsilon) \leqslant \lambda^d(\mathrm{B}_{\mathbb{R}^{d-1}}) \frac{1}{\varepsilon(d-1)}\, \mathrm{e}^{-\frac{\varepsilon^2(d-1)}{2}},$$

from which finally

$$\frac{\lambda^d(W_\varepsilon)}{\lambda^d(\mathrm{B}_{\mathbb{R}})} = \frac{2 \cdot (\lambda^d(H_0) - \lambda^d(H_\varepsilon))}{2 \cdot \lambda^d(H_0)} = 1 - \frac{\lambda^d(H_\varepsilon)}{\lambda^d(H_0)}$$

$$\geqslant 1 - \frac{2\sqrt{d-1}}{\varepsilon(d-1)}\, \mathrm{e}^{-\frac{\varepsilon^2(d-1)}{2}} = 1 - \frac{2}{\varepsilon\sqrt{d-1}}\, \mathrm{e}^{-\frac{\varepsilon^2(d-1)}{2}}$$

follows as desired. □

At the end of Chapter 8, we pointed out that the interpretation of results for $d \to \infty$ requires caution. We now illustrate why that is so using the waist concentration theorem. Its name already suggests that in a *high*-dimensional space, the volume of the unit ball is concentrated *near* its waist. The last two size specifications are represented in the theorem by (small) ε and (large) d. Caution is required in the following sense.

Remark 9.3. (i) If we denote the constant in Theorem 9.2 by

$$K_{\varepsilon,d} = 1 - \frac{2}{\varepsilon\sqrt{d-1}}\, \mathrm{e}^{-\frac{\varepsilon^2(d-1)}{2}} \in (-\infty, 1),$$

then $K_{\varepsilon,d} \xrightarrow{d\to\infty} 1$ for each fixed $\varepsilon > 0$. In this sense, the above interpretation of the theorem is to be understood.

(ii) However, if we fix the dimension d, then $K_{\varepsilon,d} \xrightarrow{\varepsilon\to 0} -\infty$. In this way, the theorem can*not* be interpreted.

(iii) If we consider $\varepsilon > 0$ and $d \in \mathbb{N}$ as fixed, the estimate $\lambda^d(W_\varepsilon) \geqslant K_{\varepsilon,d}\lambda^d(\mathrm{B}_{\mathbb{R}^d})$ from Theorem 9.2 is non-trivial only if $K_{\varepsilon,d} > 0$. By substituting $a := \varepsilon\sqrt{d-1}$, it can be seen that there exists an $a_0 \in (1,2)$ such that

$$\forall\, \varepsilon > 0,\ d \in \mathbb{N}_{\geqslant 2} \colon \varepsilon > \frac{a_0}{\sqrt{d-1}} \implies K_{\varepsilon,d} > 0$$

holds, compare Exercise 9.3. Therefore, it is indeed possible to consider arbitrarily small ε's in Theorem 9.2 — if this goes along with using a suitably large d.

(iv) Also in the surface concentration theorem, the constant $1-\mathrm{e}^{-\varepsilon d}$ depends on ε and d, but it is always positive. Nevertheless, the theorem is only of value when $1-\mathrm{e}^{-\varepsilon d}$ is close to 1, and it can be seen that in this case too a trade-off between small ε and large d is possible.

The above remarks explain very precisely why we speak of the *blessing of high dimensionality*: If the dimension d is large enough, the parameter $\varepsilon > 0$ can be chosen small enough, such that the two theorems above produce useful statements. If d is too small, however, the statements are either trivial or not useful, cf. Exercise 9.4.

Before we get to a probabilistic interpretation of the measure concentration theorems, we note the following mere reformulation of Theorem 9.2.

Corollary 9.4. *Let $d \in \mathbb{N}_{\geqslant 3}$. Then*

$$\frac{\lambda^d(\{(x_1,\dots,x_d)\in \mathrm{B}_{\mathbb{R}^d} \mid |x_1|>\varepsilon\})}{\lambda^d(\mathrm{B}_{\mathbb{R}^d})} \leqslant \frac{2}{\varepsilon\sqrt{d-1}}\,\mathrm{e}^{-\frac{\varepsilon^2(d-1)}{2}}$$

holds for every $\varepsilon > 0$. □

In the rest of the chapter, we will use the two preceding theorems and the above corollary to derive statements about uniformly distributed random vectors on the unit ball, i.e., $X\colon \Omega \to \mathbb{R}^d$ with $X \sim \mathcal{U}(\mathrm{B}_{\mathbb{R}^d})$, compare Definition 8.1. For such a random vector, we immediately obtain

$$\mathrm{P}\big[\|X\| \geqslant 1-\varepsilon\big] = \frac{\lambda^d(\overline{\mathrm{B}}_1(0)\backslash \mathrm{B}_{1-\varepsilon}(0))}{\lambda^d(\overline{\mathrm{B}}_1(0))} \geqslant 1-\mathrm{e}^{-\varepsilon d} \tag{9.2}$$

which means that a point randomly chosen from $\mathrm{B}_{\mathbb{R}^d}$ is likely to be close to the ball's surface. For the following theorem, we will choose not just one point randomly, but several points simultaneously.

Theorem 9.5. *Let $(\Omega, \Sigma, \mathrm{P})$ be a probability space, let $d \in \mathbb{N}_{\geqslant 3}$ and for $n \in \mathbb{N}_{\geqslant 2}$ let $X^{(1)},\dots,X^{(n)}\colon \Omega \to \mathbb{R}^d$ be independent with $X^{(i)} \sim \mathcal{U}(\mathrm{B}_{\mathbb{R}^d})$. Then we have*

$$\mathrm{P}\big[\|X^{(i)}\| \geqslant 1-\tfrac{2\log n}{d} \textit{ for all } i=1,\dots,n\big] \geqslant 1-\tfrac{1}{n}.$$

Proof. We put $\varepsilon = \frac{2\log n}{d}$ and fix $1 \leqslant i \leqslant n$. From Theorem 9.1 it follows

$$\mathrm{P}\big[\|X^{(i)}\| < 1-\varepsilon\big] = \frac{\lambda^d(\mathrm{B}_{1-\varepsilon}(0))}{\lambda^d(\overline{\mathrm{B}}_1(0))} \leqslant \mathrm{e}^{-\varepsilon d} = \mathrm{e}^{-2\log n} = \frac{1}{n^2}.$$

From this, we obtain

$$\begin{aligned}\mathrm{P}\big[\forall\, i\colon \|X^{(i)}\| \geqslant 1-\varepsilon\big] &= 1-\mathrm{P}\big[\exists\, i\colon \|X^{(i)}\| \geqslant 1-\varepsilon\big]\\ &= 1-\sum_{i=1}^{n} 1/n^2 = 1-\frac{1}{n}\end{aligned}$$

as claimed. □

Remark 9.6. Colloquially, Theorem 9.5 says that n-many points randomly selected uniformly from the unit ball of $\mathbb{R}^d$ will be close to its surface with high probability. We wish to highlight the following in particular:

(i) The norm estimate applies to all points — not just to all outside a certain small "exception set".

(ii) The lower bound in the norm estimate increases with the dimension d and decreases with the number of points n.

(iii) The lower bound in the probability estimate increases with the number of points n and is independent of the dimension d.

(iv) Because n enters the norm estimate logarithmically in the numerator and the probability estimate linearly in the denominator, an increase in the sample size leads to a much stronger improvement in the probability estimate than deterioration of the norm estimate.

For large d and moderately large n, we thus obtain $1 - \frac{2\log n}{d} \approx 1$ and at the same time $1 - \frac{1}{n} \approx 1$. In this sense, the interpretation given at the beginning of this remark is to be understood.

Next, we want to consider the angle between points randomly chosen from $\mathrm{B}_{\mathbb{R}^d}$. As by Theorem 9.5 the norms of such points are with high probability close to 1, we will drop normalization and only consider pairwise scalar products below. Compare however Exercise 9.3.

Theorem 9.7. *Let* $(\Omega, \Sigma, \mathrm{P})$ *be a probability space, let* $d \in \mathbb{N}_{\geqslant 3}$ *and for* $n \in \mathbb{N}_{\geqslant 2}$ *let* $X^{(1)}, \dots, X^{(n)} \colon \Omega \to \mathbb{R}^d$ *be independent with* $X^{(i)} \sim \mathcal{U}(\mathrm{B}_{\mathbb{R}^d})$. *Then we have*

$$\mathrm{P}\big[\,|\langle X^{(j)}, X^{(k)}\rangle| \leqslant \frac{\sqrt{6\log n}}{\sqrt{d-1}} \text{ for all } j \neq k\,\big] \geqslant 1 - \frac{1}{n}.$$

Proof. We first explain the idea behind the proof in an illustrative way and limit ourselves to the case where two points $x, y \in \mathrm{B}_{\mathbb{R}^d}$ are chosen uniformly and independently. According to the surface concentration theorem, we then have $\|x\| \approx 1$ and $\|y\| \approx 1$. The points x and y sit therefore just below the unit sphere. If we change the coordinate system such that x points in "north" direction, then the point y, chosen independently of x, must lie close to the equator according to the waist concentration theorem:

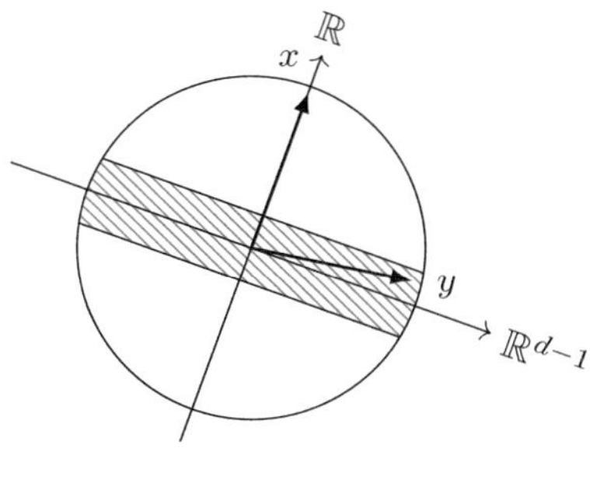

Consequently, the angle between the two points will be approximately 90°, and therefore the scalar product $\langle x, y\rangle \approx 0$ will be negligible. We will now first formalize this heuristic for two points and then, using basic probability theory, prove the estimate claimed in the theorem.

We put $\varepsilon = \frac{\sqrt{6\log n}}{\sqrt{d-1}}$ and consider two random variables

$$X, Y \colon \Omega \to \mathbb{R}^d \text{ with } X, Y \sim \mathcal{U}(\mathrm{B}_{\mathbb{R}^d}).$$

Let $\omega \in \Omega$ be such that $x := X(\omega) \neq 0$. We put $b_1 := \frac{x}{\|x\|}$ and extend to an orthonormal basis $\{b_1, \dots, b_d\}$ of $\mathbb{R}^d$. Further, we denote by $\{\mathrm{e}_1, \dots, \mathrm{e}_d\}$ the standard basis of $\mathbb{R}^d$. Then there exists precisely one linear mapping

$$T \colon \mathbb{R}^d \to \mathbb{R}^d \text{ with } Tb_i = \mathrm{e}_i \text{ for } i = 1, \dots, d$$

and with the abbreviation $y := Y(\omega)$ it follows

$$\begin{aligned} |\langle X(\omega), Y(\omega)\rangle| = |\langle x, y\rangle| &\underset{\substack{\uparrow\\ T \text{ orthog.}}}{=} |\langle Tx, Ty\rangle| \underset{\substack{\uparrow\\ x=\|x\|b_1}}{=} |\langle \|x\| Tb_1, Ty\rangle| \\ &= \|x\| \cdot |\langle \mathrm{e}_1, Ty\rangle| = \|X(\omega)\| \cdot |\langle \mathrm{e}_1, TX(\omega)\rangle|. \end{aligned}$$

Since the unit sphere is invariant under orthogonal transformations, it follows

$$\begin{aligned} \mathrm{P}\big[|\langle \mathrm{e}_1, TY\rangle| > \varepsilon\big] &= \frac{\lambda^d(\{y \in \mathrm{B}_{\mathbb{R}^d} \,|\, |\langle \mathrm{e}_1, Ty\rangle| > \varepsilon\})}{\lambda^d(\mathrm{B}_{\mathbb{R}^d})} \\ &= \frac{\lambda^d(\{T^{-1}z \in \mathrm{B}_{\mathbb{R}^d} \,|\, |\langle \mathrm{e}_1, z\rangle| > \varepsilon\})}{\lambda^d(\mathrm{B}_{\mathbb{R}^d})} \\ &= \frac{\lambda^d(\{(z_1, \dots, z_d) \in \mathrm{B}_{\mathbb{R}^d} \,|\, |z_1| > \varepsilon\})}{\lambda^d(\mathrm{B}_{\mathbb{R}^d})} \\ &\underset{\substack{\uparrow\\ \text{Corollary}\\ 9.4}}{\leqslant} \frac{2}{\varepsilon\sqrt{d-1}}\, \mathrm{e}^{-\frac{\varepsilon^2(d-1)}{2}} \\ &= \frac{2}{\sqrt{6\log n}}\, \mathrm{e}^{-\frac{6\log n}{2}} \underset{\substack{\uparrow\\ \frac{2}{\sqrt{6\log n}} \leqslant 1}}{\leqslant} \frac{1}{n^3}, \end{aligned}$$

where we point out that the mapping T depends on our preselected $\omega \in \Omega$ with $X(\omega) \neq 0$, while the upper bound that we established above is independent of ω. With the latter in mind, we calculate

$$\begin{aligned} \mathrm{P}\big[|\langle X, Y\rangle| > \varepsilon\big] &= \mathrm{P}(\{\omega \in \Omega \,|\, |\langle X(\omega), Y(\omega)\rangle| > \varepsilon\}) \\ &\underset{\substack{\uparrow\\ X(\omega)\neq 0\\ \text{automatically}}}{=} \mathrm{P}(\{\omega \in \Omega \,|\, \|X(\omega)\| \cdot |\langle \mathrm{e}_1, TY(\omega)\rangle| > \varepsilon\}) \\ &\underset{\substack{\uparrow\\ X(\omega)\in \mathrm{B}_{\mathbb{R}^d}}}{\leqslant} \mathrm{P}(\{\omega \in \Omega \,|\, |\langle \mathrm{e}_1, TY(\omega)\rangle| > \varepsilon\}) \underset{\substack{\uparrow\\ \text{as}\\ \text{above}}}{\leqslant} \frac{1}{n^3}. \end{aligned}$$

Let now finally $X^{(1)},\dots,X^{(n)}\sim\mathcal{U}(\mathrm{B}_{\mathbb{R}^d})$ be given. Then it follows

$$\begin{aligned}\mathrm{P}\big[\forall\, j\neq k\colon |\langle X^{(j)},X^{(k)}\rangle|\leqslant\varepsilon\big] &= 1-\mathrm{P}\big[\exists\, j\neq k\colon |\langle X^{(j)},X^{(k)}\rangle|>\varepsilon\big]\\ &= 1-\tbinom{n}{2}\cdot\mathrm{P}\big[\,|\langle X,Y\rangle|>\varepsilon\big]\\ &\geqslant 1-\frac{n^2-n}{n^3}\\ &\geqslant 1-\tfrac{1}{n}\end{aligned}$$

as claimed. □

For $n=2$, Theorem 9.7 bounds the probability of the scalar product of two random vectors being less than or equal to $\frac{\sqrt{6\log 2}}{\sqrt{d-1}}$ only by $1/2$ from below. The proof however shows that even

$$\mathrm{P}\big[\,|\langle X,Y\rangle|\leqslant\varepsilon\big]\geqslant 1-\frac{2}{\varepsilon\sqrt{d-1}}\,\mathrm{e}^{-\frac{\varepsilon^2(d-1)}{2}} \tag{9.3}$$

holds for $X,Y\sim\mathcal{U}(\mathrm{B}_{\mathbb{R}^d})$.

Remark 9.8. Combining Theorems 9.5 and 9.7 leads to the interpretation that n-many points chosen uniformly at random from the d-dimensional unit sphere are highly likely to be pairwise orthogonal: Theorem 9.5 provides $\|x^{(j)}\|\approx 1$ and $\|x^{(k)}\|\approx 1$, from whence we conclude with Theorem 9.7 that $\langle\frac{x^{(j)}}{\|x^{(j)}\|},\frac{x^{(k)}}{\|x^{(k)}\|}\rangle\approx 0$ holds, compare Exercise 9.3 for a formal proof. Regarding the question under which configuration of n and d, or in the case of only two points, d and ε, this interpretation is valid, we encourage the reader to proceed analogous to Remark 9.6. In particular, we recommend to work out concrete numerical examples, see Exercise 9.4.

Next we consider the distance of points chosen uniformly at random from $\mathrm{B}_{\mathbb{R}^d}$. We limit ourselves here to two points and leave it to the reader to derive results for n-many points similar to the previous theorems.

Theorem 9.9. *Let $(\Omega,\Sigma,\mathrm{P})$ be a probability space, let $X,Y\colon\Omega\to\mathbb{R}^d$ be independent random vectors with $X,Y\sim\mathcal{U}(\mathrm{B}_{\mathbb{R}^d})$. Then we have*

$$\mathrm{P}\big[\,\big|\|X-Y\|-\sqrt{2}\,\big|\leqslant\varepsilon\big]\geqslant 1-2\,\mathrm{e}^{-\varepsilon d/5}-\frac{9}{\varepsilon\sqrt{d-1}}\,\mathrm{e}^{-\frac{\varepsilon^2(d-1)}{36}}$$

for $0<\varepsilon<1$ and $d\in\mathbb{N}_{\geqslant 3}$.

Proof. We first reverse the distance estimate by considering the opposite event and then we multiply with the inequality $\|X-Y\|+\sqrt{2}\geqslant\sqrt{2}$. In this way we obtain

$$\mathrm{P}\big[\,\big|\|X-Y\|-\sqrt{2}\,\big|\leqslant\varepsilon\big]=1-\mathrm{P}\big[\,\big|\|X-Y\|-\sqrt{2}\,\big|\geqslant\varepsilon\big]$$

$$
\begin{aligned}
&\geqslant 1-\mathrm{P}\big[\,\big|\|X-Y\|-\sqrt{2}\big|(\|X-Y\|+\sqrt{2})\geqslant \varepsilon\sqrt{2}\,\big]\\
&=1-\mathrm{P}\big[\,\big|\|X-Y\|^2-2\,\big|\geqslant \varepsilon\sqrt{2}\,\big]\\
&=\mathrm{P}\big[-\varepsilon\sqrt{2}\leqslant \|X\|^2+\|Y\|^2-2\langle X,Y\rangle-2\leqslant \varepsilon\sqrt{2}\,\big]\\
&\geqslant \mathrm{P}\big[\,\big|\|X\|^2-1\,\big|\leqslant \tfrac{\varepsilon\sqrt{2}}{3}\,\big]^2\cdot \mathrm{P}\big[\,|\langle X,Y\rangle|\leqslant \tfrac{\varepsilon\sqrt{2}}{6}\,\big].
\end{aligned}
$$

Next, we estimate the first term using Bernoulli's inequality and (9.2), which leads to

$$
\begin{aligned}
\mathrm{P}\big[\,\big|\|X\|^2-1\,\big|\leqslant \tfrac{\varepsilon\sqrt{2}}{3}\,\big] &= \mathrm{P}\big[\,1-\|X\|^2\leqslant \tfrac{\varepsilon\sqrt{2}}{3}\,\big]\\
&= \mathrm{P}\big[\,\|X\|\geqslant (1-\tfrac{\varepsilon\sqrt{2}}{3})^{1/2}\,\big]\\
&\geqslant \mathrm{P}\big[\,\|X\|\geqslant 1-\tfrac{1}{2}\cdot\tfrac{\varepsilon\sqrt{2}}{3}\,\big]\\
&\geqslant 1-\mathrm{e}^{-\frac{\varepsilon\sqrt{2}}{6}d},
\end{aligned}
$$

where we may handle the second term with (9.3) and obtain

$$
\mathrm{P}\big[\,|\langle X,Y\rangle|\leqslant \tfrac{\varepsilon\sqrt{2}}{6}\,\big]\geqslant 1-\frac{6\sqrt{2}}{\varepsilon\sqrt{d-1}}\,\mathrm{e}^{-\frac{\varepsilon^2(d-1)}{36}}\,.
$$

This leads to

$$
\begin{aligned}
\mathrm{P}\big[\,\big|\|X-Y\|-\sqrt{2}\big|\leqslant\varepsilon\big] &\geqslant \big(1-\mathrm{e}^{-\frac{\varepsilon\sqrt{2}}{6}d}\big)^2\Big(1-\frac{6\sqrt{2}}{\varepsilon\sqrt{d-1}}\,\mathrm{e}^{-\frac{\varepsilon^2(d-1)}{36}}\Big)\\
&= \big(1-2\,\mathrm{e}^{-\frac{\varepsilon\sqrt{2}}{6}d}+\mathrm{e}^{-\frac{\varepsilon\sqrt{2}}{3}d}\big)\Big(1-\frac{6\sqrt{2}}{\varepsilon\sqrt{d-1}}\,\mathrm{e}^{-\frac{\varepsilon^2(d-1)}{36}}\Big)\\
&\geqslant 1-2\,\mathrm{e}^{-\frac{\varepsilon\sqrt{2}}{6}d}-\frac{6\sqrt{2}}{\varepsilon\sqrt{d-1}}\,\mathrm{e}^{-\frac{\varepsilon^2(d-1)}{36}}
\end{aligned}
$$

from whence the desired result follows, if we use $\frac{\sqrt{2}}{6}\geqslant\frac{1}{5}$ and $6\sqrt{2}\leqslant 9$. □

Remark 9.10. The results above suggest that for $X,Y\sim\mathcal{U}(\mathrm{B}_{\mathbb{R}^d})$

$$
\mathrm{E}(\|X\|)\approx 1,\quad \mathrm{E}(\langle X,Y\rangle)\approx 0\quad\text{and}\quad \mathrm{E}(\|X-Y\|)\approx\sqrt{2}
$$

hold — similarly to the Theorems 8.2 and 8.4 for standard normally distributed random vectors. Note, however, that we did not prove the above, and that our statements "with high probability" are quantitative, and therefore in a certain sense even better, see also the discussion at the end of Chapter 8.

References

This chapter is mainly based on [BHK20].

Exercises

Exercise 9.1. Let λ^d be the Lebesgue measure and $\mathrm{B}_{\mathbb{R}^d}$ the d-dimensional unit ball.

(i) Show that $\lambda^d(\mathrm{B}_{\mathbb{R}^d}) = \frac{2\pi}{d} \cdot \lambda^{d-2}(\mathrm{B}_{\mathbb{R}^{d-2}})$ holds for $d \geqslant 3$.

(ii) Use the recursion formula from (i) and the known cases $d = 1, 2$ to calculate $\lambda^d(\mathrm{B}_{\mathbb{R}^d})$ for $d = 3, \ldots, 10$ and sketch the function $d \mapsto \lambda^d(\mathrm{B}_{\mathbb{R}^d})$.

(iii) Calculate $\lim_{d\to\infty} \lambda^d(\mathrm{B}_{\mathbb{R}^d})$.

(iv) Show that $\lambda^d(\mathrm{B}_{\mathbb{R}^d}) = \frac{\pi^{d/2}}{\Gamma(d/2+1)}$ holds, where $\Gamma\colon (0,\infty) \to \mathbb{R}$ denotes the Gamma function.

Hint: For (i), first employ Fubini's theorem to get

$$\lambda^d(\mathrm{B}_{\mathbb{R}^d}) = \int_{x_1^2+x_2^2\leqslant 1} \Big(\int_{x_3^2+\cdots+x_d^2\leqslant 1-x_1^2-x_2^2} 1 \,\mathrm{d}\lambda(x_3,\ldots,x_d) \Big) \,\mathrm{d}\lambda(x_1,x_2).$$

Next, write the inner integral above as a multiple of $\lambda^{d-2}(\mathrm{B}_{\mathbb{R}^{d-2}})$. Then it remains to solve a 2-dimensional integral, which can be done by transformation into polar coordinates.

Exercise 9.2. Show that there exists $a_0 \in (1,2)$ such that

$$\forall\, \varepsilon > 0,\ d \in \mathbb{N}_{\geqslant 2}\colon \varepsilon > \tfrac{a_0}{\sqrt{d-1}} \implies 1 - \tfrac{2}{\varepsilon\sqrt{d-1}}\, \mathrm{e}^{-\frac{\varepsilon^2(d-1)}{2}} > 0,$$

compare Remark 9.3(iii). In fact, one can even find $a_0 < \sqrt{5} - 1$.

Hint: Consider $f\colon (0,\infty) \to \mathbb{R}$, $f(a) = \frac{2}{a}\,\mathrm{e}^{a^2/2}$, and first show that there exists an $a_0 > 0$ such that $f(a) < 1$ holds for all $a > a_0$. Then infer that $a_0 \in (1,2)$ must hold. Finally, substitute $a = \varepsilon\sqrt{d-1}$.

Exercise 9.3. Let $(\Omega, \Sigma, \mathrm{P})$ be a probability space, let $d \in \mathbb{N}_{\geqslant 3}$ and $n \in \mathbb{N}$ be such that $2\log(n) \leqslant d$, let $X^{(1)}, \ldots, X^{(n)}\colon \Omega \to \mathbb{R}^d$ be independent with $X^{(j)} \sim \mathcal{U}(\mathrm{B}_{\mathbb{R}^d})$. Show that the following holds:

$$\mathrm{P}\Big[\big|\big\langle \tfrac{X^{(j)}}{\|X^{(j)}\|}, \tfrac{X^{(k)}}{\|X^{(k)}\|} \big\rangle\big| \leqslant \frac{\sqrt{6\log n}}{\sqrt{d-1}} \text{ for all } j \neq k \Big] \geqslant 1 - \tfrac{1}{n}.$$

Hint: Use $1 \leqslant \frac{d^2}{(d-2\log n)^2}$ and the law of total probability.

Exercise 9.4. Go through a few concrete numerical examples to build an intuition for which n, d and ε the theorems in this chapter actually allow for the colloquially stated interpretations. E.g.: For which choice of n and d does Theorem 9.5 actually mean that $\|X^{(i)}\| \approx 1$ holds with a probability close to 1, etc.

10 Gaussian Random Vectors in High Dimensions

We now turn our attention to Gaussian random vectors. For simplicity, we will consider most of the time $X \sim \mathcal{N}(0,1,\mathbb{R}^d)$. In Theorem 8.2 we have already shown that then $|\mathrm{E}(X) - \sqrt{d}\,| \leqslant \frac{1}{\sqrt{d}}$ and $\mathrm{V}(\|X\|) \leqslant 2$ hold. Experiments, cf. page 119, have further corroborated that if we draw samples of X, then almost all of them will be located in a thin shell-shaped region with radius $\sqrt{d}$. At first glance this may be surprising as one is used to a different picture from low dimensions and even in high dimensions the Gaussian density function attains large values around zero and quickly decreases if we move away from zero:

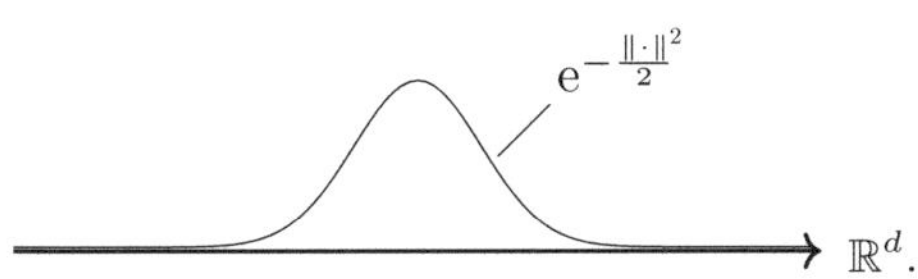

If we, on a second glance, picture the $\mathbb{R}^d$ partitioned into thin shells of equal thickness but with growing radii r, and recall Chapter 9, we see that the measure of the shells grows from inside to outside. If we now integrate the Gaussian density over such a shell, the integrand is large at small radii, but the measure of the shell is small, and therefore also the integral. At large radii, it is the other way around. At medium radii, both effects balance each other out, so that here—as we will show, in the region $r \approx \sqrt{d}$—the probability will concentrate.

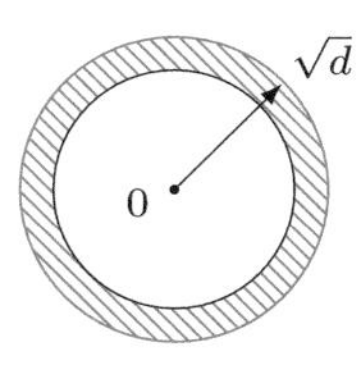

Our first goal in this chapter is to quantify and prove the above. This leads to the so-called *Gaussian annulus theorem*, whose name is based on 2-dimensional

S. A. Wegner, *Mathematical Introduction to Data Science*,
https://doi.org/10.1007/978-3-662-69426-8_10

nomenclature (annulus instead of shell). As the latter is standard in the literature, we will adopt from now on this notation. Bear in mind that by a *thin annulus*, or *shell*, we mean a set $\{x \in \mathbb{R}^d \mid r_1 \leqslant \|x\| \leqslant r_2\}$ where the two radii $r_2 > r_1 > 0$ are such that their difference $r_2 - r_1$ is small in comparison to r_1 and r_2 themselves. We refer to $r_2 - r_1$ as the *width* of the annulus.

To prove the Gaussian annulus theorem we will need some preparations and in particular two important properties of Gaussian random vectors. We note these below as facts; detailed proofs can be found in Appendix A.2.

Fact 10.1. *Let* $(\Omega, \Sigma, \mathrm{P})$ *be a probability space.*

(i) A random vector $X\colon \Omega \to \mathbb{R}^d$ *satisfies* $X \sim \mathcal{N}(0, 1, \mathbb{R}^d)$ *if and only if its coordinates* $X_1, \dots, X_d$ *are independent and* $X_i \sim \mathcal{N}(0, 1)$ *holds for all* $i = 1, \dots, d$.

(ii) If $X_1, \dots, X_n\colon \Omega \to \mathbb{R}$ *are independent random variables with distribution* $X_i \sim \mathcal{N}(0, 1)$ *for all* $i = 1, \dots, n$ *and* $\lambda_1, \dots, \lambda_n \in \mathbb{R}$, *then we have* $\lambda_1 X_1 + \cdots + \lambda_n X_n \sim \mathcal{N}(0, \lambda_1^2 + \cdots + \lambda_n^2)$.

We now start with the first of two preparatory results, which is based on the so-called *Chernoff method*.

Lemma 10.2. (Bernstein tailbound) *Let* $(\Omega, \Sigma, \mathrm{P})$ *be a probability space and let* $Y_1, \dots, Y_d\colon \Omega \to \mathbb{R}$ *be independent random variables with* $\mathrm{E}(Y_i) = 0$ *and* $|\mathrm{E}(Y_i^k)| \leqslant k!/2$ *for* $i = 1, \dots, d$ *and* $k \geqslant 2$. *Then for* $a > 0$ *we have*

$$\mathrm{P}\big[|Y_1 + \cdots + Y_d| \geqslant a\big] \leqslant 2\,\mathrm{e}^{-\frac{1}{4}\min(\frac{a^2}{d}, a)}.$$

Proof. We put $Y = Y_1 + \cdots + Y_n$ and for reasons that we will see shortly, we consider for $0 < t \leqslant 1/2$:

$$\begin{aligned}\mathrm{P}\big[Y \geqslant a\big] &\underset{t>0}{=} \mathrm{P}\big[\mathrm{e}^{tY} \geqslant \mathrm{e}^{ta}\big] \underset{\text{Markov}}{\leqslant} \frac{\mathrm{E}(\mathrm{e}^{tY})}{\mathrm{e}^{ta}} \\ &= \mathrm{e}^{-ta}\,\mathrm{E}(\mathrm{e}^{t(Y_1+\cdots+Y_d)}) \underset{Y_i \text{ independent}}{=} \mathrm{e}^{-ta}\prod_{i=1}^{n}\mathrm{E}(\mathrm{e}^{tY_i}).\end{aligned} \tag{10.1}$$

Next, we estimate the factors from above individually as follows:

$$\begin{aligned}\mathrm{E}(\mathrm{e}^{tY_i}) &= \big|\mathrm{E}\big(1 + tY_i + \sum_{k=2}^{\infty}\tfrac{(tY_i)^k}{k!}\big)\big| \leqslant 1 + \sum_{k=2}^{\infty}\frac{t^k|\mathrm{E}(Y_i^k)|}{k!} \underset{|\mathrm{E}(Y_i^k)| \leqslant k!/2}{\leqslant} 1 + \sum_{k=2}^{\infty}\frac{t^k}{2} \\ &= 1 + \tfrac{1}{2}\big(\sum_{k=0}^{\infty}t^k - t - 1\big) \underset{t<1 \text{ and geom. series}}{=} 1 + \tfrac{1}{2}\big(\tfrac{1}{1-t} - \frac{(t+1)(t-1)}{t-1}\big)\end{aligned}$$

$$= 1+\tfrac{1}{2}\frac{t^2}{1-t} \underset{\frac{1}{1-t}\leqslant 2}{\overset{}{\leqslant}} 1+t^2 \leqslant \mathrm{e}^{t^2}.$$

Since the above holds for all $0 < t \leqslant 1/2$, and in (10.1) the left side does not depend on t, we now minimize the right side over $t \in (0, 1/2]$ and obtain

$$\mathrm{P}\big[Y \geqslant a\big] \leqslant \inf_{0<t\leqslant 1/2} \mathrm{e}^{-ta} \prod_{i=1}^{d} \mathrm{e}^{t^2} \leqslant \inf_{0<t\leqslant 1/2} \mathrm{e}^{-ta+dt^2} \tag{10.2}$$

where the infimum is in fact a minimum: Consider the function $f\colon \mathbb{R} \to \mathbb{R}$, $f(t) = \mathrm{e}^{-ta+dt^2}$. Differentiating and zeroing yields

$$\tfrac{\mathrm{d}}{\mathrm{d}t} f(t) = (-a+2dt)\,\mathrm{e}^{-ta+dt^2} \overset{!}{=} 0 \implies t = \tfrac{a}{2d},$$

and since $f(t) \to \infty$ for $t \to \pm\infty$, $t_0 := \frac{a}{2d} > 0$ is the only minimizer of f. If now $t_0 \leqslant 1/2$, then $f|_{(0,1/2]}$ attains the infimum at this point. If $t_0 > 1/2$, then $f|_{(0,1/2]}$ is monotonically decreasing and the infimum is attained at $t_1 := 1/2$. In particular, in the latter case we have $a > d$ and we get for the minimizers

$$-t_0a + dt_0^2 = -\tfrac{a^2}{2d} + \tfrac{da^2}{4d^2} = -\tfrac{a^2}{4d} \quad \text{or} \quad -t_1a + dt_1^2 = -\tfrac{a}{2} + \tfrac{d}{4} \underset{d<a}{<} -\tfrac{a}{4},$$

respectively. With this, we can continue the estimate in (10.2) as follows

$$\inf_{0<t\leqslant 1/2} \mathrm{e}^{-ta+dt^2} = \mathrm{e}^{-ta+dt^2}\Big|_{t=\min(\frac{a}{2d},\frac{1}{2})} \leqslant \mathrm{e}^{-\min(\frac{a^2}{4d},\frac{a}{4})} \tag{10.3}$$

and obtain

$$\mathrm{P}\big[Y_1 + \cdots + Y_d \geqslant a\big] \leqslant \mathrm{e}^{-\frac{1}{4}\min(\frac{a^2}{d},a)}.$$

If we repeat the above, but replace everywhere Y_i with $-Y_i$, the same calculations lead to

$$\mathrm{P}\big[Y_1 + \cdots + Y_d \leqslant -a\big] = \mathrm{P}\big[-Y_1 - \cdots - Y_d \geqslant a\big] \leqslant \mathrm{e}^{-\frac{1}{4}\min(\frac{a^2}{d},a)}$$

and both together conclude the proof. □

The second preparatory result we will need in order to compute the k-th moment of a Gaussian random variable.

Lemma 10.3. *For $k \in \mathbb{N}_{\geqslant 1}$ we have* $\displaystyle\frac{1}{\sqrt{2\pi}}\int_{\mathbb{R}} t^{2k}\,\mathrm{e}^{-t^2/2}\,\mathrm{d}t = \frac{(2k)!}{2^k k!}$.

Proof. ① Using the dominated convergence theorem, we can interchange differentiation and integration in the following calculation:

$$\frac{\mathrm{d}^k}{\mathrm{d}a^k}\Big(\int_{\mathbb{R}} \mathrm{e}^{-at^2}\,\mathrm{d}t\Big)\Big|_{a=\frac{1}{2}} = \Big(\int_{\mathbb{R}} \frac{\mathrm{d}^k}{\mathrm{d}a^k}\,\mathrm{e}^{-at^2}\,\mathrm{d}t\Big)\Big|_{a=\frac{1}{2}} = (-1)^k \int_{\mathbb{R}} t^{2k}\,\mathrm{e}^{-t^2/2}\,\mathrm{d}t.$$

② By substituting $u := \sqrt{a}t$ for $a > 0$ we get:

$$\int_{\mathbb{R}} \mathrm{e}^{-at^2}\,\mathrm{d}t = \frac{1}{\sqrt{a}}\int_{\mathbb{R}} \mathrm{e}^{-u^2}\,\mathrm{d}u = \sqrt{\frac{\pi}{a}}.$$

③ If we first read ① backwards and then sub in ②, we obtain:

$$\int_{\mathbb{R}} t^{2k}\,\mathrm{e}^{-t^2/2}\,\mathrm{d}t = (-1)^k\frac{\mathrm{d}^k}{\mathrm{d}a^k}\Big(\int_{\mathbb{R}} \mathrm{e}^{-at^2}\,\mathrm{d}t\Big)\Big|_{a=\frac{1}{2}} = (-1)^k\frac{\mathrm{d}^k}{\mathrm{d}a^k}\Big(\frac{\sqrt{\pi}}{\sqrt{a}}\Big)\Big|_{a=\frac{1}{2}}.$$

Next, we calculate the derivative on the right-hand side:

$$\begin{aligned}\frac{\mathrm{d}^k}{\mathrm{d}a^k}(a^{-1/2}) &= (-1/2)\cdot(-1/2-1)\cdots(-1/2-(k-1))\cdot a^{-1/2-k}\\ &= \frac{(-1)^k}{2^k}\cdot\big(1\cdot 3\cdots(2k-3)\cdot(2k-1)\big)\cdot a^{-1/2-k}\\ &= \frac{(-1)^k}{2^k}\cdot\frac{1\cdot 2\cdot 3\cdots(2k-3)\cdot(2k-2)\cdot(2k-1)\cdot(2k)}{2\cdot 4\cdots(2k-2)\cdot(2k)}\cdot a^{-1/2-k}\\ &= \frac{(-1)^k}{2^k}\cdot\frac{(2k)!}{2^k k!}\cdot a^{-1/2-k}.\end{aligned}$$

④ Subbing in the derivative and dividing by $\sqrt{2\pi}$ leads to

$$\frac{1}{\sqrt{2\pi}}\int_{\mathbb{R}} t^{2k}\,\mathrm{e}^{-t^2/2}\,\mathrm{d}t = \frac{(-1)^k}{\sqrt{2}}\frac{(-1)^k}{2^k}\frac{(2k)!}{2^k k!}\,a^{-1/2-k}\Big|_{a=\frac{1}{2}} = \frac{1}{\sqrt{2}}\frac{1}{2^k}\frac{(2k)!}{2^k k!}\,2^{1/2+k},$$

which simplifies, after cancelling, to the expression given in the lemma. □

Now we are ready to prove the Gaussian annulus theorem. Before we begin, we point out that in Theorem 10.4, the right-hand side is equal to zero for $\varepsilon = 4\sqrt{\log 2} \approx 3.3$. The inequality thus becomes non-trivial only for larger ε, compare Example 10.5. In contrast to some of the results of Chapter 9 it is here not possible to accomodate for smaller ε's by increasing the dimension. Intuitively, this means that the "absolute width" of the annulus in the picture on page 139 is constant. Recall, that we speak of a thin annulus, if the width *relative* to the radius is small.

Theorem 10.4. (Gaussian annulus theorem) *Let* $(\Omega, \Sigma, \mathrm{P})$ *be a probability space, let* $X\colon \Omega \to \mathbb{R}^d$ *be a random vector with* $X \sim \mathcal{N}(0, 1, \mathbb{R}^d)$ *and* $0 \leqslant \varepsilon \leqslant \sqrt{d}$. *Then*

$$\mathrm{P}\big[\,\big|\|X\| - \sqrt{d}\,\big| \leqslant \varepsilon\big] \geqslant 1 - 2\,\mathrm{e}^{-\varepsilon^2/16}.$$

Proof. We multiply by the inequality $\|X\| + \sqrt{d} \geqslant \sqrt{d}$ and obtain

$$\begin{aligned}
\mathrm{P}\big[\,\big|\|X\| - \sqrt{d}\,\big| \geqslant \varepsilon\big] &\leqslant \mathrm{P}\big[\,\big|\|X\| - \sqrt{d}\big| \cdot (\|X\| + \sqrt{d}) > \varepsilon \cdot \sqrt{d}\big] \\
&= \mathrm{P}\big[\,|X_1^2 + \cdots + X_d^2 - d| > \varepsilon\sqrt{d}\big] \\
&= \mathrm{P}\big[\,|(X_1^2 - 1) + \cdots + (X_d^2 - 1)\,| > \varepsilon\sqrt{d}\big] \\
&= \mathrm{P}\big[\,\big|\tfrac{X_1^2-1}{2} + \cdots + \tfrac{X_d^2-1}{2}\big| > \tfrac{\varepsilon\sqrt{d}}{2}\big] \\
&= \mathrm{P}\big[\,|Y_1 + \cdots + Y_d\,| > a\big],
\end{aligned}$$

where in the last step we have abbreviated $Y_i := \frac{X_i^2-1}{2}$ and $a := \frac{\varepsilon\sqrt{d}}{2}$. Next we want to apply the tailbound from Lemma 10.2. For this, we first check the three conditions required there.

① The Y_i are pairwise independent according to Fact 10.1(i).

② It holds that $\mathrm{E}(Y_i) = \frac{1}{2}(\mathrm{E}(X_i^2) - 1) = \frac{1}{2}(\mathrm{E}((X_i - \mathrm{E}(X_i))^2) - 1) = 0$, where we first used $\mathrm{E}(X_i) = 0$ and then subbed in $\mathrm{E}((X_i - \mathrm{E}(X_i))^2) = \mathrm{V}(X_i) = 1$.

③ For $k \geqslant 2$ it holds

$$|\mathrm{E}(Y_i^k)| = |\mathrm{E}((\tfrac{X_i^2-1}{2})^k)| = \tfrac{1}{2^k}|\mathrm{E}((X_i^2 - 1)^k)| \leqslant \tfrac{1}{2^k}\,\mathrm{E}(X_i^{2k} + 1),$$

where the inequality follows from the monotonicity of the expectation and from the (pointwise) estimate $|X_i^2 - 1|^k \leqslant X_i^{2k} + 1$: If $|X_i| \leqslant 1$, then we have $0 \leqslant X_i^2 \leqslant 1$ and thus $|X_i^2 - 1|^k \leqslant 1$. On the other hand, if $|X_i| > 1$, then we have $X_i^2 - 1 > 0$, from whence it follows $|X_i^2 - 1|^k = (X_i^2 - 1)^k \leqslant (X_i^2)^k = X_i^{2k}$. Now we use the law of the unconscious statistician, see Proposition A.8, as well as Lemma 10.3, to calculate the $2k$-th moment of $X_i \sim \mathcal{N}(0, 1)$. We get

$$\begin{aligned}
\tfrac{1}{2^k}\,\mathrm{E}(X_i^{2k} + 1) &= \tfrac{1}{2^k}\Big(\frac{1}{\sqrt{2\pi}}\int_{\mathbb{R}} t^{2k}\,\mathrm{e}^{-t^2/2}\,\mathrm{d}t + 1\Big) = \tfrac{1}{2^k}\Big(\frac{(2k)!}{2^k k!} + 1\Big) \\
&= \frac{(2k)\cdot(2k-1)\cdot(2k-2)\cdot(2k-3)\cdots 5\cdot 4\cdot 3\cdot 2\cdot 1}{(2k)^2\cdot(2k-2)^2\cdots(2\cdot 3)^2\cdot(2\cdot 2)^2\cdot(2\cdot 1)^2}\cdot k! + \frac{1}{2^k} \\
&= \frac{(2k-1)\cdot(2k-3)\cdots 5\cdot 3\cdot 1}{(2k)\cdot(2k-2)\cdots 6\cdot 4\cdot 2}\cdot k! + \frac{1}{2^k} \\
&\leqslant 1\cdot 1\cdots 1\cdot\frac{3}{4}\cdot\frac{1}{2}\cdot k! + \frac{k!}{4\cdot 2} \\
&= \big(\tfrac{3}{4} + \tfrac{1}{4}\big)\frac{k!}{2},
\end{aligned}$$

from whence $|\mathrm{E}(Y_i^k)| \leqslant k!/2$ follows and thus also the last requirement of Lemma 10.2 is fulfilled.

Applying Lemma 10.2, together with our preliminary work at the beginning of this proof, yields

$$\begin{aligned}
\mathrm{P}\big[\big|\|X\| - \sqrt{d}\big| \geqslant \varepsilon\big] &\leqslant \mathrm{P}\big[\big|Y_1 + \cdots + Y_d\big| \geqslant \tfrac{\varepsilon\sqrt{d}}{2}\big] \\
&\leqslant 2\,\mathrm{e}^{-\frac{1}{4}\min\left(\frac{(\varepsilon\sqrt{d}/2)^2}{d},\, \frac{\varepsilon\sqrt{d}}{2}\right)} \\
&\leqslant 2\,\mathrm{e}^{-\frac{1}{4}\min\left(\frac{\varepsilon^2}{4},\, \frac{\varepsilon^2}{2}\right)} \\
&= 2\,\mathrm{e}^{-\varepsilon^2/16},
\end{aligned}$$

where we have used the assumption $\varepsilon \leqslant \sqrt{d}$ to estimate $\frac{\varepsilon\sqrt{d}}{2} \geqslant \frac{\varepsilon^2}{2}$. □

We continue our remarks from before the theorem and illustrate the above result with a numerical example.

Example 10.5. Let $\varepsilon = 10$, $d \geqslant 100$ and denote by

$$R_d := \overline{\mathrm{B}}_{\sqrt{d}+10}(0) \backslash \mathrm{B}_{\sqrt{d}-10}(0)$$

the annulus with center zero and radii $\sqrt{d} \pm 10$. By Theorem 10.4 and for $X \sim \mathcal{N}(0,1,\mathbb{R}^d)$ we then have

$$\mathrm{P}\big[X \in R_d\big] = \mathrm{P}\big[\sqrt{d} - 10 \leqslant \|X\| \leqslant \sqrt{d} + 10\big] \geqslant 0.99.$$

Notice, that $R_d = \overline{\mathrm{B}}_{20}(0)$ is a full ball for $d = 100$. For $d > 100$, R_d then becomes a proper annulus which is thin in the sense discussed above provided that d is large enough: Indeed R_d's width is constant 20, while its radii grow like $\sqrt{d}$.

With the same arguments that were used in the proof of Theorem 10.4, we obtain an estimate for $\mathrm{P}\big[\big|\|X\|^2 - d\big| \geqslant \varepsilon\big]$, compare our calculations of expectation (8.1) and and variance (8.2) on page 120. We leave the proof, as well as the question of for which ε and d the estimates in Corollary 10.6 are non-trivial, as Exercise 10.2 to the reader.

Corollary 10.6. *Let $(\Omega, \Sigma, \mathrm{P})$ be a probability space, $X\colon \Omega \to \mathbb{R}^d$ be a random vector with $X \sim \mathcal{N}(0,1,\mathbb{R}^d)$. Then*

$$\mathrm{P}\big[\big|\|X\|^2 - d\big| \leqslant \varepsilon\big] \geqslant 1 - 2\,\mathrm{e}^{-\frac{1}{8}\min\left(\frac{\varepsilon^2}{2d},\, \varepsilon\right)}$$

holds for $0 < \varepsilon \leqslant \sqrt{d}$. □

Next, we consider angles between standard normally distributed independent random variables. We advise the reader to recall our results on expected value and variance from Theorem 8.4, as well as the simulations from Exercise 8.2. We again point out that the bound in the following Gaussian orthogonality theorem is non-trivial for $\varepsilon > \frac{2}{\sqrt{d-7}}$. In contrast to the annulus theorem, a tradeoff is possible here in the sense that smaller ε's can be considered the larger the dimension is.

Theorem 10.7. (Gaussian orthogonality theorem) *Let* $(\Omega, \Sigma, \mathrm{P})$ *be a probability space and let* $X, Y\colon \Omega \to \mathbb{R}^d$ *be independent random vectors with* $X, Y \sim \mathcal{N}(0, 1, \mathbb{R}^d)$. *For* $d \in \mathbb{N}$ *and* $\varepsilon > 0$ *we have*

$$\mathrm{P}\big[\big|\big\langle \tfrac{X}{\|X\|}, \tfrac{Y}{\|Y\|}\big\rangle\big| \leqslant \varepsilon\big] \geqslant 1 - \frac{2/\varepsilon + 7}{\sqrt{d}}.$$

Here, we tacitly understand the left expression as a conditional probability under the condition that $\|X\|, \|Y\| \neq 0$ *holds — which of course happens with probability one.*

Proof. Since X and Y are independent, it is enough to consider $Y \equiv y \in \mathbb{R}^d\backslash\{0\}$. We put $\lambda_i := \frac{y_i}{\|y\|}$, and define the random variable $U\colon \Omega \to \mathbb{R}$ by

$$U := \big\langle X, \tfrac{y}{\|y\|}\big\rangle = \sum_{i=1}^{d} \tfrac{y_i}{\|y\|} X_i = \sum_{i=1}^{d} \lambda_i X_i.$$

With Fact 10.1(ii) it follows $U \sim \mathcal{N}(0, \lambda_1^2 + \cdots + \lambda_d^2) = \mathcal{N}(0,1)$. Now we first use the theorem of total probability, but we do not even write down the second summand and instead estimate it from below with zero. This may seem very rough, but because of the annulus theorem we actually lose almost nothing, since $\|X\| \leqslant \frac{\sqrt{d}}{2}$ only occurs with small probability provided that d is suitably large. We get

$$\begin{aligned}
&\mathrm{P}\big[\big|\big\langle \tfrac{X}{\|X\|}, \tfrac{y}{\|y\|}\big\rangle\big| \leqslant \varepsilon\big]\\
&\quad= \mathrm{P}\big[\big|\big\langle \tfrac{X}{\|X\|}, \tfrac{y}{\|y\|}\big\rangle\big| \leqslant \varepsilon \,\big|\, \tfrac{\sqrt{d}}{2} \leqslant \|X\|\big] \cdot \mathrm{P}\big[\tfrac{\sqrt{d}}{2} \leqslant \|X\|\big] + \cdots\\
&\quad\geqslant \mathrm{P}\big[|U| \leqslant \varepsilon\|X\| \,\big|\, \tfrac{\sqrt{d}}{2} \leqslant \|X\|\big] \cdot \mathrm{P}\big[\sqrt{d} - \tfrac{\sqrt{d}}{2} \leqslant \|X\| \leqslant \sqrt{d} + \tfrac{\sqrt{d}}{2}\big]\\
&\quad\geqslant \mathrm{P}\big[|U| \leqslant \tfrac{\varepsilon\sqrt{d}}{2}\big] \cdot \mathrm{P}\big[\big|\|X\| - \sqrt{d}\big| \leqslant \tfrac{\sqrt{d}}{2}\big]\\
&\quad\geqslant \mathrm{P}\big[|U| \leqslant \tfrac{\varepsilon\sqrt{d}}{2}\big] \cdot \big(1 - 2\,\mathrm{e}^{-(\frac{\sqrt{d}}{2})^2/16}\big),
\end{aligned}$$

where the last estimate of the second factor is Theorem 10.4 with $\varepsilon = \frac{\sqrt{d}}{2}$. Since $U \sim \mathcal{N}(0,1)$, we get for the first factor

$$\begin{aligned}
\mathrm{P}\big[|U| \leqslant \tfrac{\varepsilon\sqrt{d}}{2}\big] &= \frac{1}{\sqrt{2\pi}} \int_{-\frac{\varepsilon\sqrt{d}}{2}}^{\frac{\varepsilon\sqrt{d}}{2}} \mathrm{e}^{-t^2/2}\,\mathrm{d}t \geqslant 1 - \frac{2}{\sqrt{2\pi}} \int_{-\frac{\varepsilon\sqrt{d}}{2}}^{\infty} \tfrac{1}{t^2}\,\mathrm{d}t\\
&= 1 - \frac{2}{\sqrt{2\pi}} \Big(-\tfrac{1}{t}\Big|_{-\frac{\varepsilon\sqrt{d}}{2}}^{\infty}\Big) \geqslant 1 - 1 \cdot \frac{2}{\varepsilon\sqrt{d}},
\end{aligned}$$

and if we sub this in above, we get

$$\mathrm{P}\big[|\langle \tfrac{X}{\|X\|}, \tfrac{y}{\|y\|}\rangle| \leqslant \varepsilon\big] \geqslant \big(1-\tfrac{2}{\varepsilon\sqrt{d}}\big)\big(1-2\,\mathrm{e}^{-\frac{d}{64}}\big) = 1-\frac{2}{\varepsilon\sqrt{d}} - 2\,\mathrm{e}^{-\frac{d}{64}} + \frac{4\,\mathrm{e}^{-\frac{d}{64}}}{\varepsilon\sqrt{d}}.$$

Now we estimate the positive summand at the end with zero, and furthermore use $2\,\mathrm{e}^{-\frac{d}{64}} \leqslant \frac{7}{\sqrt{d}}$. This provides the desired lower bound. □

We note the following variant, where Y is constant and where the normalization of X is omitted. We will need this in Chapter 12.

Corollary 10.8. *Let $(\Omega, \Sigma, \mathrm{P})$ be a probability space, $Z\colon \Omega \to \mathbb{R}^d$ a random vector with $Z \sim \mathcal{N}(0,1,\mathbb{R}^d)$ and let $\xi \in \mathbb{R}^d$ be fixed. For $d \in \mathbb{N}$ and $\varepsilon > 0$ we have*

$$\mathrm{P}\big[|\langle Z, \xi\rangle| \leqslant \varepsilon\big] \geqslant 1 - \frac{\|\xi\|}{\varepsilon}.$$

Proof. For $\xi = 0$ the statement is trivial. Otherwise, we calculate with $U := \langle Z, \frac{\xi}{\|\xi\|}\rangle \sim \mathcal{N}(0,1)$ as in the previous proof

$$\begin{aligned}\mathrm{P}\left[|\langle Z, \xi\rangle| \leqslant \varepsilon\right] &= \mathrm{P}\left[|\langle Z, \tfrac{\xi}{\|\xi\|}\rangle| \leqslant \tfrac{\varepsilon}{\|\xi\|}\right] = \mathrm{P}\left[|U| \leqslant \tfrac{\varepsilon}{\|\xi\|}\right]\\ &\geqslant 1 - \frac{2}{\sqrt{2\pi}}\int_{-\frac{\varepsilon}{\|\xi\|}}^{\infty} \frac{1}{t^2}\,\mathrm{d}t \geqslant 1 - 1\cdot\frac{\|\xi\|}{\varepsilon}.\end{aligned}$$ □

We now consider a numerical example to build an intuition for which dimensions the Gaussian orthogonality theorem provides useful estimates.

Example 10.9. Let $\varepsilon = 0.1$. For $X, Y \sim \mathcal{N}(0,1,\mathbb{R}^d)$ we then have

$$\mathrm{P}\Big[\big|\langle \tfrac{X}{\|X\|}, \tfrac{Y}{\|Y\|}\rangle\big| \leqslant 0.1\Big] \geqslant 0.9,$$

provided that $d \geqslant 100\,000$. The latter means that the angle $\measuredangle(X,Y)$ is $90^\circ \pm 6^\circ$ with a probability of more than 0.9. Note that we need significantly higher dimensions to apply the orthogonality theorem than for the annulus theorem. The following table contains the means of the pairwise scalar products of 100 randomly chosen vectors with respect to the standard normal distribution in $\mathbb{R}^d$, compare Exercise 8.2.

d	1	10	100	1 000	10 000	100 000
$\frac{1}{100\cdot 99}\sum_{i\neq j}\langle \frac{x^{(i)}}{\|x^{(i)}\|}, \frac{x^{(j)}}{\|x^{(j)}\|}\rangle$	-0.0097	0.0007	-0.0023	-0.0006	0.0001	-0.00004
Variance	0.9999	0.1018	0.0098	0.0010	0.0001	0.00001

The simulation suggests that even in lower dimensions, we may end up with almost orthogonal vectors.

Finally, let us consider to the distance of independent random vectors.

Theorem 10.10. (First Gaussian distance theorem) *Let* $(\Omega, \Sigma, \mathrm{P})$ *be a probability space, let* $X, Y\colon \Omega \to \mathbb{R}^d$ *be independent random vectors with* $X, Y \sim \mathcal{N}(0, 1, \mathbb{R}^d)$ *and let* $0 < \varepsilon \leqslant \sqrt{d}$. *Then we have*

$$\mathrm{P}\big[\,\big|\|X - Y\| - \sqrt{2d}\,\big| \leqslant \varepsilon\big] \geqslant 1 - 4\,\mathrm{e}^{-\varepsilon^2/72} - \frac{6\sqrt{2}}{\varepsilon d} - \frac{7}{\sqrt{d}}.$$

Proof. We proceed analogously to the proof of the annulus theorem, respectively to the proof of Theorem 9.9. That is, we switch to the opposite probability and then multiply with $\|X - Y\| + \sqrt{2d} \geqslant \sqrt{2d}$. This leads to

$$\begin{aligned}
&\mathrm{P}\big[\,\big|\|X - Y\| - \sqrt{2d}\,\big| \leqslant \varepsilon\big]\\
&\qquad = 1 - \mathrm{P}\big[\,\big|\|X - Y\| - \sqrt{2d}\,\big| \geqslant \varepsilon\big]\\
&\qquad \geqslant 1 - \mathrm{P}\big[\,\big|\|X - Y\| - \sqrt{2d}\,\big|\big(\|X - Y\| + \sqrt{2d}\big) \geqslant \varepsilon\sqrt{2d}\,\big]\\
&\qquad = 1 - \mathrm{P}\big[\,\big|\|X - Y\|^2 - 2d\,\big| \geqslant \varepsilon\sqrt{2d}\,\big]\\
&\qquad = \mathrm{P}\big[-\varepsilon\sqrt{2d} \leqslant \|X\|^2 + \|Y\|^2 - 2\langle X, Y\rangle - 2d \leqslant \varepsilon\sqrt{2d}\,\big]\\
&\qquad \geqslant \mathrm{P}\big[\,\big|\|X\|^2 - d\,\big| \leqslant \tfrac{\varepsilon\sqrt{2d}}{3}\big]^2 \cdot \mathrm{P}\big[\,|\langle X, Y\rangle| \leqslant \tfrac{\varepsilon\sqrt{2d}}{6}\big]
\end{aligned}$$

and now we estimate both factors individually. Then we get

$$\mathrm{P}\big[\,\big|\|X\|^2 - d\,\big| \leqslant \tfrac{\varepsilon\sqrt{2d}}{3}\big] \underset{\text{Cor 10.6}}{\overset{}{\geqslant}} 1 - 2\,\mathrm{e}^{-\frac{1}{8}\min(\frac{\varepsilon^2}{9}, \frac{\varepsilon\sqrt{2d}}{3})} \underset{\varepsilon \leqslant \sqrt{d}}{\geqslant} 1 - 2\,\mathrm{e}^{-\varepsilon^2/72}$$

and

$$\mathrm{P}\big[\,|\langle X, Y\rangle| \leqslant \tfrac{\varepsilon\sqrt{2d}}{6}\big] \underset{\text{Thm 10.7}}{\geqslant} 1 - \frac{\frac{12}{\varepsilon\sqrt{2d}} + 7}{\sqrt{d}} = 1 - \big(\tfrac{6\sqrt{2}}{\varepsilon d} + \tfrac{7}{\sqrt{d}}\big).$$

Expanding and further estimating yields

$$\begin{aligned}
&\big(1 - 2\,\mathrm{e}^{-\varepsilon^2/72}\big)^2\big(1 - \big(\tfrac{6\sqrt{2}}{\varepsilon d} + \tfrac{7}{\sqrt{d}}\big)\big)\\
&\qquad = \big(1 - 4\,\mathrm{e}^{-\varepsilon^2/72} + 4\,\mathrm{e}^{-\varepsilon^2/36}\big)\big(1 - \big(\tfrac{6\sqrt{2}}{\varepsilon d} + \tfrac{7}{\sqrt{d}}\big)\big)\\
&\qquad \geqslant 1 - 4\,\mathrm{e}^{-\varepsilon^2/72} - \frac{6\sqrt{2}}{\varepsilon d} - \frac{7}{\sqrt{d}}
\end{aligned}$$

as claimed. □

We leave it to the reader to check for which ε and d the theorem actually allows for the interpretation that with high probability $\|X - Y\| \approx \sqrt{2d}$ holds, as well as the comparison with the experimental data in the following table

d	1	10	100	1 000	10 000	100 000
$\frac{1}{100\cdot 99}\sum_{i\neq j}\lVert x^{(i)} - x^{(j)}\rVert$	1.06	4.41	14.05	44.65	141.50	447.35
$\sqrt{2d}$	1.41	4.47	14.14	44.72	141.42	447.21
Variance	0.74	1.10	0.92	0.89	0.96	1.08

in which again for 100 points randomly chosen with respect to normal distribution in $\mathbb{R}^d$ their pairwise average distances are given.

In conclusion, we formulate the following corollary about the squares of the distances of randomly Gaussian distributed points.

Corollary 10.11. *We have*

$$\mathrm{P}\big[\,\big|\lVert X - Y\rVert^2 - 2d\,\big| \leqslant \varepsilon\,\big] \geqslant 1 - 4\,\mathrm{e}^{-\varepsilon^2/72} - \frac{6\sqrt{2}}{\varepsilon d} - \frac{7}{\sqrt{d}}$$

under the assumptions of Theorem 10.10. □

References

This chapter is mainly based on [BHK20] and [Ver18], but also contains some reformulations and small extensions of the results discussed there.

Exercises

Exercise 10.1. (Classic example of the Chernoff method) Let $Y_1, \dots, Y_n$ be independent Bernoulli random variables with $\mathrm{P}[Y_i = 1] = p \in [0, 1]$ and let $Y := Y_1 + \cdots + Y_n$. Let $\delta > 0$.

(i) Show that $\mathrm{E}(\mathrm{e}^{tY_i}) \leqslant \mathrm{e}^{p(\mathrm{e}^t - 1)}$ holds for $t > 0$.

(ii) Proceed as in the proof of Lemma 10.2 to show the following estimate:

$$\mathrm{P}\big[X \geqslant (1+\delta)np\big] \leqslant \Big(\frac{\mathrm{e}^{\delta}}{(1+\delta)^{1+\delta}}\Big)^{np}.$$

Hint: When applying the Chernoff method it is often not even necessary to explicitly calculate the infimum, as done in Lemma 10.2. In the current example, it is sufficient to choose $t = \log(1 + \delta)$.

(iii) We now consider an experiment in which a fair dice is rolled n-many times. Apply (ii) to estimate the probability that a six is rolled in at least 70% of the n-many rolls.

(iv) Compare the bound from (iii) with the bounds obtained by applying the Markov or Chebyshev inequalities. Simulate the dice game and test how sharp the three theoretical bounds are.

Exercise 10.2. Prove Corollary 10.6.

Exercise 10.3. According to the Gaussian orthogonality theorem, two points randomly selected in high-dimensional space according to a normal distribution are almost orthogonal with high probability. On the other hand, it is clear that the probability that the points are exactly orthogonal is zero. Therefore, one may expect that the probability for the scalar product being "very small" will be small again. Quantify this by showing that

$$\mathrm{P}\big[\big|\big\langle \tfrac{X}{\|X\|}, \tfrac{Y}{\|Y\|}\big\rangle\big| \leqslant \tfrac{\varepsilon}{2\sqrt{d}}\big] \leqslant 2\,\mathrm{e}^{-d/16} + \varepsilon$$

holds for $\varepsilon > 0$ and $X, Y \sim \mathcal{N}(0,1)$. Also find a numerical example that illustrates what is meant by "very small scalar product" in the preceding text.

Hint: Bring the random variable U as in the proof of Theorem 10.7 into play and then use the fact that for random variables A, B and constants a, b the inequality $\mathrm{P}[AB \leqslant ab] \leqslant \mathrm{P}[A \leqslant a \text{ or } B \leqslant b]$ holds.

11

Dimensionality Reduction à la Johnson-Lindenstrauss

At the beginning of Chapter 8 we discussed situations where the dimension d in a dataset D is very large and may even exceed the number $n = \#D$ of data points. Especially in the latter case, the question arises whether all dimensions are actually necessary to represent the information contained in the dataset. Strategies to reduce the dimensionality (formally: $\dim \operatorname{span} D$) without, or at least with a controllable loss of information, are known as *dimensionality reduction*. An obvious idea is to *project* the dataset onto a k-dimensional subspace $V \subset \mathbb{R}^d$, where k is significantly smaller than d.

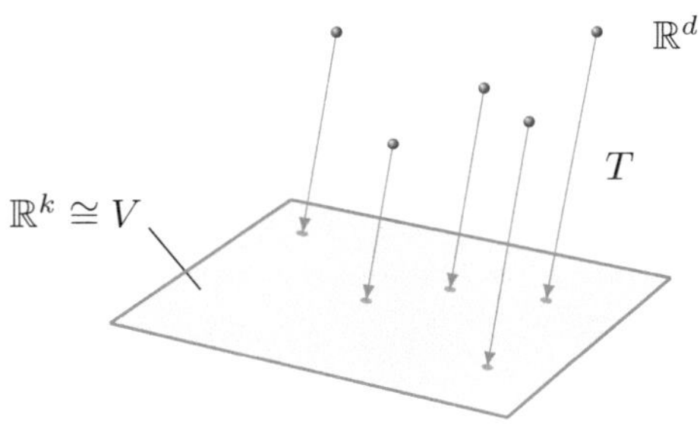

Here, one is at first inclined to look for a projection $T\colon \mathbb{R}^d \to \mathbb{R}^d$ in the sense of linear algebra, i.e., $T \in \mathrm{L}(\mathbb{R}^d)$ with $T^2 = T$, and then to put $V := \operatorname{ran} T$. Moreover, it would be good if T is independent of the given dataset and ideally the pairwise distances between data points should be invariant under T.

One quickly realizes that this wish list of properties for T is a bit too much: if we want to preserve all distances, regardless of the dataset, then T needed to be an isometry and would consequently be injective, which is not compatible with the idea of dimensionality *reduction*.

On the other hand we know from linear algebra that an isometry is given by multiplication with an orthogonal matrix. It thus seems natural to try an

S. A. Wegner, *Mathematical Introduction to Data Science*,
https://doi.org/10.1007/978-3-662-69426-8_11

orthogonal projection, or in other words, to consider

$$T_A\colon \mathbb{R}^d \to \mathbb{R}^k,\ x \mapsto Ax,$$

where $A \in \mathbb{R}^{k\times d}$ has orthonormal rows. The latter guarantees $AA^\intercal = \mathrm{id}_{\mathbb{R}^k}$ and therefore we may understand T_A as well as an orthogonal projection on the k-dimensional subspace $V := \operatorname{ran} A^\intercal \subset \mathbb{R}^d$, compare the remarks on p. 103. We now need a *concrete* matrix A. Let $(\Omega, \Sigma, \mathrm{P})$ be a probability space. We consider a measurable mapping

$$U\colon \Omega \to \mathbb{R}^{k\times d},\ \ U = \begin{bmatrix} u_{11} & \cdots & u_{1d} \\ \vdots & & \vdots \\ u_{k1} & \cdots & u_{kd} \end{bmatrix}$$

with $U \sim \mathcal{N}(0,1,\mathbb{R}^{k\times d})$. The latter means that the entries $u_{ij} \sim \mathcal{N}(0,1)$ are independent random variables and we can therefore easily obtain realizations of U. We note the following properties of such a *random matrix*.

Remark 11.1. Let $(\Omega, \Sigma, \mathrm{P})$ be a probability space and for $k < d$ let $U\colon \Omega \to \mathbb{R}^{k\times d}$ be a random matrix with $U \sim \mathcal{N}(0,1,\mathbb{R}^{k\times d})$.

(i) If we denote the rows of U by $u_i = [u_{i1} \cdots u_{id}]$, then these, as well as their transposes, are independent $\mathbb{R}^d$-valued random vectors with $u_i \sim \mathcal{N}(0,1,\mathbb{R}^d)$ for $i = 1, \dots, k$.

(ii) For $i \neq j$ and $\varepsilon > 0$, the Gaussian orthogonality theorem provides

$$\mathrm{P}\Big[\big|\big\langle \tfrac{u_i}{\|u_i\|}, \tfrac{u_j}{\|u_j\|}\big\rangle\big| \leqslant 0.1\Big] \geqslant 1 - \frac{2/\varepsilon + 7}{\sqrt{d}}.$$

We thus obtain that the rows of U are almost orthogonal with high probability. If we were to now normalize them, we would get $UU^\intercal \approx \mathrm{id}_{\mathbb{R}^k}$ and thus had come very close to a matrix A with the properties we initially wished for.

Theorem 11.2. (Random projection theorem) *Let $(\Omega, \Sigma, \mathrm{P})$ be a probability space and for $k < d$ let $U\colon \Omega \to \mathbb{R}^{k\times d}$ be a random matrix with $U \sim \mathcal{N}(0,1,\mathbb{R}^{k\times d})$. For $x \in \mathbb{R}^d\backslash\{0\}$ and $0 < \varepsilon \leqslant 1$ we have*

$$\mathrm{P}\big[\big|\|Ux\| - \sqrt{k}\,\|x\|\big| \leqslant \varepsilon\sqrt{k}\,\|x\|\big] \geqslant 1 - 2\,\mathrm{e}^{-k\varepsilon^2/16}.$$

Proof. We first note that for any $x \in \mathbb{R}^d$ the application of U to x, i.e.,

$$Ux = \begin{bmatrix} u_{11} & \cdots & u_{1d} \\ \vdots & & \vdots \\ u_{k1} & \cdots & u_{kd} \end{bmatrix} \begin{bmatrix} x_1 \\ \vdots \\ x_d \end{bmatrix} = \begin{bmatrix} \langle u_1, x\rangle \\ \vdots \\ \langle u_k, x\rangle \end{bmatrix},$$

just yields the above vector of scalar products. Now we fix $x \in \mathbb{R}^d\backslash\{0\}$ and consider the k-dimensional random vector

$$U(\cdot)\tfrac{x}{\|x\|}\colon \Omega \to \mathbb{R}^k.$$

Its i-th coordinate function is given by

$$\big(U(\cdot)\tfrac{x}{\|x\|}\big)_i = \big\langle u_i, \tfrac{x}{\|x\|}\big\rangle = \sum_{j=1}^{d} u_{ij}\tfrac{x_j}{\|x\|}$$

and with Fact 10.1(ii), we obtain that

$$\big(U(\cdot)\tfrac{x}{\|x\|}\big)_i \sim \mathcal{N}\big(0, \frac{x_1^2}{\|x\|^2} + \cdots + \frac{x_d^2}{\|x\|^2}\big) = \mathcal{N}(0,1)$$

for all $i = 1, \dots, k$, and therefore $U(\cdot)\tfrac{x}{\|x\|} \sim \mathcal{N}(0,1,\mathbb{R}^k)$ by applying Fact 10.1(i). It follows

$$\begin{aligned}\mathrm{P}\big[\,\big|\|Ux\| - \sqrt{k}\,\|x\|\big| \leqslant \varepsilon\sqrt{k}\,\|x\|\,\big] &= \mathrm{P}\big[\,\big|\|U\tfrac{x}{\|x\|}\| - \sqrt{k}\big| \leqslant \varepsilon\sqrt{k}\,\big]\\ &\leqslant 1 - 2\,\mathrm{e}^{-(\varepsilon\sqrt{k})^2/16}\\ &= 1 - 2\,\mathrm{e}^{-\varepsilon^2 k/16},\end{aligned}$$

where we have used the Theorem 10.4 for the estimate, but with k instead of d and $\varepsilon\sqrt{k}$ instead of ε. The necessary condition $0 < \varepsilon\sqrt{k} \leqslant \sqrt{k}$ holds due to our assumption $\varepsilon \leqslant 1$. □

Since we specify k beforehand and then keep it fixed, it makes sense to divide by $\sqrt{k}$ and then "absorb" $\frac{1}{\sqrt{k}}$ into the matrix U.

Definition 11.3. Let $(\Omega, \Sigma, \mathrm{P})$ be a probability space and for $k < d$ let $U\colon \Omega \to \mathbb{R}^{k\times d}$ be a random matrix with $U \sim \mathcal{N}(0,1,\mathbb{R}^{k\times d})$. Let further $\omega \in \Omega$. Then the map

$$T_{U(\omega)}\colon \mathbb{R}^d \to \mathbb{R}^k,\ T_{U(\omega)}x := \tfrac{1}{\sqrt{k}}U(\omega)x$$

is called the *Johnson-Lindenstrauss projection.*

It should be noted that the term "projection" is not formally consistent with its meaning in the sense of linear algebra — it is however "almost" consistent, namely if we identify $\mathbb{R}^k$ with a subspace of $\mathbb{R}^d$ via $U(\omega)^{\mathsf{T}}$ and ignore that $U(\omega)^{\mathsf{T}}$ may not have full rank.

Remark 11.4. (i) With the above notation, Theorem 11.2 implies

$$\forall\, x \in \mathbb{R}^d\backslash\{0\},\ 0 < \varepsilon \leqslant 1\colon\ \mathrm{P}\big[\,\big|\|T_U x\| - \|x\|\big| \leqslant \varepsilon\|x\|\,\big] \geqslant 1 - 2\,\mathrm{e}^{-k\varepsilon^2/16},$$

where $U \sim \mathcal{N}(0,1,\mathbb{R}^{k\times d})$. The heuristic is now of course that, by the above,

$$\left|\frac{\|T_U x\|}{\|x\|} - 1\right| \leqslant \varepsilon, \quad \text{respectively,} \quad \frac{\|T_U x\|}{\|x\|} \approx 1$$

hold with high probability.

(ii) Notice that in (i) only the relative change in length is close to one, while the absolute change may be quite large. However, if we choose our points from a bounded set $B \subseteq \overline{\mathrm{B}}_r(0)$ with fixed $r > 0$, we may achieve by suitable choice of ε and k that

$$\big|\|T_U x\| - \|x\|\big| \approx 0$$

is highly likely as well. Consider, for example, a dataset in the unit ball, i.e., $r = 1$. If we choose $\varepsilon = 0.2$ and $k = 1000$, then $\mathrm{P}\big[\big|\|T_U x\| - \|x\|\big| < 0.2\big] \geqslant 0.8$ holds for every $x \in \mathbb{R}^d$ with $\|x\| \leqslant 1$—and even independently of $d \geqslant 1000$.

Next, we want to investigate what happens to pairwise distances when we project data points using T_U.

Theorem 11.5. (Johnson-Lindenstrauss lemma) *Let $(\Omega, \Sigma, \mathrm{P})$ be a probability space and let further $0 < \varepsilon < 1$, $n \geqslant 1$ and $k \geqslant \frac{48}{\varepsilon^2}\log n$. Let $U\colon \Omega \to \mathbb{R}^{k\times d}$ be a random matrix with $U \sim \mathcal{N}(0,1,\mathbb{R}^{k\times d})$. Then for every system $x^{(1)},\dots,x^{(n)} \in \mathbb{R}^d$ of n-many points the following estimate holds:*

$$\mathrm{P}\big[\forall\, i,j\colon (1-\varepsilon)\|x_i - x_j\| \leqslant \|T_U x_i - T_U x_j\| \leqslant (1+\varepsilon)\|x_i - x_j\|\big] \geqslant 1 - \tfrac{1}{n}.$$

Proof. For $i \neq j$ we define $x := x_i - x_j$ and calculate via the previous Theorem 11.2:

$$\begin{aligned}
P_{ij} &:= \mathrm{P}\big[(1-\varepsilon)\|x_i - x_j\| \leqslant \|T_U x_i - T_U x_j\| \leqslant (1+\varepsilon)\|x_i - x_j\|\big]\\
&= \mathrm{P}\big[(1-\varepsilon)\|x\| \leqslant \|T_U x\| \leqslant (1+\varepsilon)\|x\|\big]\\
&= \mathrm{P}\big[\big|\|T_U x\| - \|x\|\big| \leqslant \varepsilon\|x\|\big]\\
&= \mathrm{P}\big[\big|\|Ux\| - \sqrt{k}\,\|x\|\big| \leqslant \varepsilon\sqrt{k}\,\|x\|\big]\\
&\geqslant 1 - 2\,\mathrm{e}^{-k\varepsilon^2/16}.
\end{aligned}$$

Since there are $\binom{n}{2} \leqslant \frac{n^2}{2}$ possible choices for $1 \leqslant i,j \leqslant n$ with $i \neq j$, we get for the probability stated in the theorem

$$\begin{aligned}
\mathrm{P}\big[\cdots\big] &= 1 - \mathrm{P}\big[\exists\, i \neq j\colon \|T_U x_i - T_U x_j\| \notin \big((1-\varepsilon)\|x_i - x_j\|, (1+\varepsilon)\|x_i - x_j\|\big)\big]\\
&= 1 - \sum_{i\neq j}(1 - P_{ij}) \geqslant 1 - \sum_{i\neq j} 2\,\mathrm{e}^{-k\varepsilon^2/16} \geqslant 1 - \frac{n^2}{2}\cdot 2\,\mathrm{e}^{-k\varepsilon^2/16}\\
&\geqslant 1 - n^2\,\mathrm{e}^{-(\frac{48}{\varepsilon^2}\log n)\,\varepsilon^2/16} \geqslant 1 - n^2\,\mathrm{e}^{\log(n^{-3})}
\end{aligned}$$

$$= 1 - \frac{1}{n}$$

as claimed. □

Let us point out the following remarkable observations in connection with the Johnson-Lindenstrauss lemma (Theorem 11.5).

(i) The dimension d of the output space does neither appear in the norm estimate nor in the probability estimate.

(ii) The number n of points enters the probability estimate as $1/n$, but appears only inside a logarithm within the dimension k of the low-dimensional space.

(iii) The norm estimate holds for *all* pairs of points, with high probability — not with high probability for each pair of points.

Now we return to the discussion we started in Remark 11.4(ii).

Remark 11.6. We consider a dataset

$$D = \{(x^{(i)}, y^{(i)}) \mid i = 1, \dots, n_1\} \subseteq \mathbb{R}^d \times \mathbb{R},$$

where the $x^{(i)}$ all lie in a bounded set $B \subseteq \mathbb{R}^d$, and we assume that we are interested in determining a predictor defined on B. To this end, we choose n_2-many unlabeled points (test data) in B, project the labeled points (training data) as well as the unlabeled points onto $\mathbb{R}^k$, and assign labels to the unlabeled points, by executing a suitable algorithm on the low-dimensional data. By choosing a suitably large test or training dataset, we achieve that $n = n_1 + n_2$ is large and thus the probability in Theorem 11.5 will be close to one. Since n only enters k logarithmically, k practically only depends on ε. Since all data points come from a bounded set, we can achieve as in Remark 11.4(ii) that

$$\big| \|T_U x^{(i)} - T_U x^{(j)}\| \big| \approx 0 \quad \text{and} \quad k \ll d$$

holds for all points $x^{(i)}$, $x^{(j)}$ from the training set and from the test set.

In Exercise 11.3 we will treat randomly generated data with the Johnson-Lindenstrauss projection, calculate the actual error of the distances and compare it with the bound from the Johnson-Lindenstrauss lemma. For $n = 300$ independent samples from $X \sim \mathcal{N}(0, 1, \mathbb{R}^{1000})$ the following picture emerges for the actual multiplicative error (solid line) compared to the bound $\varepsilon = \sqrt{48 \log(n)/k}$ (dashed line) provided by Johnson-Lindenstrauss.

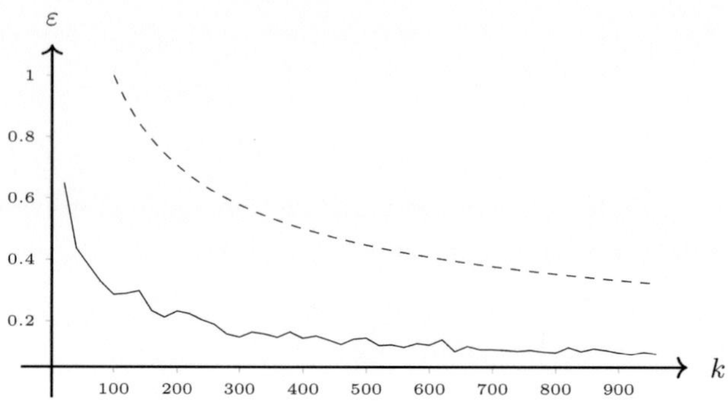

References

This chapter is again based on [BHK20]. We also refer to the original work [JL84]. Some of the exercises in this and the previous chapters are taken from [For20]. In addition, we like to point out that on the internet there are versions of the Johnson-Lindenstrauss lemma in which the constant is 8 (instead of 48); it is unknown to the author whether these versions are correct.

Exercises

Exercise 11.1. Let $X = X_1 + \cdots + X_d$ with $X_i \sim \mathcal{N}(0,1)$ be given.

(i) Show $\operatorname{E}(e^{tX_i}) = (1-2t)^{-d/2}$ for $t \in (0, \frac{1}{2})$.

(ii) Show $\operatorname{P}\big[X \geqslant a\big] \leqslant \inf\limits_{t\in(0,\frac{1}{2})} \frac{e^{-ta}}{(1-2t)^{d/2}}$ for $a > 0$.

Hint: For (i) use the law of the unconscious statistician

$$\operatorname{E}(f(X_i)) = \frac{1}{\sqrt{2\pi}} \int_{\mathbb{R}} f(t) e^{-\frac{t^2}{2}} \,\mathrm{d}t.$$

Exercise 11.2. Use Exercise 11.1 to provide an alternative proof for the Johnson-Lindenstrauss lemma, which does *not* require the Gaussian annulus theorem.

Exercise 11.3. Let $d > k$ be given.

(i) Implement the Johnson-Lindenstrauss projection $T_U \colon \mathbb{R}^d \to \mathbb{R}^k$ and test it for small d and k.

(ii) Put $d = 1000$, generate 300 points in $\mathbb{R}^d$ at random, according to a distribution of your choice, project these via T_U onto $\mathbb{R}^k$ and calculate for different k the largest occurring distance change

$$\varepsilon := \max\Bigg(1 - \min_{x\neq y} \frac{\|T_U x - T_U y\|}{\|x-y\|}, \max_{x\neq y} \frac{\|T_U x - T_U y\|}{\|x-y\|} - 1\Bigg).$$

(iii) Compare via a plot the experimental values for ε with the bound for ε that the Johnson-Lindenstrauss lemma provides. This should lead to a picture as the one on page 155.

12

Separation and Fitting of High-Dimensional Gaussians

After having studied Gaussian distributions in high-dimensional space in Chapter 10, we are going to apply our findings now to datasets like the one in the following picture, consisting of data points sampled from a superposition of several Gaussian distributions.

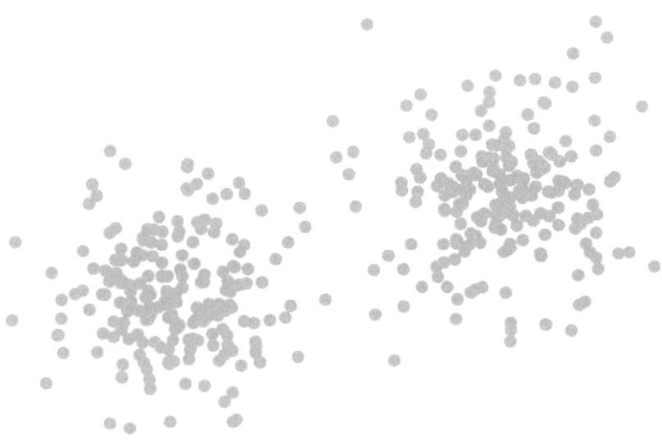

We formalize this as follows.

Definition 12.1. Let $(\Omega, \Sigma, \mathrm{P})$ be a probability space.

(i) Let $X\colon \Omega \to \mathbb{R}^d$ be a random vector with $X \sim \mathcal{N}(\mu, \sigma^2, \mathbb{R}^d)$ and let $\omega_1, \ldots, \omega_n \in \Omega$. Then we call the set $G = \{X(\omega_i) \mid i = 1, \ldots, n\} \subseteq \mathbb{R}$ a *(d-dimensional) Gaussian*.

(ii) Let $X_1, \ldots, X_m\colon \Omega \to \mathbb{R}^d$ be independent random vectors such that $X_i \sim \mathcal{N}(\mu_i, \sigma_i^2, \mathbb{R}^d)$, and let further points $\omega_1, \ldots, \omega_n \in \Omega$ be given. Then we call the Gaussians $G_j = \{X_j(\omega_i) \mid i = 1, \ldots, n\}$ for $j = 1, \ldots, m$ *independent*.

S. A. Wegner, *Mathematical Introduction to Data Science*,
https://doi.org/10.1007/978-3-662-69426-8_12

Let independent Gaussians $G_1, \ldots, G_m$ be given. We put

$$D := G_1 \cup \cdots \cup G_m \subseteq \mathbb{R}^d.$$

The picture on the previous page 159 shows the union $D = G_1 \cup G_2$ of two Gaussians with different means and 200 points each. As this monochrome picture already suggests, we now pretend to have forgotten which data point comes from which Gaussian. The first naturally arising task is to retrieve this information based on D, i.e., to *separate* or *disentangle* the Gaussians. Once this is done, the second natural task is to determine the parameters μ_i and σ_i for each Gaussian — assuming that we have forgotten these as well. This is referred to as *parameter estimation* or *fitting*.

Before we start with the separation task, we note that in practice, both tasks of course cannot be solved in a "definitive" sense: In the above picture, e.g., it may happen that a point, although it comes from the Gaussian on the bottom left, still happens to lie top right. Any sensible separation algorithm will then classify this point as belonging to the right Gaussian. If in practice some dataset is given, we have no way of detecting this misclassification and therefore do not really strive for a "correct" classification but rather a classification with "high plausibility". The latter will be made precise later on, see the explanations before Definition 12.9, using the concept of conditional probability.

12.1 Separation of Gaussians

We start with the separation task described above and limit ourselves in the following to two Gaussians. For simplicity we let both variances be one and refer the reader to Remark 12.13.

The picture on page 159 suggests the following idea for constructing a classifier: Firstly, we randomly select a point $z \in D$, and note that this point will with high probability be close to one of the means of the two Gaussians. Then we assign the label 1 to z and determine for all other points in D their distances to z. If such a distance turns out to be small, then we label the point with 1. Otherwise, we assign the label 2. Provided that the distance of the means $\Delta := \|\mu_1 - \mu_2\|$ of the Gaussians is large enough, it is to be expected that after calculating all distances $\|z - x\|$, $x \in D$, it makes sense to speak of "small" and "large" distances. We continue to consider the dataset D from page 159 and test the above procedure.

Example 12.2. In the following pictures, the point z picked in the first step is marked by a black dot. Below we then see the distribution of distances $\|z - x\|$, $x \in D$.

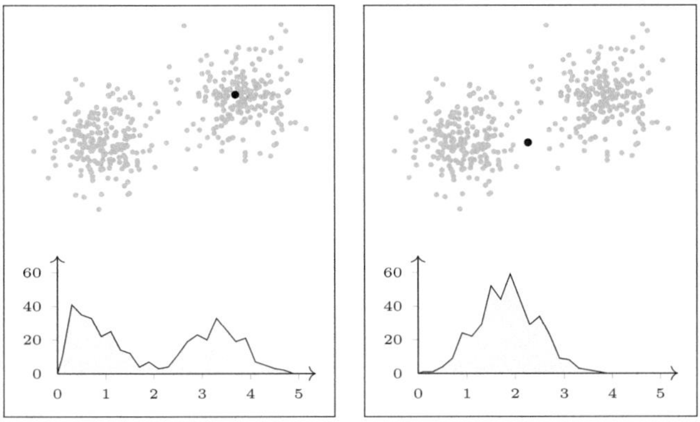

As we have already mentioned, a situation like on the right, where z is "halfway" between the Gaussians, only occurs with low probability, cf. Exercise 12.3.

An canonical strategy to get from the distances to the classifier is the following: Given a dataset with n-many points, we label the $n/2$-nearest neighbors of z with 1 and the rest with 2. We formulate this in pseudocode.

Algorithm 12.3. *Let a dataset* $D = \{x^{(1)}, \ldots, x^{(n)}\} \subseteq \mathbb{R}^d$ *be given, which is the union of two independent Gaussians, each consisting of* $n/2$*-many points.*

1: **function** SEPARATION (D)
2: $\quad z \leftarrow$ random point of D
3: $\quad$ **for** $i \leftarrow 1$ to n **do**
4: $\quad\quad d_i \leftarrow \|z - x^{(i)}\|$
5: $\quad I \leftarrow (i_1, \ldots, i_n)$ with $d_{i_j} \leqslant d_{i_{j+1}}$ for $j = 1, \ldots, n$
6: $\quad$ **for** $j \leftarrow 1$ to $n/2$ **do**
7: $\quad\quad \ell_{i_j} \leftarrow 1$
8: $\quad$ **for** $j \leftarrow n/2 + 1$ to n **do**
9: $\quad\quad \ell_{i_j} \leftarrow 2$
10: $\quad$ **return** $(\ell_1, \ldots, \ell_n)$

Here, a suitable sorting algorithm can be used in line 5 for sorting the data points by their distances from z.

In Exercises 12.1 and 12.4 we will implement Algorithm 12.3 for both low-dimensional and high-dimensional datasets, respectively. We note that we may of course also read the vector ℓ as a mapping $\ell\colon D \to \{1,2\}$, $\ell(x^{(i)}) := \ell_i$. Furthermore, we note that our current method is an instance of unsupervised learning.

We now turn our attention to the situation of high-dimensional data. Here we know that a Gaussian does not look like in the picture on page 159, but that a large part of its data points will accumulate in a thin annulus with radius $\sqrt{d}$ around the center μ of the Gaussian. If we stick with two Gaussians, we need to assume, similarly to the low-dimensional case, that their centers μ_1 and μ_2 are far enough apart for a separation to be successful. It would certainly be sufficient to require that $\Delta := \|\mu_1 - \mu_2\|$ is significantly larger than $2\sqrt{d}$—because then we could expect a similar distribution of distances from a randomly chosen point as in Example 12.2. As it turns out, we can even manage with a smaller distance, as long as it is not too small compared to $\sqrt{d}$. The following three pictures

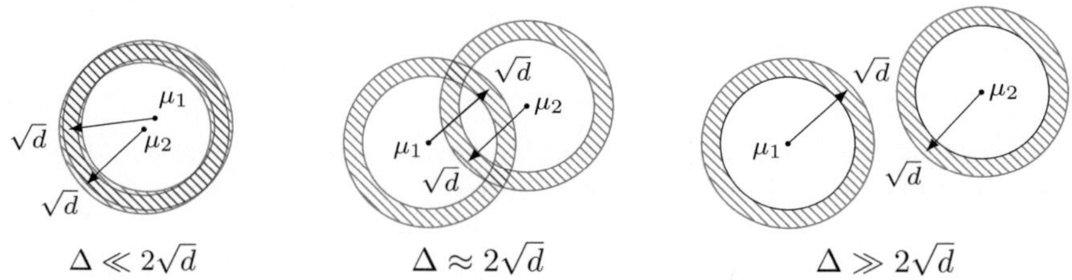

of two Gaussians suggest that even if Δ is "somewhat smaller" than $2\sqrt{d}$, the two annuli, in which the main part of the points of the Gaussians are located, only have a relatively small intersection. In the following remark we explain, how the symbols "$\ll$", "$\approx$" and "$\gg$" are to be understood in the current context.

Remark 12.4. In order to describe "how large" the distance between the centers, i.e., $\Delta = \|\mu_1 - \mu_2\|$, of the two Gaussians needs to be, we are interested, analogous to the previous Chapters 8–10, in statements that hold *for all suitably large dimensions*. Below, we thus think of $\Delta = \Delta(d)$ as a function of the dimension.

(i) If Δ grows too slowly compared to $\sqrt{d}$, we will, for large d, eventually end up in a situation where the annuli overlap heavily, see the far left picture before this remark. This is what we mean by $\Delta \ll 2\sqrt{d}$.

(ii) If Δ grows in such a way that for large d always $\Delta(d) > 2\sqrt{d} + \varepsilon$ holds with a fixed ε that is larger than the width of the annuli, then a separation should be possible. Recall that the width of the annuli is constant according to the Gaussian annulus theorem, compare also Example 10.5. This is how $\Delta \gg 2\sqrt{d}$ is to be understood.

(iii) If Δ grows as fast as $2\sqrt{d}$, then the middle picture before this remark suggests that a separation is possible. We will see in the following that this remains the case even if $\Delta(d)$ grows slightly slower than $2\sqrt{d}$. This is what is meant by $\Delta \approx 2\sqrt{d}$.

Our goal now is to find out *how much slower* $\Delta(d)$ *can grow compared to* $2\sqrt{d}$ such that a separation via Algorithm 12.3 can still be achieved. We consider

$x, y \in D$. According to Chapter 10 we expect that $\|x-y\| \approx \sqrt{2d}$ holds if the two points belong to the same Gaussian, and we therefore have to find out of which magnitude $\|x-y\|$ will be if x and y belong to different Gaussians. In order to do so, let us assume that x and y belong to different Gaussians. Then we get the following picture.

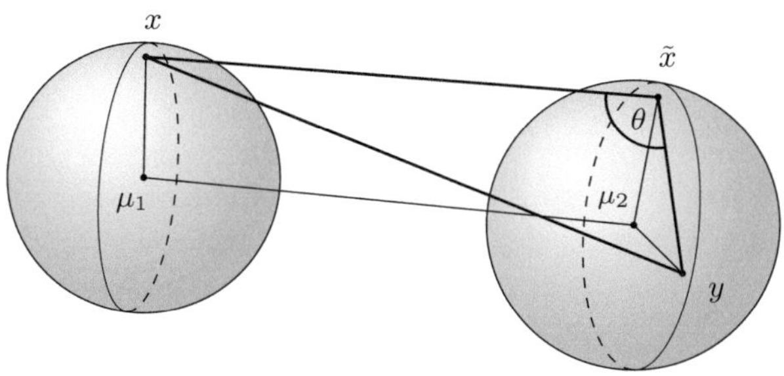

Indeed:

1. With high probability, the point x will be close to the surface of a sphere around μ_1 with radius $\sqrt{d}$ (Gaussian annulus theorem) and we will have, again with high probability, that $x-\mu_1 \perp \mu_2-\mu_1$ almost holds (Gaussian orthogonality theorem with $Y \equiv \mu_2 - \mu_1$).
2. We obtain the point $\tilde{x} := \mu_2 + x - \mu_1 = x + (\mu_2 - \mu_1)$ by parallel shift, such that, according to the above, with high probability it will almost hold $\tilde{x} - \mu_2 \perp \mu_1 - \mu_2$.
3. The point y will with high probability be close to the surface of a sphere around μ_2 with radius $\sqrt{d}$ (again annulus theorem). At the same time, the distance between $\tilde{x}$ and y will be approximately $\sqrt{2d}$ (first Gaussian distance theorem), and finally, we will almost have $y - \mu_2 \perp \mu_1 - \mu_2$ (again orthogonality theorem).

In the resulting triangle, $\theta \approx 90°$ holds and it follows

$$\|x-y\|^2 \approx \|x-\tilde{x}\|^2 + \|\tilde{x}-y\|^2 \approx \Delta^2 + 2d$$

with the Pythagorean theorem. The following formalizes the above making use of expectation and variance and employing the law of cosines.

Theorem 12.5. *Let* $(\Omega, \Sigma, \mathrm{P})$ *be a probability space, let* $\mu_1, \mu_2 \in \mathbb{R}^d$ *and let* $X, Y\colon \Omega \to \mathbb{R}^d$ *be independent random vectors with* $X \sim \mathcal{N}(\mu_1, 1, \mathbb{R}^d)$ *and* $Y \sim \mathcal{N}(\mu_2, 1, \mathbb{R}^d)$. *Let* $\Delta = \|\mu_1 - \mu_2\|$. *Then:*

(i) $\mathrm{E}(\|X-Y\|^2) = \Delta^2 + 2d$,

(ii) $\mathrm{V}(\|X-Y\|^2) \leqslant 3d + 8\Delta^2 + 5\sqrt{d}\Delta$.

Proof. (i) Let $X, Y : \Omega \to \mathbb{R}^d$ be as above and let $\tilde{X} := \mu_2 + X - \mu_1$. That is, $\tilde{X} \sim \mathcal{N}(\mu_2, 1, \mathbb{R}^d)$ and $\tilde{X}$ and Y are independent. We apply the law of cosines pointwise with $\theta = \measuredangle(X - \tilde{X}, Y - \tilde{X})$ and obtain

$$\begin{aligned}
\|X - Y\|^2 &= \|X - \tilde{X}\|^2 + \|Y - \tilde{X}\|^2 - 2\|X - \tilde{X}\|\|Y - \tilde{X}\| \cos(\theta) \\
&= \|\mu_1 - \mu_2\|^2 + \|Y - \tilde{X}\|^2 - 2\langle X - \tilde{X}, Y - \tilde{X}\rangle \\
&= \Delta^2 + \|Y - \tilde{X}\|^2 - 2\big(\langle \mu_1 - \mu_2, Y - \mu_2\rangle + \langle \mu_2 - \mu_1, X - \mu_1\rangle\big).
\end{aligned}$$

Shifting by μ_2 and applying (8.1) from page 120 yields $\mathrm{E}(\|Y - \tilde{X}\|^2) = 2d$. Application of Theorem 8.4 further shows that the expected values of the two scalar products above are each zero. We thus obtain

$$\mathrm{E}(\|X_1 - X_2\|^2) = 2d + \Delta^2 + 2(0 - 0) = \Delta^2 + 2d.$$

(ii) In order to estimate the variance of $\|X - Y\|^2$, we have to take into account that the three summands that we get by expanding $\|X - Y\|^2$ are not independent, see Proposition A.6. Therefore, we have

$$\begin{aligned}
\mathrm{V}(\|X - Y\|^2) &= \mathrm{V}(\Delta^2 + \|Y - \tilde{X}\|^2 - 2\langle X - \tilde{X}, Y - \tilde{X}\rangle) \\
&= \mathrm{V}(\|Y - \tilde{X}\|^2) + 4\,\mathrm{V}(\langle X - \tilde{X}, Y - \tilde{X}\rangle) \\
&\qquad + \mathrm{Cov}(\|Y - \tilde{X}\|^2 - 2\langle X - \tilde{X}, Y - \tilde{X}\rangle) \\
&\leqslant \mathrm{V}(\|Y - \tilde{X}\|^2) + 4\,\mathrm{V}(\langle X - \tilde{X}, Y - \tilde{X}\rangle) \\
&\qquad + \sqrt{\mathrm{V}(\|Y - \tilde{X}\|^2) \cdot 4\,\mathrm{V}(\langle X - \tilde{X}, Y - \tilde{X}\rangle)},
\end{aligned}$$

where we can estimate $\mathrm{V}(\|Y - \tilde{X}\|^2) \leqslant 3d$ according to the hint in Exercise 8.3. For the variance of the scalar product, we get

$$\begin{aligned}
\mathrm{V}(\langle X - \tilde{X}, Y - \tilde{X}\rangle) &= \mathrm{V}(\langle X - \mu_1, \mu_2 - \mu_1\rangle + \langle Y - \mu_2, \mu_1 - \mu_2\rangle) \\
&= \mathrm{V}(\langle X - \mu_1, \mu_2 - \mu_1\rangle) + \mathrm{V}(\langle Y - \mu_2, \mu_1 - \mu_2\rangle) \\
&\leqslant 2\|\mu_1 - \mu_2\|^2,
\end{aligned}$$

where we have applied Theorem 8.4 with $Z := X_i - \mu_i \sim \mathcal{N}(0, 1, \mathbb{R}^d)$ and $\xi := \pm(\mu_1 - \mu_2)$ to conclude $\mathrm{V}(\langle Z, \xi\rangle) = \|\xi\|^2$. Finally, we obtain

$$\begin{aligned}
\mathrm{V}(\|X - Y\|^2) &\underset{\substack{\uparrow \\ \text{see} \\ \text{above}}}{\leqslant} \mathrm{V}(\|Y - \tilde{X}\|^2) + 4\,\mathrm{V}(\langle X - \tilde{X}, Y - \tilde{X}\rangle) + \sqrt{\cdots} \\
&\leqslant 3d + 8\|\mu_1 - \mu_2\|^2 + \sqrt{3d \cdot 8\|\mu_1 - \mu_2\|^2} \\
&\leqslant 3d + 8\Delta^2 + 5\sqrt{d}\Delta
\end{aligned}$$

as claimed. □

With a bit more work, we will now obtain estimates for the expected value and variance of the distance itself, i.e., without the square.

Theorem 12.6. *Let* $(\Omega, \Sigma, \mathrm{P})$ *be a probability space, let* $\mu_1, \mu_2 \in \mathbb{R}^d$ *and let* $X, Y : \Omega \to \mathbb{R}^d$ *be independent random vectors with* $X \sim \mathcal{N}(\mu_1, 1, \mathbb{R}^d)$ *and* $Y \sim \mathcal{N}(\mu_2, 1, \mathbb{R}^d)$. *Let* $\Delta = \|\mu_1 - \mu_2\|$. *Then:*

(i) $\forall\, d \in \mathbb{N}\colon |\mathrm{E}(\|X-Y\|) - \sqrt{\Delta^2+2d}\,| \leqslant 5/\sqrt{d}$.

(ii) $\forall\, d \in \mathbb{N}\colon\ \mathrm{V}(\|X-Y\|) \leqslant 5/\sqrt{d} + 14 \leqslant 19$.

Proof. (i) We proceed as in the proof of Theorem 8.2 and decompose the term for which the expected value is sought as follows:

$$\|X-Y\|-\sqrt{\Delta^2+2d} = \underbrace{\frac{\|X-Y\|^2-(\Delta^2+2d)}{2\sqrt{\Delta^2+2d}}}_{=:S_d} - \underbrace{\frac{(\|X-Y\|^2-(\Delta^2+2d))^2}{2\sqrt{\Delta^2+2d}(\|X-Y\|+\sqrt{\Delta^2+2d})^2}}_{=:R_d}.$$

According to Theorem 12.5(i) we have $\mathrm{E}(S_d) = 0$. In R_d, we estimate in the denominator $\|X-Y\| \geqslant 0$. In the numerator we first apply Theorem 12.5(i) to rewrite the expected value into a variance and then Theorem 12.5(ii) to estimate the variance. This leads to

$$0 \leqslant \mathrm{E}(R_d) \leqslant \frac{\mathrm{V}(\|X-Y\|^2)}{2(\Delta^2+2d)^{3/2}} \leqslant \frac{3d+8\Delta^2+5\sqrt{d}\Delta}{2(\Delta^2+2d)^{3/2}} =: r_d,$$

which we will now estimate further by case analysis. Let first $\Delta \geqslant \sqrt{2d}$. Then it follows

$$r_d = \frac{3d+8\Delta^2+5\sqrt{d}\,\Delta}{2(\Delta^2+2d)^{3/2}} \underset{0\leqslant d\leqslant \frac{1}{2}\Delta^2}{\overset{\uparrow}{\leqslant}} \frac{\frac{3}{2}\Delta^2+8\Delta^2+\frac{5}{\sqrt{2}}\Delta^2}{2(\Delta^2)^{3/2}} \leqslant \frac{5}{\sqrt{d}}.$$

If, on the other hand, $\Delta < \sqrt{2d}$, then we have

$$r_d = \frac{3d+8\Delta^2+5\sqrt{d}\,\Delta}{2(\Delta^2+2d)^{3/2}} \underset{0\leqslant \Delta^2\leqslant 2d}{\overset{\uparrow}{\leqslant}} \frac{3d+16d+5\sqrt{2}\,d}{2(2d)^{3/2}} \underset{1\leqslant 2^{3/2}}{\overset{\uparrow}{\leqslant}} \frac{5}{\sqrt{d}}$$

and thus $\mathrm{E}(R_d) \leqslant \frac{5}{\sqrt{d}}$ for all $d \in \mathbb{N}$.

(ii) To handle the variance, we first calculate

$$\mathrm{V}(\|X-Y\|) = \big|\mathrm{E}(\|X-Y\|^2) - [\mathrm{E}(\|X-Y\|)]^2\big|$$

$$
\begin{aligned}
&\underset{\substack{\uparrow\\ \text{Thm 12.5(i)}}}{=} \big|(\Delta^2+2d) - [\,\mathrm{E}(\|X-Y\| - \sqrt{\Delta^2+2d}\,) + \sqrt{\Delta^2+2d}\,]^2\big| \\
&= \big|(\Delta^2+2d) - \big[\big(\mathrm{E}(\|X-Y\| - \sqrt{\Delta^2+2d}\,)\big)^2 \\
&\qquad + 2\sqrt{\Delta^2+2d}\cdot \mathrm{E}\big(\|X-Y\| - \sqrt{\Delta^2+2d}\,\big) + (\Delta^2+2d)\big]\big| \\
&\underset{\substack{\uparrow\\ \text{see}\\ \text{above}}}{\leqslant} r_d + 2\sqrt{\Delta^2+2d}\cdot r_d.
\end{aligned}
$$

The first term can be estimated by $5/\sqrt{d}$ and we treat the second term again via case analysis. If $\Delta \geqslant \sqrt{2d}$, then

$$
2\sqrt{\Delta^2+2d}\cdot r_d = \frac{3d+8\Delta^2+5\sqrt{d}\,\Delta}{\Delta^2+2d} \underset{\substack{\uparrow\\ 0\,\leqslant\, d\,\leqslant\, \frac{1}{2}\Delta^2}}{\leqslant} \frac{\frac{3}{2}\Delta^2+8\Delta^2+\frac{5}{\sqrt{2}}\Delta^2}{\Delta^2} \leqslant 14
$$

holds and if $\Delta < \sqrt{2d}$, we obtain similarly

$$
2\sqrt{\Delta^2+2d}\cdot r_d = \frac{3d+8\Delta^2+5\sqrt{d}\,\Delta}{\Delta^2+2d} \underset{\substack{\uparrow\\ 0\,\leqslant\, \Delta^2\,\leqslant\, 2d}}{\leqslant} \frac{3d+16d+5\sqrt{2}\,d}{2d} \leqslant 14.
$$

This conclude the proof. □

Theorem 12.6 tells us that we have to expect a distance of $\|x-y\| \approx \sqrt{\Delta^2+2d}$ between randomly chosen points from different Gaussians and that for large d the distribution of these distances, regardless of d, will concentrate tightly around this value. We formalize this further through an explicit probability estimate.

Theorem 12.7. (Second Gaussian distance theorem) *Let $(\Omega,\Sigma,\mathrm{P})$ be a probability space, let μ_1, $\mu_2 \in \mathbb{R}^d$ and let $X,Y\colon \Omega \to \mathbb{R}^d$ be independent random vectors with $X \sim \mathcal{N}(\mu_1,1,\mathbb{R}^d)$ and $Y \sim \mathcal{N}(\mu_2,1,\mathbb{R}^d)$. Let $\Delta = \|\mu_1-\mu_2\|$. For $\varepsilon \leqslant \sqrt{d}$ we have*

$$
\mathrm{P}\big[\,\big|\|X-Y\| - \sqrt{\Delta^2+2d}\,\big| \leqslant \varepsilon\,\big] \geqslant 1 - 4\,\mathrm{e}^{-\varepsilon^2/648} - \frac{26}{\varepsilon d} - \frac{7}{\sqrt{d}} - \frac{12}{\varepsilon}.
$$

Proof. We use the same technique as in the proof of the Gaussian annulus theorem (Theorem 10.4), i.e., we estimate

$$
P := \mathrm{P}\big[\,\big|\|X-Y\| - \sqrt{\Delta^2+2d}\,\big| \geqslant \varepsilon\,\big]
$$

from above by multiplying the estimate in $\mathrm{P}[\cdots]$ with the trivial inequality $\|X-Y\| + \sqrt{\Delta^2+2d} \geqslant \sqrt{\Delta^2+2d}$. Then, as in the proof of Theorem 12.5, we introduce the random variable $\tilde{X} := \mu_2 + X - \mu_1 \sim \mathcal{N}(\mu_2,1,\mathbb{R}^d)$ and apply the law of cosines. This yields

$$\begin{aligned}
P &\leqslant \mathrm{P}\big[\,\big|\|X-Y\|^2-(\Delta^2+2d)\big|\geqslant \varepsilon\sqrt{\Delta^2+2d}\,\big]\\
&= 1-\mathrm{P}\big[-\varepsilon\sqrt{\Delta^2+2d}\leqslant \Delta^2+\|Y-\tilde{X}\|^2-2\big(\langle\mu_1-\mu_2,Y-\mu_2,\rangle\\
&\qquad\qquad +\langle\mu_2-\mu_1,X-\mu_1\rangle\big)-(\Delta^2+2d)\leqslant \varepsilon\sqrt{\Delta^2+2d}\,\big]\\
&\leqslant 1-\mathrm{P}\big[-\tfrac{\varepsilon\sqrt{\Delta^2+2d}}{3}\leqslant \|Y-\tilde{X}\|^2-2d\leqslant \tfrac{\varepsilon\sqrt{\Delta^2+2d}}{3}\,\big]\\
&\qquad\qquad \cdot\mathrm{P}\big[-\tfrac{\varepsilon\sqrt{\Delta^2+2d}}{3}\leqslant -2\langle\mu_1-\mu_2,Y-\mu_2\rangle\leqslant \tfrac{\varepsilon\sqrt{\Delta^2+2d}}{3}\,\big]\\
&\qquad\qquad \cdot\mathrm{P}\big[-\tfrac{\varepsilon\sqrt{\Delta^2+2d}}{3}\leqslant -2\langle\mu_2-\mu_1,X-\mu_1\rangle\leqslant \tfrac{\varepsilon\sqrt{\Delta^2+2d}}{3}\,\big]\\
&\leqslant 1-\mathrm{P}\big[\,\big|\|Y-\tilde{X}\|^2-2d\,\big|\leqslant \tfrac{\varepsilon\sqrt{2d}}{3}\,\big]\cdot\mathrm{P}\big[\,|\langle Z,\xi\rangle|\leqslant \tfrac{\varepsilon\sqrt{\Delta^2}}{6}\,\big]^2,
\end{aligned}$$

where $Z:=X_i-\mu_i\sim\mathcal{N}(0,1,\mathbb{R}^d)$ and $\xi:=\mu_1-\mu_2$. The two probabilities at the end of the above calculation can be estimated using Corollary 10.11 by

$$\mathrm{P}\big[\,\big|\|Y-\tilde{X}\|^2-2d\,\big|\leqslant \tfrac{\varepsilon\sqrt{2d}}{3}\,\big]\geqslant 1-4\,\mathrm{e}^{-\varepsilon^2/648}-\frac{18\sqrt{2}}{\varepsilon d}-\frac{7}{\sqrt{d}}$$

and with Theorem 10.8 by

$$\mathrm{P}\big[\,|\langle Z,\xi\rangle|\leqslant \tfrac{\varepsilon\sqrt{\Delta^2}}{6}\,\big]\geqslant 1-\frac{6\|\xi\|}{\varepsilon\Delta}=1-\frac{6}{\varepsilon},$$

respectively. Since we are actually interested in $1-P$, we now consider

$$\begin{aligned}
1-P &\underset{\substack{\uparrow\\ \text{see}\\ \text{above}}}{\geqslant} \mathrm{P}\big[\,\big|\|Y-\tilde{X}\|^2-2d\,\big|\leqslant \tfrac{\varepsilon\sqrt{2d}}{3}\,\big]\cdot\mathrm{P}\big[\,|\langle Z,\xi\rangle|\leqslant \tfrac{\varepsilon\sqrt{\Delta^2}}{6}\,\big]^2\\
&\geqslant \Big(1-4\,\mathrm{e}^{-\varepsilon^2/648}-\frac{18\sqrt{2}}{\varepsilon d}-\frac{7}{\sqrt{d}}\Big)\big(1-\tfrac{6}{\varepsilon}\big)^2\\
&\geqslant 1-4\,\mathrm{e}^{-\varepsilon^2/648}-\frac{26}{\varepsilon d}-\frac{7}{\sqrt{d}}-\frac{12}{\varepsilon},
\end{aligned}$$

where for the last inequality we first expanded the product and then estimated all positive terms from below with zero. □

We point out that $\varepsilon>100$ is necessary for the lower bound in the probability estimate of Theorem 12.7 to become positive. This is due to the two terms on the right-hand side which are independent of d. Similar to Example 10.9, we now consider a simulation and compare its outcome with Theorem 12.7.

Example 12.8. The following picture shows the distribution of distances $\|x-y\|$ for $(x,y)\in G_1\times G_2$. Here, G_1 is a Gaussian in $\mathbb{R}^{100}$ with variance one and center $\mu_1=0$, and G_2 is a Gaussian in $\mathbb{R}^{100}$ with variance one and center $\mu_2=(\Delta,0,\dots,0)$ and $\Delta=10$.

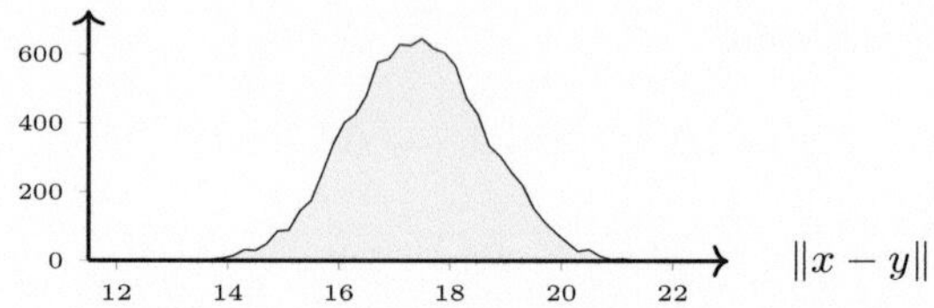

We get that $\sqrt{\Delta^2+2d} \approx 17.32$ indeed holds and that in our simulation even for *all* pairs $(x,y) \in G_1 \times G_2$ the estimate $\big|\|x-y\| - \sqrt{\Delta^2+2d}\,\big| < 5$ is valid.

If we have randomly chosen a point $z \in D = G_1 \cup G_2$ as in Algorithm 12.3, we know that the distances $\|z-x\|$ for $x \in D$ will concentrate around $\sqrt{2d}$ and $\sqrt{\Delta^2+2d}$. If Δ is large enough, we thus expect a picture of the following form.

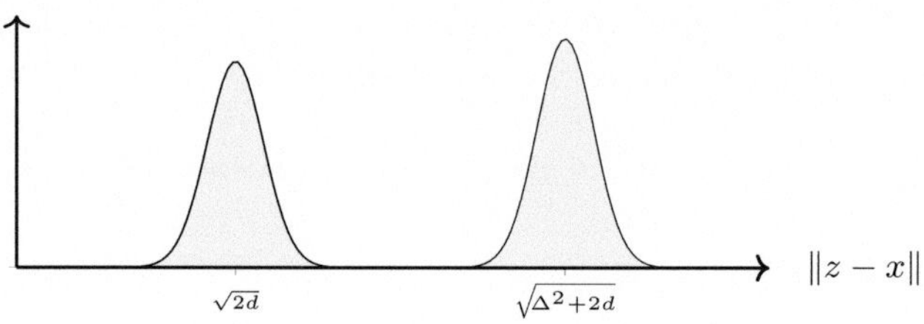

Instead of sorting all points according to their distances as in Algorithm 12.3, in high dimensions we may simply label all x with $\sqrt{2d}-\varepsilon \leqslant \|z-x\| \leqslant \sqrt{2d}+\varepsilon$ with 1. Here, one can determine ε by trial and error or even estimate it *independently of the dataset* by choosing ε for instance in such a way that

$$\mathrm{P}\big[\,\big|\|X-Y\| - \sqrt{2d}\,\big| \leqslant \varepsilon\big] \underset{\substack{\uparrow \\ \text{Thm 10.10}}}{\geqslant} 1 - 4\,\mathrm{e}^{-\varepsilon^2/72} - \frac{6\sqrt{2}}{\varepsilon d} - \frac{7}{\sqrt{d}} \geqslant 0.9$$

applies. For this, knowledge of d is sufficient and no further information about D is necessary. We leave it as Exercise 12.4 to formulate and test these variants of the separation algorithm.

It is clear that we can only be successful with any version of the separation algorithm if $\Delta := \|\mu_1 - \mu_2\|$ is large enough. We will now answer the question of "how large" by an asymptotic growth condition on $\Delta = \Delta(d)$, understood as a *function of the dimension.* For this, we need to quantify the probability that two points come from the same Gaussian, *assuming* that their distance is close to $\sqrt{2d}$. In simulations, say with the union of two Gaussians $D = G_1 \cup G_2$, the corresponding *frequency* can be determined by forming the quotient:

$$L := \frac{\#\{(x,y) \in D \times D \,\big|\, \big|\|x-y\| - \sqrt{2d}\,\big| < \varepsilon \wedge (x,y \in G_1 \vee x,y \in G_2)\}}{\#\{(x,y) \in D \times D \,\big|\, \big|\|x-y\| - \sqrt{2d}\,\big| < \varepsilon\}}.$$

For a fixed dataset $D = G_1 \cup G_2$ as above, we consider two discrete (!) and independent random vectors $A, B\colon \Omega \to \mathbb{R}^d$, which are uniformly distributed on D, i.e., $A, B \sim \mathcal{U}(G_1 \cup G_2)$. We denote by the abbreviation $x \sim y$, for $x, y \in \mathbb{R}^d$, if x and y both belong to G_1 or both belong to G_2. Then it follows that

$$L \underset{\substack{\uparrow \\ \text{expand} \\ \text{with } \#D^2}}{=} \frac{\mathrm{P}\big[\,\big|\|A-B\| - \sqrt{2d}\,\big| < \varepsilon \,\wedge\, A \sim B\big]}{\mathrm{P}\big[\,\big|\|A-B\| - \sqrt{2d}\,\big| < \varepsilon\big]} = \mathrm{P}\big[A \sim B \,\big|\, \big|\|A-B\| - \sqrt{2d}\,\big| < \varepsilon\big]$$

is the conditional probability that two points randomly chosen from D belong to the same Gaussian, *given* that their distance is ε-close to $\sqrt{2d}$. With $\#G_1 = \#G_2$ it further follows

$$\begin{aligned} L &\underset{\substack{\uparrow \\ \text{Bayes}}}{=} \frac{\mathrm{P}\big[\,\big|\|A-B\| - \sqrt{2d}\,\big| < \varepsilon \,\big|\, A \sim B\big] \cdot \mathrm{P}\big[A \sim B\big]}{\mathrm{P}\big[\,\big|\|A-B\| - \sqrt{2d}\,\big| < \varepsilon\big]} \\ &\underset{\substack{\uparrow \\ \text{Thm} \\ \text{of total} \\ \text{probability}}}{=} \frac{\mathrm{P}\big[\,\big|\|A-B\| - \sqrt{2d}\,\big| < \varepsilon \,\big|\, A \sim B\big] \cdot \frac{1}{2}}{\mathrm{P}\big[\,\big|\|X-Y\| - \sqrt{2d}\,\big| |\varepsilon \,\big|\, A \sim B\big] \cdot \frac{1}{2} + \mathrm{P}\big[\,\big|\|A-B\| - \sqrt{2d}\,\big| < \varepsilon \,\big|\, A \not\sim B\big] \cdot \frac{1}{2}}. \end{aligned}$$

In the latter formula, we can express each of the three conditional probabilities involving the discrete random vectors A and B, including the factor $1/2$, as (unconditional) probabilities involving Gaussian random vectors: This is possible, since $x \not\sim y$ for $x, y \in D$ by definition means that $x \in G_1$ and $y \in G_2$ or vice versa.

Definition 12.9. We define $L\colon \mathbb{N} \times [0, \infty) \times (0, \infty) \to [0, 1]$ by

$$L(d, \Delta, \varepsilon) := \frac{\mathrm{P}\big[\,\big|\|X_1 - Y_1\| - \sqrt{2d}\,\big| < \varepsilon\big]}{\mathrm{P}\big[\,\big|\|X_1 - Y_1\| - \sqrt{2d}\,\big| < \varepsilon\big] + \mathrm{P}\big[\,\big|\|X_2 - Y_2\| - \sqrt{2d}\,\big| < \varepsilon\big]},$$

where $X_1, Y_1 \sim \mathcal{N}(0, 1, \mathbb{R}^d)$ and $X_2 \sim \mathcal{N}(\mu_1, 1, \mathbb{R}^d)$, $Y_2 \sim \mathcal{N}(\mu_2, 1, \mathbb{R}^d)$ are independent random vectors with $\Delta = \|\mu_1 - \mu_2\|$.

We point out that in the definition, for each d we consider new random vectors with the specified properties. However, we may assume that these are all defined on a single probability space $(\Omega, \Sigma, \mathrm{P})$, cf. Theorem A.13.

Theorem 12.10. (Asymptotic separation theorem) *Let $(\Omega, \Sigma, \mathrm{P})$ be a probability space. For each $d \geqslant 1$, let $\mu_1, \mu_2 \in \mathbb{R}^d$ and independent random vectors X_1, Y_1, X_2, $Y_2\colon \Omega \to \mathbb{R}^d$ with $X_1, Y_1 \sim \mathcal{N}(0, 1, \mathbb{R}^d)$ and $X_2 \sim \mathcal{N}(\mu_1, 1, \mathbb{R}^d)$, $Y_2 \sim \mathcal{N}(\mu_2, 1, \mathbb{R}^d)$ be given. Furthermore, let $\Delta(d) := \|\mu_1 - \mu_2\|$ (read as a function of the dimension!) be such that $\sqrt{(\Delta(d))^2 - 2d} - \sqrt{2d} \to \infty$ for*

$d \to \infty$. *Then for fixed* $\varepsilon > 0$ *we have*

$$\liminf_{d\to\infty} L(d, \Delta(d), \varepsilon) \geqslant 1 - 4\,\mathrm{e}^{-\varepsilon^2/72}$$

where L *is the function from Definition 12.9.*

Proof. Let $\varepsilon > 0$ be fixed and let $d_0 \in \mathbb{N}$ be such that $\varepsilon < \sqrt{d_0}$. Using the first Gaussian distance theorem (Theorem 10.10), we can estimate the numerator of L, independently of Δ, by

$$\mathrm{P}\big[\,\big|\|X_1 - Y_1\| - \sqrt{2d}\,\big| < \varepsilon\big] \geqslant 1 - 4\,\mathrm{e}^{-\varepsilon^2/72} - \frac{6\sqrt{2}}{\varepsilon d} - \frac{7}{\sqrt{d}}$$

from below. In the denominator, we estimate the first summand from above with one. Notice that by Theorem 10.10 we are not giving much away here, provided that d is sufficiently large. We put

$$\delta(d) := \min\big(\sqrt{d}, \sqrt{\Delta^2 + 2d} - \sqrt{2d} - \varepsilon\big) \xrightarrow{d\to\infty} \infty,$$

which tends to infinity by assumption. By increasing d_0 if necessary, we achieve $\delta(d) > 0$ for $d \geqslant d_0$. For each $d \geqslant d_0$ we thus have $\sqrt{2d} + \varepsilon \leqslant \sqrt{\Delta^2 + 2d} - \delta(d)$ and we obtain for the second summand in the denominator of L the estimate

$$\begin{aligned}\mathrm{P}\big[\,\big|\|X_2 - Y_2\| - \sqrt{2d}\,\big| < \varepsilon\big] &= 1 - \mathrm{P}\big[\,\big|\|X_2 - Y_2\| - \sqrt{2d}\,\big| \geqslant \varepsilon\big]\\ &\leqslant 1 - \mathrm{P}\big[\,\big|\|X_2 - Y_2\| - \sqrt{\Delta^2 + 2d}\,\big| \leqslant \delta(d)\big]\\ &\leqslant 4\,\mathrm{e}^{-\delta(d)^2/648} - \frac{26}{\delta(d)d} - \frac{7}{\sqrt{d}} - \frac{12}{\delta(d)}\end{aligned}$$

employing the second Gaussian distance theorem (Theorem 12.7). Combining what we have proven so far it follows

$$L(d, \Delta, \varepsilon) \geqslant \frac{1 - 4\,\mathrm{e}^{-\varepsilon^2/72} - \dfrac{6\sqrt{2}}{\varepsilon d} - \dfrac{7}{\sqrt{d}}}{1 + 4\,\mathrm{e}^{-\delta(d)^2/648} - \dfrac{26}{\delta(d)d} - \dfrac{7}{\sqrt{d}} - \dfrac{12}{\delta(d)}} \xrightarrow{d\to\infty} 1 - 4\,\mathrm{e}^{-\varepsilon^2/72}$$

for $d \geqslant d_0$ and thus the statement about the limit inferior is established. □

The following proposition explicitly states which growth rates for $\Delta = \Delta(d)$ are allowed in Theorem 12.10.

Proposition 12.11. *For* $\Delta\colon \mathbb{N} \to [0, \infty)$ *TFAE:*

(i) $\displaystyle\lim_{d\to\infty} \frac{\Delta(d)}{\sqrt[4]{d}} = \infty$.

(ii) $\displaystyle\lim_{d\to\infty} \sqrt{\Delta(d)^2 - 2d} - \sqrt{2d} = \infty$.

Proof. (i) $\Longrightarrow$ (ii) Squaring in condition (i) leads to

$$\forall\, R > 0\; \exists\, d_0 \in \mathbb{N}\; \forall\, d \geqslant d_0 \colon \Delta(d)^2 \geqslant R \cdot \sqrt{d}.$$

If the latter is given, then (ii) follows from

$$\sqrt{\Delta(d)^2 - 2d} - \sqrt{2d} \geqslant \sqrt{R\sqrt{d} - 2d} - \sqrt{2d} \xrightarrow{d\to\infty} \frac{R}{2\sqrt{2}}.$$

(ii) $\Longrightarrow$ (i) Multiplication with $\sqrt{\Delta(d)^2 - 2d} + \sqrt{2d}$ in condition (ii) yields

$$\forall\, R > 0\; \exists\, d_0 \in \mathbb{N}\; \forall\, d \geqslant d_0 \colon \Delta(d)^2 \geqslant R \cdot \big(\sqrt{\Delta(d)^2 - 2d} + \sqrt{2d}\big) \geqslant R \cdot \sqrt{d},$$

from which we infer (i) via division by $\sqrt{d}$ and taking roots. □

In combination, Theorem 12.10 and Proposition 12.11 state that, for $d \to \infty$, a separation of two Gaussians is possible if the distance Δ between the centers grows "strictly faster" than $\sqrt[4]{d}$. In the following example we will illustrate this, but also highlight some shortcomings of this merely asymptotic statement.

Example 12.12. (i) In the following, we consider the distance $\Delta = \Delta(d)$ as a function of the dimension, that is, we consider a sequence of datasets $D_d \subseteq \mathbb{R}^d$, each consisting of the union of two Gaussians, whose centers have the distance $\Delta = cd^\alpha$ for various $\alpha > 0$ and $c > 0$. This yields the following graphs.

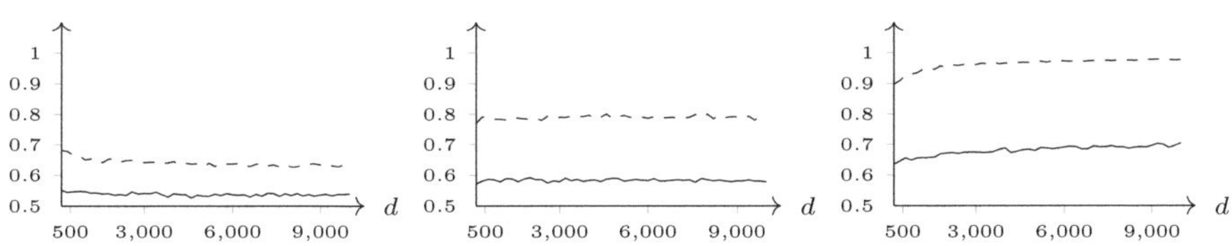

$\Delta = 2d^{0.20}$ (dashed) $\Delta = d^{0.20}$ (solid)
$\Delta = 2d^{0.25}$ (dashed) $\Delta = d^{0.25}$ (solid)
$\Delta = 2d^{0.30}$ (dashed) $\Delta = d^{0.30}$ (solid)

Specifically, we used the dimensions $d = 200, 400, 600, \ldots, 10000$ above and Gaussians G_1, G_2 each with 100 points around 0 and around $(\Delta, 0, \ldots, 0)$. After choosing a random point $z \in G_1 \cup G_2$, we assigned label 1 to the 100-nearest neighbors of z and label 2 to the remaining points. Then we calculated the rate of correctly classified points. To increase the accuracy, the simulation was performed 100 times per dimension; the plot above shows the averages of the correctly labeled parts.

We would like to point out that the constant 2, which is irrelevant for the asymptotic separation Theorem 12.10, significantly improves the quality of the separation in the right picture. To achieve a "good" separation without the constant but with exponent 0.30, one has to increase the dimension much further, see Exercise 12.5. At the same time, however, one can also see that $\Delta = cd^{0.25}$, for both $c = 1$ and $c = 2$, is the threshold of separability.

(ii) Now we fix the dimension $d = 10\,000$ and carry out the separation experiment with $\Delta = 30$. This yields the following distribution of pairwise distances:

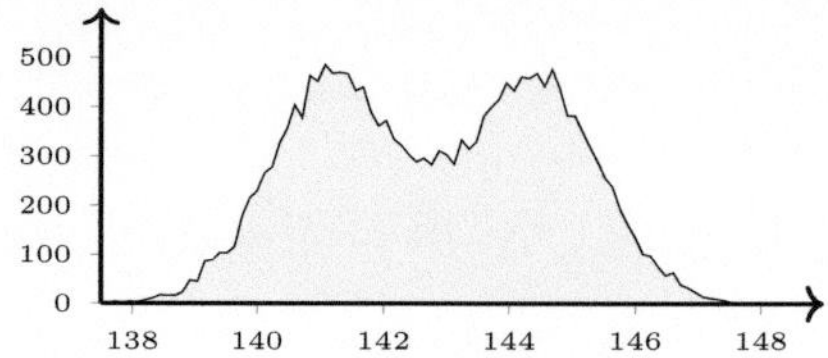

$\Delta = 30$ (96%)

Although the two "humps" already merge into each other, compared to the abstract drawing on page 168, the separation Algorithm 12.3 correctly classifies 96% of the points. Next, we consider $\Delta = 20$ and $\Delta = 10$ and again look at the distribution of pairwise distances and the rate of correctly classified points:

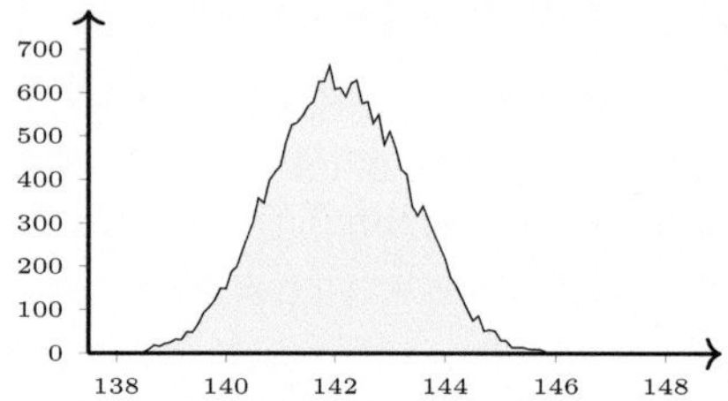

$\Delta = 20$ (81%)

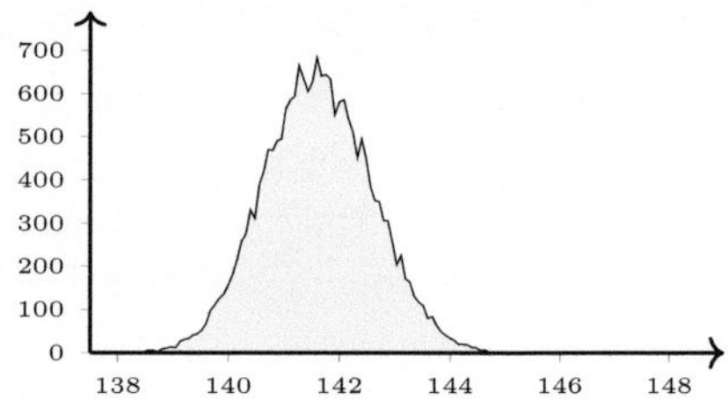

$\Delta = 10$ (54%)

At $\Delta = 20$, the formerly two "humps" seem to have completely merged into one. However, the separation Algorithm 12.3 still provides quite usable results. At $\Delta = 10 = \sqrt[4]{d}$, the distance of the Gaussians is then too small and their union cannot be disentangled anymore.

We add the following remarks, firstly on the case where the variances of the Gaussians are not both one, and secondly addressing the treatment of more than two Gaussians.

Remark 12.13. (i) Given $X \sim \mathcal{N}(\mu, \sigma^2, \mathbb{R}^d)$, one can easily calculate that $\mathrm{E}(\|X\|^2) = \sigma^2 d$ holds. It is therefore not far-fetched to assume that our Algorithm 12.3 will also apply to a union $D = G_1 \cup G_2$ of Gaussians with variances not equal to one. If the variances of G_1 and G_2 are different and d is large enough, one may even expect that Gaussians can be separated—given that their centers $\mu_1 \approx \mu_2$ are close together. Indeed, if the centers are the same, the Gaussian annulus theorem suggests that the samples will be located within two thin annuli, which are now concentric but have different radii.

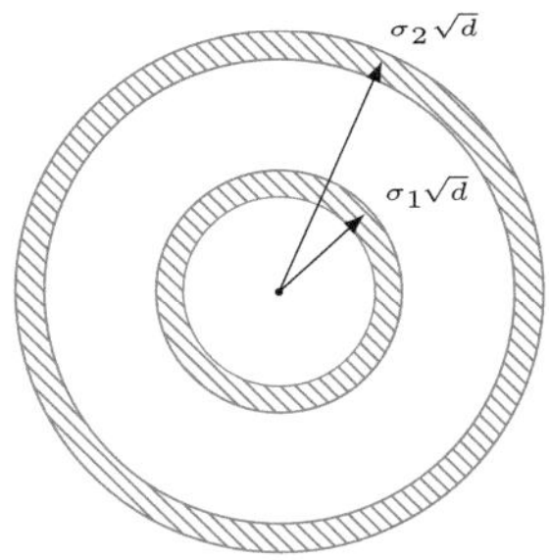

We leave it to the reader as Exercise 12.6 to verify this experimentally.

(ii) If a given dataset $D = G_1 \cup \cdots \cup G_m$ consists of several Gaussians, with n-many points per Gaussian, one may attempt the separation by first labeling the n-nearest neighbors of a random $z \in D$ with 1. Then these points are removed from D, a new point z is chosen and its n-nearest neighbors are labeled with 2, etc.

12.2 Parameter Estimation for Gaussians

After the separation of Gaussians, the second natural task is to estimate mean and variance for each individual Gaussian. For simplicity, we only consider the 1-dimensional situation here, i.e., a Gaussian

$$G = \{x_1, \dots, x_n\} \subseteq \mathbb{R}, \tag{12.1}$$

in which the x_i are realizations of a Gaussian distributed random variable $X\colon \Omega \to \mathbb{R}$ on a probability space $(\Omega, \Sigma, \mathrm{P})$, and we pretend to have forgotten mean and variance. It is then natural to try estimating the parameters by considering the mean and variance of the data points:

$$\mu_{\mathrm{s}} := \tfrac{1}{n}\sum_{i=1}^{n} x_i \;\text{ und }\; \sigma_{\mathrm{s}}^2 := \tfrac{1}{n}\sum_{i=1}^{n}(x_i - \mu_{\mathrm{s}})^2$$

Above, the "s" stands for *sample mean* or for *sample variance*. In order to formalize in what sense this is actually a good estimate, we use the maximum likelihood method. If a set (12.1) is given, we define the likelihood function

$$L\colon \mathbb{R} \times (0, \infty) \to \mathbb{R},\;\; L(\mu, \sigma) := \prod_{i=1}^{n} \tfrac{1}{\sqrt{2\pi}\sigma}\, \mathrm{e}^{-\frac{(x_i-\mu)^2}{2\sigma^2}}\,. \tag{12.2}$$

If $X_1, \dots, X_n$ are independent copies of $X \sim \mathcal{N}(\mu, \sigma^2)$, then it holds

$$\mathrm{P}\big[\forall\, i\colon X_i \in [x_i - \varepsilon, x_i + \varepsilon]\big] = \big(\tfrac{1}{\sqrt{2\pi}\sigma}\big)^n \int_{x_1-\varepsilon}^{x_1+\varepsilon} \mathrm{e}^{-\frac{(x-\mu)^2}{2\sigma^2}}\, \mathrm{d}x \cdots \int_{x_n-\varepsilon}^{x_n+\varepsilon} \mathrm{e}^{-\frac{(x-\mu)^2}{2\sigma^2}}\, \mathrm{d}x.$$

For small and fixed $\varepsilon > 0$, the above probability becomes large if and only if the integrands, each evaluated at x_i, are large. This leads to the task of maximizing the function L, and to the heuristic that, for a maximizer (μ^*, σ^*) of L, it is most likely that the data points in G come from a Gaussian distribution with mean μ^* and variance ${\sigma^*}^2$, compared to all other Gaussian distributions.

Theorem 12.14. (Fitting theorem) *Let $G = \{x_1, \ldots, x_n\} \subseteq \mathbb{R}$ be given such that not all x_i are equal. Then the associated likelihood function L, defined as in* (12.2)*, has precisely one maximizer, given by*

$$\mu^* = \tfrac{1}{n} \sum_{i=1}^{n} x_i \quad \textit{und} \quad {\sigma^*}^2 = \tfrac{1}{n} \sum_{i=1}^{n} (x_i - \mu^*)^2.$$

Proof. As in earlier proofs, we consider instead of L the log-likelihood function $\ell\colon \mathbb{R} \times (0, \infty) \to \mathbb{R}$, $\ell(\mu, \sigma) := \log L$, for which

$$\begin{aligned} \ell(\mu, \sigma) &= \log\Big(\prod_{i=1}^{n} \tfrac{1}{\sqrt{2\pi}\sigma}\, \mathrm{e}^{-\frac{(x_i-\mu)^2}{2\sigma^2}} \Big) \\ &= \log\Big(\big(\tfrac{1}{\sqrt{2\pi}\sigma}\big)^n \mathrm{e}^{-\frac{(x_1-\mu)^2}{2\sigma^2} - \cdots - \frac{(x_n-\mu)^2}{2\sigma^2}} \Big) \\ &= n(0 - \log\sqrt{2\pi} - \log\sigma) - \frac{1}{2\sigma^2} \sum_{i=1}^{n} (x_i - \mu)^2 \end{aligned}$$

holds. Setting the partial derivatives to zero yields

$$\frac{\partial \ell}{\partial \mu} = \frac{1}{2\sigma^2} \sum_{i=1}^{n} 2(x_i - \mu) \cdot 1 \stackrel{!}{=} 0 \iff n\mu - \sum_{i=1}^{n} x_i = 0 \iff \mu = \tfrac{1}{n} \sum_{i=1}^{n} x_i$$

and

$$\begin{aligned} \frac{\partial \ell}{\partial \sigma} = -\tfrac{n}{\sigma} - \frac{-2}{2\sigma^3} \sum_{i=1}^{n} (x_i - \mu)^2 \stackrel{!}{=} 0 &\iff n = \frac{1}{\sigma^2} \sum_{i=1}^{n} (x_i - \mu)^2 \\ &\iff \sigma^2 = \tfrac{1}{n} \sum_{i=1}^{n} (x_i - \mu)^2 \end{aligned}$$

which means that there is precisely one critical point. To see that this point is the uniquely determined maximizer, we show that the log-likelihood function diverges to $-\infty$ if we approach a boundary point of $\mathbb{R} \times (0, \infty)$ or infinity. To simplify our notation, we omit the $(\frac{1}{\sqrt{2\pi}})^n$-term and consider

$$f\colon \mathbb{R} \times (0, \infty) \to \mathbb{R},\ f(\mu, \sigma) := \tfrac{1}{\sigma^n}\, \mathrm{e}^{-\frac{1}{2\sigma^2} g(\mu)}$$

with

$$g\colon \mathbb{R} \to \mathbb{R},\ g(\mu) := \sum_{i=1}^{n} (x_i - \mu)^2.$$

① First, we look at points in $\{0\} \times (0, \infty)$. Let $(\mu, \sigma) \to (\mu_0, 0)$ with $\mu_0 \in \mathbb{R}$. Since $\lim_{|\mu|\to\infty} g(\mu) = \infty$ and g is continuous and non-zero, $c := \min_{\mu\in\mathbb{R}} g(\mu) > 0$. It follows

$$f(\mu, \sigma) = \tfrac{1}{\sigma^n}\, \mathrm{e}^{-\frac{1}{2\sigma^2} g(\mu)} \leqslant \tfrac{1}{\sigma^n}\, \mathrm{e}^{-\frac{1}{2\sigma^2} c} \xrightarrow{\sigma\to 0} 0$$

and thus $\ell(\mu, \sigma) = \log\big((\frac{1}{\sqrt{2\pi}})^n f(\mu, \sigma)\big) \to -\infty$.

② To see that $\lim_{\|(\mu,\sigma)\|\to\infty} f(\mu, \sigma) = 0$, we estimate the values of f outside a box as in the following picture.

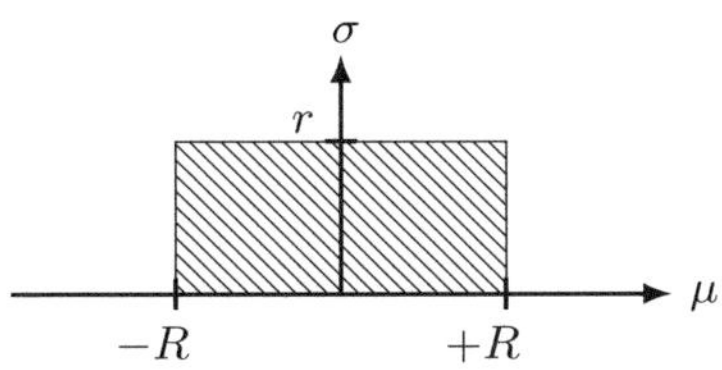

Formally, we claim

$$\forall\, \varepsilon > 0\; \exists\, r,\, R > 0\; \forall\, (\mu, \sigma) \in \mathbb{R} \times (0, \infty)\colon \big[|\mu| \geqslant R \text{ or } \sigma \geqslant r\big] \Longrightarrow f(\mu, \sigma) < \varepsilon.$$

The condition in the square bracket can be equivalently written as $\|(\mu, \sigma)\|_\infty \geqslant \min(R, r)$, and, because $\|\cdot\|_\infty$ and $\|\cdot\|_2$ on $\mathbb{R}^2$ are equivalent norms, we then get the desired statement. Let thus now $\varepsilon > 0$ be given. Firstly, we choose $r > 0$ such that $\frac{1}{\sigma^n} < \varepsilon$ holds for all $\sigma \geqslant r$. With $c = \min_{\mu\in\mathbb{R}} g(\mu) > 0$ as in the first part, we next choose $\Sigma > 0$ such that $\frac{1}{\sigma^n}\, \mathrm{e}^{-\frac{c}{2\sigma^2}} < \varepsilon$ holds for all $0 < \sigma < \Sigma$. Since we have $\lim_{\mu\to\infty} g(\mu) = \infty$, we can finally choose $R > 0$ such that for all μ with $|\mu| \geqslant R$ the estimate $\frac{1}{\Sigma^n}\, \mathrm{e}^{-\frac{g(\mu)}{2r^2}} < \varepsilon$ holds.

Let now (μ, σ) be given with $|\mu| \geqslant R$ or $\sigma \geqslant r$. If $\sigma \geqslant r$, then, due to $g \geqslant 0$ and according to our previous choice of r, we have:

$$\tfrac{1}{\sigma^n}\, \mathrm{e}^{-\frac{g(\mu)}{2\sigma^2}} \leqslant \tfrac{1}{\sigma^n} < \varepsilon.$$

Otherwise, $\sigma < r$ and then $|\mu| \geqslant R$ must hold. We distinguish two subcases: If $\sigma < \Sigma$, we get

$$f(\mu, \sigma) = \tfrac{1}{\sigma^n}\, \mathrm{e}^{-\frac{g(\mu)}{2\sigma^2}} \leqslant \tfrac{1}{\sigma^n}\, \mathrm{e}^{-\frac{c}{2\sigma^2}} < \varepsilon$$

by our choice of Σ. Finally, if $\sigma \geqslant \Sigma$, and still (!) $\sigma < r$ and $|\mu| \geqslant R$, then from these three inequalities it follows

$$f(\mu, \sigma) = \tfrac{1}{\sigma^n}\, \mathrm{e}^{-\frac{g(\mu)}{2\sigma^2}} \leqslant \tfrac{1}{\Sigma^n}\, \mathrm{e}^{-\frac{g(\mu)}{2\sigma^2}} \leqslant \tfrac{1}{\Sigma^n}\, \mathrm{e}^{-\frac{g(\mu)}{2r^2}} < \varepsilon$$

according to the choice of R. This proves the assertion and we have already established that this implies $\ell(\mu, \sigma) \to -\infty$ for $\|(\mu, \sigma)\| \to \infty$. □

We add the following remarks on the so-called *maximum likelihood estimator*

(μ^*, σ^*).

Remark 12.15. (i) Theorem 12.14 induces a mapping

$$M\colon \{(x_1,\ldots,x_n)\in\mathbb{R}^n \mid \exists\, i,j\colon x_i\neq x_j\}\to\mathbb{R}^2,\ M(x_1,\ldots,x_n):=(\mu^*,\sigma^*),$$

by considering the estimator as a function of the data, which can be extended to the whole space $\mathbb{R}^n$ in view of the explicit formulas for μ^* and σ^*. We have already seen in Exercise 2.14 that M can also be considered as a random variable and in this sense $\mathrm{E}(M)$ and $\mathrm{V}(M)$ can be computed. If one does so, it may seem more natural in the formula for σ^* to replace the denominator n with $n-1$.

(ii) If $X\colon\Omega\to\mathbb{R}^d$, $X\sim\mathcal{N}(\mu,\sigma^2,\mathbb{R}^d)$ with $\mu=(\mu_1,\ldots,\mu_d)\in\mathbb{R}^d$ and $\sigma>0$, then one can show similarly to Theorem 12.14, that

$$\mu^*=\tfrac{1}{n}\sum_{i=1}^n x_i\in\mathbb{R}^d \ \text{ and } \ \sigma^{*2}=\tfrac{1}{n}\sum_{i=1}^n\|\mu^*-x_i\|^2$$

maximizes a likelihood function analogous to (12.2).

We conclude this chapter on high-dimensional spaces by pointing out that we have, throughout this and the previous four chapters, considered spherical Gaussian distributions only. In general, one considers Gaussian random variables $X\sim\mathcal{N}(\mu,\Sigma,\mathbb{R}^d)$ given by the density function

$$\rho(x)=\frac{1}{(2\pi)^{d/2}}\cdot\frac{1}{(\det\Sigma)^{1/2}}\,\mathrm{e}^{-\frac{1}{2}\langle x-\mu,\Sigma^{-1}(x-\mu)\rangle}$$

with positive definite *covariance matrix* Σ. In two dimensions, samples then look as follows, where the left picture shows the case we have dealt with, i.e., Σ is the identity matrix multiplied by a scalar $\sigma>0$.

In the middle picture, Σ is a diagonal matrix and on the right an arbitrary positive definite matrix. The treatment of Gaussians in this more general setting is unsurprisingly much more technical than what we have done in the previous chapters.

References

The basis for this chapter is again [BHK20], but we have made changes and additions. We also refer to the original work [DS07]. Sorting algorithms, as mentioned in Algorithm 12.3, can be found in [CLRS22, Part II]. For more details on the treatment of non-spherical Gaussians we refer to [Bis06, Chapter 9]. The author would like to thank Thomas Sauerwald for various discussions and suggestions for improvement regarding Chapters 8–12.

Exercises

Exercise 12.1. The following table contains the heights of Americans aged 40-49. E.g., 1.8% of women are shorter then 147.32cm; 3.4% measure between 147.32cm and 149.86cm; and so on.

	147.32	149.86	152.40	154.94	157.48	160.02	162.56	165.10	167.64	170.18
Women	1.6	3.4	5.8	9.0	11.0	15.2	12.0	14.2	10.8	8.2
Men	0	0	0	0	1.9	1.9	1.8	4.2	9.6	10.9

	172.72	175.26	177.80	180.34	182.88	185.42	187.96	190.50	193.04	195.58
Women	3.5	3.1	1.6	0.1	0	0	0	0	0.5	0
Men	10.1	14.0	15.2	9.5	8.3	5.1	5.2	1.3	0.4	0.5

Source: https://www2.census.gov/library/publications/2010/compendia/statab/130ed/tables/11s0205.pdf

Plot the two (Gaussian) distributions, then create the union set and try to separate it again with Algorithm 12.3.

Exercise 12.2. Generate two Gaussians $G_1, G_2 \subseteq \mathbb{R}^{100}$, one with center zero and the other with center $(\Delta, 0, \ldots, 0)$ for various $\Delta > 0$. Calculate and plot the distribution of pairwise distances in the union set $D = G_1 \cup G_2$.

Exercise 12.3. A weak spot of Algorithm 12.3 is that the starting point z could be an "untypical point". In two dimensions, for example, z could lie "halfway" between the Gaussians, in high dimensions, outside the two annuli. Develop strategies for choosing z that counteract this.

Exercise 12.4. (i) Apply Algorithm 12.3 to high-dimensional Gaussians (that you generate yourself) and have your program calculate the rate of points that are correctly separated.

(ii) Test the variations of the separation algorithm suggested in Example 12.8 and compare with the original version.

Exercise 12.5. Example 12.12 suggests that for $\Delta = d^{0.30}$ the quality of the separation algorithm is smaller than for $\Delta = 2d^{0.30}$, but still growing. Simulate this by examining as large dimensions as possible and verify that from $d = 10^6$ onwards it is to be expected that disentanglement will be achieved with $\geqslant 80\%$ accuracy.

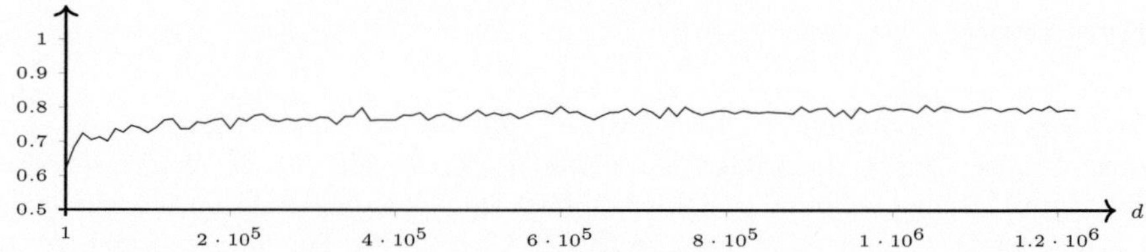

Note: On a standard computer, this can take quite a while and depending on the programming language, your simulation will give up sooner or later.

Exercise 12.6. Generate two Gaussians in $\mathbb{R}^d$ with variances $\sigma_1^2 \neq \sigma_2^2$ and center zero. Find out by trial and error from which d onwards the separation succeeds.

Exercise 12.7. Generate a Gaussian G and then estimate its parameters μ and σ using the maximum likelihood estimator.

13 Perceptron

In this chapter, we consider classification problems of the following form: Given a labeled dataset $D = \{(x_i, y_i) \mid i = 1, \ldots, n\} \subseteq \mathbb{R}^d \times \{-1, 1\}$ with binary labels, we are looking for an *affine linear classifier*, or "perceptron", i.e., a function of the form

$$h\colon \mathbb{R}^d \to \mathbb{R},\ h(x) = \operatorname{sgn}(\langle w, x\rangle + b),$$

with $0 \neq w \in \mathbb{R}^d$ and $b \in \mathbb{R}$, such that $h(x_i) = y_i$ holds for all $i = 1, \ldots, n$. Here, $\langle \cdot\,, \cdot \rangle$ denotes the standard scalar product. Of course, we cannot expect that this is possible for arbitrary datasets D. We will therefore begin with two geometric characterizations of those datasets for which a classifier h as above exists. The first characterization states that the data points with label -1 can be strictly separated from those with label $+1$ by an affine hyperplane (left picture). The second characterizing property is that the convex hulls of the points with label $+1$, respectively -1, have an empty intersection (right picture).

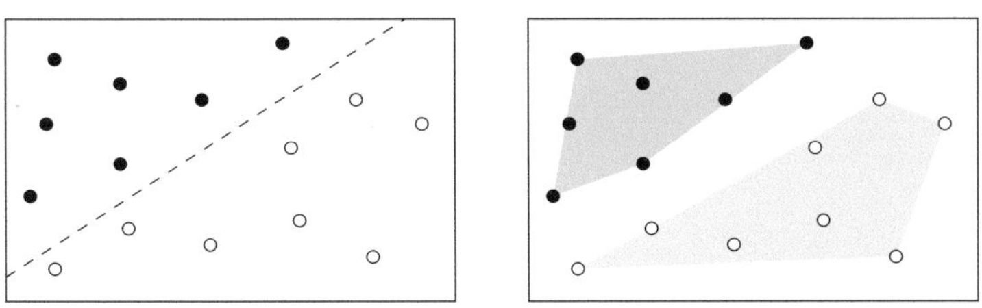

For the second characterization, we need Carathéodory's theorem about convex hulls, as well as Minkowski's theorem on the separation of (compact) convex sets via a hyperplane; both of which we state below without a proof. The reader can find detailed references at the end of the chapter.

We now use the existence of a classifier of the form described above to define the class of datasets that we are going to study in this chapter. This may seem somewhat tautological but, firstly, all three properties will later turn out to be equivalent anyway, and secondly, the existence of the classifier is the easiest of

S. A. Wegner, *Mathematical Introduction to Data Science*,
https://doi.org/10.1007/978-3-662-69426-8_13

the three properties to formulate.

Definition 13.1. We say that a dataset $D = \{(x_i, y_i) \mid i = 1, \dots, n\} \subseteq \mathbb{R}^d \times \{-1, 1\}$ is *linearly separable* if there exists a function $h\colon \mathbb{R}^d \to \mathbb{R}$, of the form $h(x) = \operatorname{sgn}(\langle w, x\rangle + b)$ with $0 \neq w \in \mathbb{R}^d$, $b \in \mathbb{R}$, that correctly classifies all data points.

Often, $w_1, \dots, w_d$ are called the *weights* and b the *bias* of the classifier, because b determines what value the weighted sum $w_1x_1 + \cdots + w_dx_d$ must exceed for the classifier to switch from value -1 to value $+1$.

We point out that we could also have allowed $w = 0$ in the previous definition. This would lead to the functions $h \equiv 0$, $h \equiv -1$ and $h \equiv +1$. In the first case $D = \emptyset$, whereas the other two cases correspond to datasets in which all data points have the same label. Apart from the fact that these are uninteresting cases, a correct affine linear classifier with $w \neq 0$ can of course also be found in all three situations. Our requirement $w \neq 0$ thus means no loss of generality, and it will become practical when we now present the first alternative characterization. In order to do so, we need the following definition.

Definition 13.2. A subset $\mathcal{H} \subseteq \mathbb{R}^d$ is called an *affine hyperplane* if $\mathcal{H}$ is of the form $\mathcal{H} = X + x_0$, where $X \subseteq \mathbb{R}^d$ is a linear subspace of dimension $d - 1$ and $x_0 \in \mathbb{R}^d$.

Next, we prove that affine hyperplanes are precisely the zero sets of affine linear classifiers.

Lemma 13.3. *A subset $\mathcal{H} \subseteq \mathbb{R}^d$ is an affine hyperplane if and only if there exist $0 \neq w \in \mathbb{R}^d$ and $b \in \mathbb{R}$ such that $\mathcal{H} = \{x \in \mathbb{R}^d \mid \langle w, x\rangle + b = 0\}$ holds. In this case, w and b are unique up to a non-zero factor.*

Proof. "$\Longrightarrow$" Let $\mathcal{H} = X + x_0$ be as in Definition 13.2. We choose $0 \neq w \in X^\perp$ and note that $X^\perp$ has dimension one. Then we put $b := -\langle w, x_0\rangle$. Now let $x = x_1 + x_0 \in X + x_0$ be given. Then we have

$$\langle w, x\rangle + b = \langle w, x_1 + x_0\rangle - \langle w, x_0\rangle = \langle w, x_1\rangle$$

and the latter is zero as $w \perp x_1$. Conversely, let $x \in \mathbb{R}^d$ be given with $\langle w, x\rangle + b = 0$. We put $x_1 := x - x_0$ and obtain $x = x_1 + x_0$. With the same calculation as above it follows $w \perp x_1$.

"$\Longleftarrow$" If $0 \neq w \in \mathbb{R}^d$ and $b \in \mathbb{R}$ are given, we put $X := \{w\}^\perp$ and choose x_0 such that $\langle w, x_0\rangle = -b$ holds.

To prove the uniqueness statement, let $(w_1, b_1), (w_2, b_2) \in \mathbb{R}^{d+1}$ be given with $w_1 \neq 0 \neq w_2$. We show that

$$\mathcal{H} := \{x \in \mathbb{R}^d \mid \langle w_1, x\rangle + b_1 = 0\} = \{x \in \mathbb{R}^d \mid \langle w_2, x\rangle + b_2 = 0\}$$

holds if and only if there exists $\alpha \neq 0$ with $w_1 = \alpha w_2$ and $b_1 = \alpha b_2$.

"$\Longrightarrow$" According to Definition 13.2, there exist a subspace $X \subseteq \mathbb{R}^d$, with $\dim X = d - 1$, and a vector $x_0 \in \mathbb{R}^d$ such that $\mathcal{H} = X + x_0$. In view of the previous parts of the proof we have $w_1, w_2 \perp X$. In other words, w_1 and w_2 both belong to the one-dimensional subspace $X^\perp$. Since both vectors are non-zero, there exists $\alpha \neq 0$ with $w_1 = \alpha w_2$. Because further $x_0 \in X + x_0$ holds, we get

$$0 = \langle w_1, x_0 \rangle + b_1 = \langle \alpha w_2, x_0 \rangle + b_1 = \alpha \langle w_2, x_0 \rangle + b_1 = -\alpha b_2 + b_1$$

and thus $b_1 = \alpha b_2$ as claimed.

"$\Longleftarrow$" Follows from the bilinearity of the scalar product. □

Remark 13.4. If $(w, b) \in \mathbb{R}^{d+1}$ with $w \neq 0$ is given, then the associated hyperplane $\mathcal{H}$ divides $\mathbb{R}^d$ into two *affine half-spaces*

$$\mathcal{H}_+ := \{x \in \mathbb{R}^d \mid \langle w, x \rangle + b > 0\} \text{ and } \mathcal{H}_- := \{x \in \mathbb{R}^d \mid \langle w, x \rangle + b < 0\}.$$

Note that $\mathcal{H}_\pm$, in contrast to $\mathcal{H}$, depend on the choice of (w, b) to the extend that under the transition to $(\alpha w, \alpha b)$ with $\alpha < 0$, the half-spaces swap places. If we fix $x_0 \in \mathcal{H}$, we can illustrate $\mathcal{H}$, $\mathcal{H}_+$ and $\mathcal{H}_-$ via the following picture:

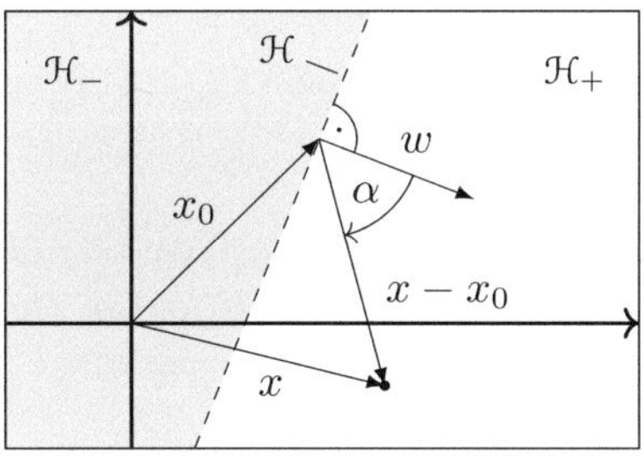

Here we see that $\langle w, x \rangle + b = \langle w, x - x_0 \rangle > 0$ holds if and only if the angle $\alpha = \arccos(\langle \frac{w}{\|w\|}, \frac{x - x_0}{\|x - x_0\|} \rangle)$ between w and $x - x_0$ lies in the interval $(-\pi/2, \pi/2)$. This again explains the term "half-space".

Theorem 13.5. *A dataset $D = \{(x_i, y_i) \mid i = 1, \ldots, n\} \subseteq \mathbb{R}^d \times \{-1, 1\}$ is linearly separable if and only if a hyperplane $\mathcal{H}$ exists such that all points with label $+1$ belong to one of the two half-spaces and all points with label -1 belong to the other.*

Proof. "$\Longrightarrow$" If D is linearly separable then, by definition, there exists $(w, b) \in \mathbb{R}^{d+1}$ with $w \neq 0$, such that $y_i = h(x_i) = \operatorname{sgn}(\langle w, x_i \rangle + b)$ holds for all $i = 1, \ldots, n$. This means that whenever $y_i = +1$ we must have $\langle w, x_i \rangle + b > 0$, or, in other words, $x_i \in \mathcal{H}_+$. Similarly, $x_i \in \mathcal{H}_-$ holds whenever $y_i = -1$.

"$\Longleftarrow$" If the condition in the theorem holds and the hyperplane is given by $(w, b) \in \mathbb{R}^{d+1}$, then either all points with label $+1$ are in $\mathcal{H}_+$ and all points with label -1 are in $\mathcal{H}_-$, or it is exactly the other way round. In the first case, we put $h = \operatorname{sgn}(\langle w, \cdot \rangle + b)$, and in the second case $h = \operatorname{sgn}(\langle -w, \cdot \rangle - b)$. □

Before we come to the second characterization, let us note that our definition has the advantage over the condition in Theorem 13.5 that it determines the sign of w and b: In simple terms, w points in the direction of the positively labeled data points.

Theorem 13.6. *Let* $D = \{(x_i, y_i) \mid i = 1, \ldots, n\} \subseteq \mathbb{R}^d \times \{-1, 1\}$ *and let* $D_\pm := \{x_i \mid y_i = \pm 1\}$. *Then the dataset* D *is linearly separable if and only if* $\operatorname{conv} D_+ \cap \operatorname{conv} D_- = \emptyset$ *holds, i.e., if the convex hulls of the data points with label* $+1$ *and label* -1, *respectively, are disjoint.*

Proof. "$\Longrightarrow$" If D is linearly separable, then there exists $(w, b) \in \mathbb{R}^{d+1}$ with

$$D_+ \subseteq \{x \in \mathbb{R}^d \mid \langle w, x \rangle + b > 0\} =: \mathcal{H}_+.$$

As $\mathcal{H}_+$ is convex, we get $\operatorname{conv} D_+ \subseteq \mathcal{H}_+$ and analogously $\operatorname{conv} D_- \subseteq \mathcal{H}_-$. The intersection $\mathcal{H}_+ \cap \mathcal{H}_-$ is empty by construction.

"$\Longleftarrow$" We show that $K_+ := \operatorname{conv} D_+$ and $K_- := \operatorname{conv} D_-$ can be separated by an affine hyperplane (in the sense of Theorem 13.5) and use the following theorem in order to do so.

Theorem 13.7. (Minkowski's hyperplane separation theorem) *Let* $A, B \subseteq \mathbb{R}^d$ *both be non-empty, convex, and closed. If one of the sets is compact, then there exists a hyperplane* $\mathcal{H}$ *such that* A *is contained in one of the associated half-spaces and* B *in the other.* ◇

In our case, K_- and K_+ are convex by definition and we now show that both sets are also compact. We claim, more generally, that for any compact set $X \subseteq \mathbb{R}^d$, its convex hull $K := \operatorname{conv} X$ is again compact. To describe the convex hull, we use the following theorem.

Theorem 13.8. (Carathéodory's convex hull theorem) *Let* $X \subseteq \mathbb{R}^d$ *and* $z \in K := \operatorname{conv} X$. *Then there exist* $z_1, \ldots, z_{d+1} \in X$ *and* $\alpha_1, \ldots, \alpha_{d+1} \in [0, 1]$ *with* $\alpha_1 + \cdots + \alpha_{d+1} = 1$ *and* $z = \alpha_1 z_1 + \cdots + \alpha_{d+1} z_{d+1}$. ◇

In other words, we have

$$K = \Big\{ \sum_{k=1}^{d+1} \alpha_k z_k \;\Big|\; z = (z_1, \ldots, z_{d+1}) \in X^{d+1}, \alpha = (\alpha_1, \ldots, \alpha_{d+1}) \in \Sigma_d \Big\},$$

where the set

$$\Sigma_d := \big\{\alpha = (\alpha_1, \dots, \alpha_d) \in \mathbb{R}^{d+1}_{\geqslant 0} \mid \sum_{k=1}^{d+1} \alpha_k = 1\big\}$$

is obviously compact. Now we consider the map

$$\varphi\colon X^{d+1} \times \Sigma_d \to \mathbb{R}^d,\ (z, \alpha) \mapsto \sum_{k=1}^{d+1} \alpha_k z_k,$$

which is clearly continuous. As $X^{d+1} \times \Sigma_d$ is compact, it follows that $K = \operatorname{ran}\varphi$ must also be compact. This proves the claim and applying it with $X = D_{\pm}$ completes the proof. □

After having proven the two geometric characterizations of linearly separable sets, we turn to the question of how we can find an affine linear classifier, assuming that a given dataset D is linearly separable. This will lead us to the so-called *perceptron*. To formulate the corresponding Algorithm 13.11 efficiently, we will combine weights and biases into one $(d+1)$-dimensional vector. Accordingly, we will embed the data points into $\mathbb{R}^{d+1}$. The combination of weights and biases into one vector has already proven useful in Subchapter 2.4 on logistic regression. For the convenience of the reader we repeat the definition.

Definition 13.9. For data points $z = (z_1, \dots, z_d) \in \mathbb{R}^d$, we denote by $\widehat{z} := (z, 1) = (z_1, \dots, z_d, 1) \in \mathbb{R}^{d+1}$ the *(by one) extended data point.* For weights $w = (w_1, \dots, w_d) \in \mathbb{R}^d$ and biases $b \in \mathbb{R}$, we denote by $(w, b) = (w_1, \dots, w_d, b) \in \mathbb{R}^{d+1}$ the *generalized weight vector.*

Remark 13.10. Let $w \in \mathbb{R}^d$ and $b \in \mathbb{R}$ be given. With the notation introduced above, and for $z = (z_1, \dots, z_d) \in \mathbb{R}^d$, we have

$$\langle w, z\rangle + b = \sum_{i=1}^{d} w_i z_i + b \cdot 1 \ = \left\langle \begin{bmatrix} w_1 \\ \vdots \\ w_d \\ b \end{bmatrix}, \begin{bmatrix} z_1 \\ \vdots \\ z_d \\ 1 \end{bmatrix} \right\rangle = \langle w, \widehat{z}\rangle,$$

where we (abuse of notation!) denote the "normal" weight vector on the left with w and the generalized weight vector on the right, also with w.

Algorithm 13.11. *The following pseudocode outlines the so-called* perceptron algorithm. *Let $D = \{(x_i, y_i) \mid i = 1, \dots, n\} \subseteq \mathbb{R}^d \times \{-1, 1\}$ be a dataset and let $w \in \mathbb{R}^{d+1}$. Notice that the output of the algorithm is the generalized (!) weight vector $w^{(j)} \in \mathbb{R}^{d+1}$, i.e., the first d-many entries of $w^{(j)}$ are the weights, and the last entry is the bias of the classifier.*

```
1: function PERCEPTRON (D, w)
2:     w^(0) ← w
3:     for j ← 0 to ∞ do
4:         if ∃ i ∈ {1, ..., n}: y_i⟨w^(j), x̂_i⟩ ⩽ 0 then
5:             w^(j+1) ← w^(j) + y_i x̂_i
6:         else
7:             break
8:     return w^(j)
```

We will show in Theorem 13.13 that the perceptron algorithm always terminates and that it provides a correct classifier under the assumption that D is linearly separable. Before that, however, we will discuss the heuristic behind the algorithm.

Remark 13.12. We first consider the `if`-query, i.e., the condition that leads to an update of the weight vector. Here we observe

$$\begin{aligned} y_i\langle w^{(j)}, \widehat{x}_i\rangle > 0 &\iff \big[(y_i = +1 \wedge \langle w^{(j)}, \widehat{x}_i\rangle > 0) \vee (y_i = -1 \wedge \langle w^{(j)}, \widehat{x}_i\rangle < 0)\big] \\ &\iff h(x_i) \underset{\substack{\uparrow \\ \text{Rmk} \\ 13.10}}{=} \operatorname{sgn}(\langle w^{(j)}, \widehat{x}_i\rangle) = y_i \end{aligned}$$

and it follows by negation that $y_i\langle w^{(j)}, \widehat{x}_i\rangle \leqslant 0$ holds if and only if x_i is classified incorrectly. Line 4 thus finds incorrectly classified data points. The update in line 5 then "tries" to correct the misclassification by adding a correction term:

$$y_i\langle w^{(j+1)}, x_i\rangle = y_i\langle w^{(j)} + y_i x_i, x_i\rangle = \underbrace{y_i\langle w^{(j)}, x_i\rangle}_{\substack{\text{erroneously} \\ \text{less than zero}}} + \underbrace{y_i^2\,\langle x_i, x_i\rangle}_{\substack{\text{addition of a} \\ \text{positive cor-} \\ \text{rection term}}}.$$

However, there is no guarantee that $\widehat{x}_i$ will be correctly classified afterwards, cf. Exercise 13.2. In the following picture, however, this is the case:

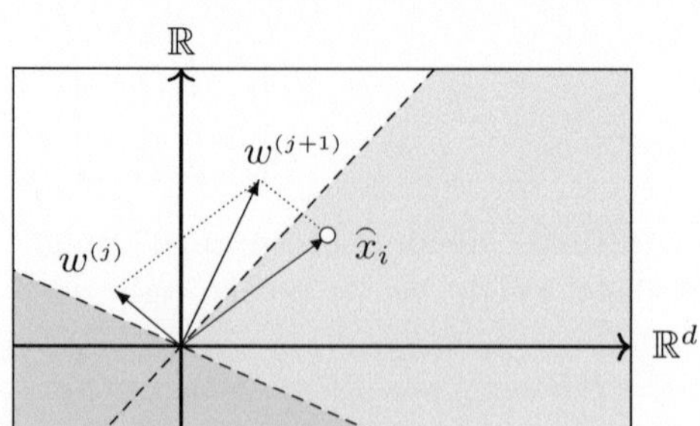

Above we have sketched the hyperplanes given by $w^{(j)}$ and $w^{(j+1)}$. It can be seen that the update rotates these in such a way that the point $\widehat{x}_i$, which was not correctly classified by $w^{(j)}$, will be correctly classified after the update, i.e., by $w^{(j+1)}$. We point out that we are working with the extended data points and generalized weight vectors in the picture. For this reason, the affine hyperplanes are indeed subspaces.

Now we will formally prove that the perceptron algorithm always terminates and in doing so we will establish a bound for its runtime.

Theorem 13.13. (Perceptron theorem) *Let* $D = \{(x_i, y_i) \mid i = 1, \ldots, n\} \subseteq \mathbb{R}^d \times \{-1, 1\}$ *be non-empty and linearly separable. Then there exists* $w^* \in \mathbb{R}^{d+1}$ *such that, for all* $i = 1, \ldots, n$, $y_i\langle w^*, \widehat{x}_i\rangle > 0$ *and* $\|w^*\| = 1$ *hold. Let* $\gamma := \min_{i=1,\ldots,n} y_i\langle w^*, \widehat{x}_i\rangle$ *(*>0 *according to the previous!) and let* $R := \max_{i=1,\ldots,n} \|x_i\|$. *Then we have:*

(i) The perceptron algorithm initialized with $w^{(0)} = 0$ *stops after at most* $\lfloor (R/\gamma)^2 \rfloor$ *iterations.*

(ii) If the perceptron algorithm stops and outputs the generalized weight vector $w^{(J)}$, *then the corresponding classifier* $h\colon \mathbb{R}^d \to \mathbb{R}$ *classifies all points of* D *correctly.*

Proof. Since D is linearly separable, there exists some $w \in \mathbb{R}^{d+1}$, such that all points will be classified correctly. This vector w cannot be the zero vector, because then $h \equiv 0$ and all data points would be classified incorrectly. We put $w^* := w/\|w\|$. Then $\|w^*\| = 1$ and $y_i\langle w^*, x_i\rangle = \frac{1}{\|w\|} y_i\langle w, x_i\rangle > 0$ hold for all $i = 1, \ldots, n$. With γ as in the theorem, it follows

$$\forall\, i = 1, \ldots, n\colon y_i\langle w^*, \widehat{x}_i\rangle \geqslant \gamma > 0.$$

Let now $0 = w^{(0)}, w^{(1)}, w^{(2)}, \ldots$ be the sequence of weight vectors that the perceptron algorithm, initialized with the zero vector, calculates. Let $w^{(j)}$ be a weight vector that does not provide a correct classification of all points. Then the algorithm performs the update $w^{(j+1)} = w^{(j)} + y_i\widehat{x}_i$, where $i \in \{1, \ldots, n\}$ is chosen such that $\widehat{x}_i$ was incorrectly classified by the weight vector $w^{(j)}$. This implies

$$\langle w^*, w^{(j+1)}\rangle = \langle w^*, w^{(j)} + y_i\widehat{x}_i\rangle = \langle w^*, w^{(j)}\rangle + y_i\langle w^*, \widehat{x}_i\rangle \geqslant \langle w^*, w^{(j)}\rangle + \gamma.$$

Iteratively, and because $w^{(0)} = 0$, we get

$$\langle w^*, w^{(j+1)}\rangle \geqslant \langle w^*, w^{(0)}\rangle + j \cdot \gamma = j \cdot \gamma.$$

On the other hand, the update, with R as defined in the theorem, provides the estimate

$$\|w^{(j+1)}\|^2 = \|w^{(j)} + y_i\widehat{x}_i\|^2 = \langle w^{(j)} + y_i\widehat{x}_i, w^{(j)} + y_i\widehat{x}_i\rangle$$

$$= \|w^{(j)}\|^2 + \underbrace{2y_i\langle w^{(j)}, \widehat{x}_i\rangle}_{\substack{<0,\text{ as } \widehat{x}_i \\ \text{is classified} \\ \text{incorrectly}}} + \underbrace{\|x_i\|^2}_{\leqslant R} < \|w^{(j)}\|^2 + R^2.$$

Again, iteratively, and because $w^{(0)} = 0$, it follows

$$\|w^{(j+1)}\|^2 \leqslant \|w^{(0)}\|^2 + j\cdot R^2 = j\cdot R^2.$$

Combining both estimates yields

$$j\cdot\gamma \leqslant \langle w^*, w^{(j+1)}\rangle \leqslant \|w^*\|^2\|w^{(j+1)}\|^2 = \|w^{(j+1)}\|^2 \leqslant \sqrt{j}\cdot R$$

and division by γ and $\sqrt{j}$ implies $\sqrt{j} = j/\sqrt{j} \leqslant R/\gamma$. Since $j \in \mathbb{N}$, this is only possible if $j \leqslant \lfloor (R/\gamma)^2\rfloor$ holds and we see that the perceptron algorithm must have stopped after at most $\lfloor (R/\gamma)^2\rfloor$ updates. This shows (i). If the perceptron algorithm stops in the J-th round, then by definition the condition

$$\exists\, i \in \{1,\dots,n\}\colon y_i\langle w^{(J)}, x_i\rangle \leqslant 0$$

is not fulfilled. But this means that all data points in D are classified correctly via the weight vector $w^{(J)}$. This shows (ii). □

Remark 13.14. (i) In line 4 of Algorithm 13.11 we did not specify how the incorrectly classified point $\widehat{x}_i$ is selected, in case that there is more than one choice. In fact, the proof of Theorem 13.13 has shown that the bound for the runtime is independent of how we handle this! However, the actual runtime does indeed depend on how we proceed here, compare Exercise 13.3.

(ii) If we start the perceptron algorithm with $w^{(0)} \neq 0$, we can adjust the above proof and get a similar result, see Exercise 13.5.

(iii) In Exercise 13.3 we will consider training data consisting of words that occurred in spam and no-spam e-mails. Based on this, we are going to use the perceptron algorithm to obtain a spam classifier. In Exercise 13.3 this will lead to

$$h(x_1,\dots,x_5) = \operatorname{sgn}(0\cdot x_1 + 2\cdot x_2 + 0\cdot x_3 - 1\cdot x_4 + 1\cdot x_5 + 0)$$

which allows us to interpret that the words “and” and “the” are neutral with respect to spam, whereas “bonus” and “contract” are more or less indicative of spam, according to their different weights. Finally, “cafeteria” is indicative of no-spam. One can thus say that the perceptron algorithm provides *explainable* results.

(iv) Instead of using labels -1 and $+1$, one can also work with 0 and 1 and use classifiers of the type $h = \mathbb{1}_{[0,\infty)}(\langle w,\cdot\rangle + b)$. The perceptron algorithm then requires two `if`-queries, since in the case $y_i = 0$, the product $y_i\langle w, \widehat{x}_i\rangle$ does not reveal anything about the sign of $\langle w, \widehat{x}_i\rangle$. In addition, with this modification it may happen that data points end up precisely on the decision boundary — this undesirable effect is automatically prevented if we use ± 1 as labels.

(v) The perceptron, at that time not an abstract algorithm but a physical machine, was developed in the 1950s by Frank Rosenblatt, who was then convinced that it was the foundation for a "machine which senses, recognizes, and responds like the human mind". At the end of the 1960s, Martin Minsky and Seymour Papert published a book in which they argued that the perceptron was not suitable to develop algorithms for complex pattern recognition, especially because a single perceptron cannot represent the logical XOR function, see Exercise 13.3. Although this is true, it can be overcome by using several "coupled" perceptrons, leading to the concept of artificial neural networks, see Chapter 16. The latter had been known since the early 1960s, however, based on Minsky and Papert's book, public opinion shifted, research funds were cut, and it was not until the 1980s that research on neural networks experienced a resurgence.

References

First, we note that our version of Minkowski's Theorem 13.7 admits an elementary proof, see, e.g., [Roc70, Corollary 11.4.2]. Alternatively, it also follows from the Hahn-Banach Theorem [Rud91, Theorem 3.4(b)]. For the elementary proof of Carathéodory's Theorem we refer to [LL16, Theorem 3.3.10]. The proof of Theorem 13.13 presented in this chapter is based on [Col12]. We further note that Rosenblatt's quote in Remark 13.14(iv) is taken from [Ros59]. We also refer to [Lef19] as well as to [Ola96] for more historical details and an analysis of the so-called "perceptron controversy".

Exercises

Exercise 13.1. Assume that the perceptron algorithm outputs the summarized weight vector $w = (2, 1, 1)$. Sketch which points in $\mathbb{R}^2$ are classified as ± 1.

Exercise 13.2. Find a dataset D and a generalized weight vector w such that a data point $(x, y) \in D$ is not correctly classified and a (single) update using this data point does *not* lead to x being correctly classified after the update.

Exercise 13.3. The following table contains five words that appeared within e-mails, and the information whether the particular e-mail was spam or not.

	"and"	"bonus"	"the"	"cafeteria"	"contract"	Spam
e-Mail 1	✓	✓	✗	✓	✓	Yes
e-Mail 2	✗	✗	✓	✓	✗	No
e-Mail 3	✗	✓	✓	✗	✗	Yes
e-Mail 4	✓	✗	✗	✓	✗	No
e-Mail 5	✓	✗	✓	✗	✓	Yes
e-Mail 6	✓	✗	✓	✓	✗	No

(i) Based on the table, create a numerical dataset of points $x_1, \ldots, x_6 \in \mathbb{R}^5$ with labels $y_1, \ldots, y_6 \in \{-1, 1\}$.

(ii) Run the perceptron algorithm cyclically on the data points from (i) until it terminates. Convince yourself that the runtime depends on the order in which you input the data points.

(iii) Classify a new e-mail, which contains the words "bonus", "contract", "the" and "cafeteria", as spam or no-spam.

Exercise 13.4. We now deal with the question of whether the perceptron algorithm is able to "learn" the logical functions *and* and *exclusive or*:

		AND
f	f	f
f	t	f
t	f	f
t	t	t

		XOR
f	f	f
f	t	t
t	f	t
t	t	f

(i) Replace in the truth table for AND each "t" with +1 and each "f" with −1. Then run the perceptron algorithm on this training set cyclically *by hand*, starting with the generalized weight vector $w^{(0)} = (1, 2, 3)$.

(ii) Do the same for XOR; recall Remark 13.14(iv).

Exercise 13.5. Let $D = \{(x_i, y_i) | i = 1, \ldots, n\} \subseteq \mathbb{R}^d \times \{-1, 1\}$ be a linearly separable dataset, in which not all labels coincide. Show that the perceptron algorithm with any initial value $w^{(0)} \in \mathbb{R}^{d+1}$ terminates in a finite number of steps, and that the runtime is bounded by $\lfloor (C/\gamma)^2 \rfloor$, where $C > 0$ is a constant depending on $w^{(0)} \in \mathbb{R}^{d+1}$.

14

Support Vector Machines

In Chapter 13, we explained how an affine linear classifier $h\colon \mathbb{R}^d \to \mathbb{R}$, $h = \operatorname{sgn}(\langle w, \cdot\rangle + b)$, can be obtained for a linearly separable dataset

$$D = \{(x_i, y_i) \mid i = 1, \dots, n\} \subseteq \mathbb{R}^d \times \{-1, 1\},$$

via the perceptron algorithm (Algorithm 13.11). We noted that, by construction, all data points have a strictly positive distance to the decision boundary $\mathcal{H} = \{x \in \mathbb{R}^d \mid h(x) = 0\}$, see Remark 13.14(iii). Since there are always multiple correct classifiers (left picture), it is natural to seek classifiers where the distance between $\mathcal{H}$ and the data points closest to $\mathcal{H}$ is as large as possible. In the dataset below, there is precisely one classifier that maximizes this distance (right picture) and it can be easily determined graphically, cf. Exercise 14.1.

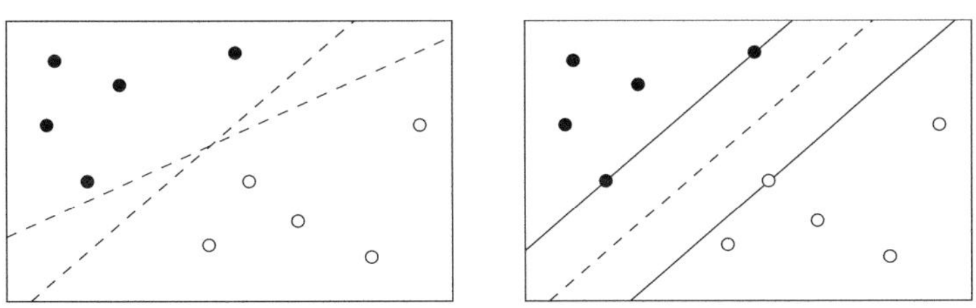

Let a linearly separable dataset be given such that not all points have the same label. In this chapter, we will first show that then there always exists a unique affine linear classifier that correctly classifies all data points *and* maximizes the minimum distance of the data points to the decision boundary. This unique classifier is called the *support vector machine* (SVM). In the second part of the chapter, we will provide a method by which the parameters of the SVM can be calculated via a quadratic optimization problem. For this, we rely on the Karush-Kuhn-Tucker theorem, which we will note without proof. We will also make a forward reference to Chapter 17, in order to use existence and uniqueness results for minimizers of convex functions.

S. A. Wegner, *Mathematical Introduction to Data Science*,
https://doi.org/10.1007/978-3-662-69426-8_14

Definition 14.1. Let $h\colon \mathbb{R}^d \to \mathbb{R}$, $h(x) = \operatorname{sgn}(\langle w, x\rangle + b)$ with $0 \neq w \in \mathbb{R}^d$ and let $b \in \mathbb{R}$ be an affine linear classifier for a dataset $D = \{(x_i, y_i) \mid i = 1, \ldots, n\} \subseteq \mathbb{R}^d \times \{-1, 1\}$, where $\mathcal{H} = \{x \in \mathbb{R}^d \mid \langle w, x\rangle + b = 0\}$ denotes the associated affine hyperplane. We call

$$\gamma(h) := \min_{i=1,\ldots,n} \operatorname{dist}(\mathcal{H}, x_i)$$

the *margin* of the classifier h, and the affine hyperplanes

$$\mathcal{R}_\pm = \left\{x \in \mathbb{R}^d \mid \operatorname{dist}(\mathcal{H}, x) = \gamma \text{ and } h(x) = \pm 1\right\}$$

its *gutters*.

The following picture suggests to associate the classifier h with a road having $\mathcal{H}$ as its center line and $\mathcal{R}_\pm$ as its gutters — which explains the origin of this notation.

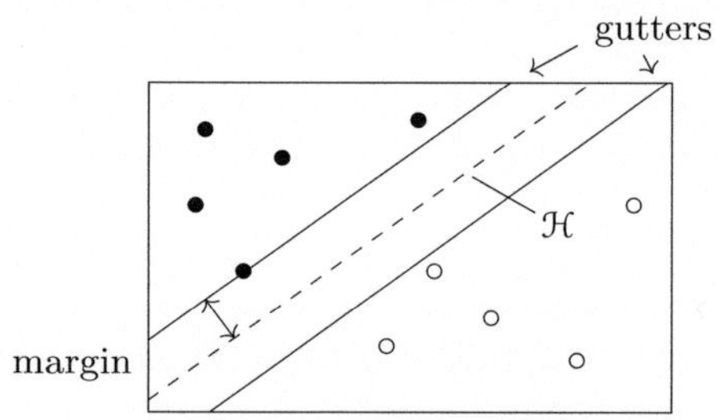

From Chapter 13, we know that the parameters (w, b) are uniquely determined by the classifier up to a scalar factor. We restate this in the following lemma for subsequent use.

Lemma 14.2. *Let $h\colon \mathbb{R}^d \to \mathbb{R}$, $h = \operatorname{sgn}(\langle w, \cdot\rangle + b)$ with $0 \neq w \in \mathbb{R}^d$, $b \in \mathbb{R}$ be given. Then $h = \operatorname{sgn}(\langle w', \cdot\rangle + b')$ holds if and only if there exists $\alpha > 0$ such that $(w', b') = \alpha \cdot (w, b)$.*

Proof. According to Lemma 13.3, $\mathcal{H} = \{x \in \mathbb{R}^d \mid \langle w, x\rangle + b = 0\}$ is determined by (w, b) up to a factor $\alpha \neq 0$. If $(w', b') = \alpha(w, b)$ with $\alpha < 0$, then $\operatorname{sgn}(\langle w', \cdot\rangle + b') = -h$. If $\alpha > 0$, then we get equality. □

We now need a way to calculate the margin $\gamma(h)$ based on the parameters (w, b). The following lemma provides a formula for determining the distance of a point from the affine hyperplane associated with h.

Lemma 14.3. *Let $0 \neq w \in \mathbb{R}^d$, $b \in \mathbb{R}$ and let $\mathcal{H} = \{x \in \mathbb{R}^d \mid \langle w, x\rangle + b = 0\}$. For $x_0 \in \mathbb{R}^d$ we have $\operatorname{dist}(\mathcal{H}, x_0) = \frac{1}{\|w\|}|\langle w, x_0\rangle + b|$.*

Proof. We put $w' := \frac{w}{\|w\|}$ and $b' := \frac{b}{\|w\|}$. Then, according to Lemma 14.2, (w', b') provides the same classifier, and the same affine hyperplane as (w, b). Hence, we need to prove

$$\min_{x\in\mathcal{H}} \|x_0 - x\| = |\langle w', x_0\rangle + b'|.$$

We define $x_1 := x_0 - (\langle w', x_0\rangle + b')w'$ and show the following two equations

$$\min_{x\in\mathcal{H}} \|x_0 - x\| \overset{(1)}{=} \|x_0 - x_1\| \overset{(2)}{=} |\langle w', x_0\rangle + b'|.$$

By subbing in, we first see

$$\begin{aligned}\langle w', x_1\rangle + b' &= \big\langle w', x_0 - (\langle w', x_0\rangle + b')w'\big\rangle + b' \\ &= \langle w', x_0\rangle - (\langle w', x_0\rangle + b')\langle w', w'\rangle + b' \;=\; 0,\end{aligned}$$

so $x_1 \in \mathcal{H}$ and we have already completed one inequality in (1). For the other, let $x \in \mathcal{H}$ be arbitrary. Then $\langle w', x_1 - x\rangle = 0$ and therefore we have

$$\begin{aligned}\|x_0 - x\|^2 &= \|x_0 - x_1 + x_1 - x\|^2 \\ &= \|x_0 - x_1\|^2 + 2\big\langle x_0 - x_1, x_1 - x\big\rangle + \|x_1 - x\|^2 \\ &\geqslant \|x_0 - x_1\|^2 + 2\big\langle(\langle w', x_0\rangle + b')w', x_1 - x\big\rangle \\ &= \|x_0 - x_1\|^2 + 2(\langle w', x_0\rangle + b')\langle w', x_1 - x\rangle \\ &= \|x_0 - x_1\|^2,\end{aligned}$$

which completes the proof of (1) by taking the square root. Again, by subbing in, we see

$$\operatorname{dist}(\mathcal{H}, x_0) = \|x_0 - x_1\| = \Big\|\big(\big\langle \tfrac{w}{\|w\|}, x_0\big\rangle + b\big)\tfrac{w}{\|w\|}\Big\| = \tfrac{1}{\|w\|}|\langle w, x_0\rangle + b|$$

and thus equation (2) is also proven. □

If we consider, instead of a single point, a whole dataset $D = \{(x_i, y_i) \mid i = 1, \ldots, n\} \subseteq \mathbb{R}^d \times \{-1, 1\}$, it follows from the above that the margin of a correct classifier $h_{w,b} = \operatorname{sgn}(\langle w, \cdot\rangle + b)$ with $0 \neq w \in \mathbb{R}^d$ and $b \in \mathbb{R}$ is given by

$$\gamma(h_{w,b}) = \min_{i=1,\ldots,n} \operatorname{dist}(\mathcal{H}, x_i) = \frac{1}{\|w\|} \min_{i=1,\ldots,n} |\langle w, x_i\rangle + b|.$$

In order to find "the best" classifier, the expression on the right has thus to be maximized over a suitable set of (w, b)'s. Due to Lemma 14.2, this set does not have to include all (w, b) that lead to correct classifiers. One might now think that the reduction to (w, b) with $\|w\| = 1$ is sensible, as w would then be the unit normal vector of the affine hyperplane associated with h. This approach, although feasible, leads however to a rather unpleasant optimization problem, see Exercise 14.3. We therefore take a different approach and first note the following, which we have essentially already observed in Remark 13.12.

Lemma 14.4. *Let* $D = \{(x_i, y_i) \mid i = 1, \ldots, n\} \subseteq \mathbb{R}^d \times \{-1, 1\}$ *be a dataset. For* $(w, b) \in \mathbb{R}^{d+1}$ *with* $w \neq 0$*, the following are equivalent.*

(i) The function $h \colon \mathbb{R}^d \to \mathbb{R}$, $h = \operatorname{sgn}(\langle w, \cdot \rangle + b)$*, classifies correctly.*

(ii) We have $y_i(\langle w, x_i \rangle + b) > 0$ *for all* $i = 1, \ldots, n$.

(iii) We have $\min\limits_{i=1,\ldots,n} y_i(\langle w, x_i \rangle + b) > 0$.

Proof. It suffices to notice that $h(x_i) = y_i$ is equivalent to $y_i \cdot h(x_i) > 0$, and then to use that $h(x_i) > 0$ holds if and only if we have $\langle w, x_i \rangle + b > 0$. This shows the equivalence of (i) and (ii). The equivalence of (ii) and (iii) is trivial. □

Datasets are by Definition 13.1 linearly separable if and only if an affine linear classifier exists that classifies all data points correctly. This also applies in the case that all points have the same label, including the pathological case of an empty dataset. We discussed that, in this case, there are correct classifiers with $w = 0$, and therefore $\mathcal{H}$ will either be empty or equal to the whole space—and therefore not an affine hyperplane. Since the aforementioned situations are clearly uninteresting, we will exclude them from now on:

Remark 14.5. In order not to get lost in pathologies, we will from now on always assume that our dataset $D = \{(x_i, y_i) \mid i = 1, \ldots, n\} \subseteq \mathbb{R}^d \times \{-1, 1\}$ is linearly separable and that not all labels coincide. The latter implies that $D \neq \emptyset$ holds, and that every correct affine linear classifier is of the form $h = \operatorname{sgn}(\langle w, \cdot \rangle + b)$ with $w \neq 0$.

In order to get rid of the redundancy within the parameter set (as described in Lemma 14.2) without excluding correct classifiers, and in order to end up with a solvable optimization problem, we restrict ourselves to a minimal set of parameters as follows.

Proposition 14.6. *Let* $D = \{(x_i, y_i) \mid i = 1, \ldots, n\} \subseteq \mathbb{R}^d \times \{-1, 1\}$ *be a linearly separable dataset and such that not all labels coincide. With*

$$\mathcal{R}(D) := \Big\{(w, b) \in \mathbb{R}^{d+1} \;\Big|\; \min_{i=1,\ldots,n} y_i(\langle w, x_i \rangle + b) = 1\Big\} \text{ and}$$

$$\mathcal{K}(D) := \Big\{h \colon \mathbb{R}^d \to \mathbb{R} \;\Big|\; h \text{ is a correct affine linear classifier for } D\Big\},$$

the following holds true.

(i) The map $\phi\colon \mathcal{R}(D) \to \mathcal{K}(D)$, $(w, b) \mapsto h_{w,b} := \operatorname{sgn}(\langle w, \cdot\rangle + b)$ *is bijective.*
(ii) We have $\gamma(h_{w,b}) = \frac{1}{\|w\|}$ *for all* $(w, b) \in \mathcal{R}(D)$.

Proof. (i) According to Lemma 14.4, every $(w, b) \in \mathcal{R}(D)$ leads to a correct classifier, so ϕ is well-defined. If $h = \operatorname{sgn}(\langle w, \cdot\rangle + b)$ is any correct classifier, then, again according to Lemma 14.4,

$$m := \min_{i=1,\ldots,n} y_i(\langle w, x_i\rangle + b) > 0$$

and $(w/m, b/m)$ belongs to $\mathcal{R}(D)$ and yields the same classifier by Lemma 14.2. Thus, ϕ is surjective. Finally, if (w, b), $(w', b') \in \mathcal{R}(D)$ are given with $h_{w,b} = h_{w',b'}$, then according to Lemma 14.2 there exists an $\alpha > 0$ with $(w', b') = (\alpha w, \alpha b)$. By the definition of $\mathcal{R}(D)$ it follows

$$\alpha = \alpha \cdot 1 = \alpha \min_{i=1,\ldots,n} y_i(\langle w, x_i\rangle + b) = \min_{i=1,\ldots,n} y_i(\langle w', x_i\rangle + b') = 1$$

and thus ϕ is injective.

(ii) Let $(w, b) \in \mathcal{R}(D)$ be arbitrary. By Lemma 14.4, we obtain $y_i(\langle w, x_i\rangle + b) > 0$ for all i, from whence we infer $|\langle w, x_i\rangle + b| = y_i(\langle w, x_i\rangle + b)$ for all i. With Lemma 14.3 it follows

$$\gamma(h_{w,b}) = \min_{i=1,\ldots,n} \operatorname{dist}(\mathcal{H}, x_i) = \frac{1}{\|w\|} \min_{i=1,\ldots,n} |\langle w, x_i\rangle + b| \underset{(w,b)\in\mathcal{R}(D)}{\overset{}{\underset{\uparrow}{=}}} \frac{1}{\|w\|} \cdot 1,$$

which concludes the proof. □

Remark 14.7. The requirement

$$\min_{i=1,\ldots,n} y_i(\langle w, x_i\rangle + b) \overset{!}{=} 1,$$

that we used to define $\mathcal{R}(D)$, is called the *gutter condition*. If a pair of parameters (w, b) satisfies the latter, then $h_{w,b}$ is a correct classifier and the gutters of the classifier are given by

$$\mathcal{R}_{\pm} = \big\{x \in \mathbb{R}^d \mid \langle w, x\rangle + b = \pm 1\big\}.$$

In fact, it holds $\operatorname{dist}(x, \mathcal{H}) = \frac{1}{\|w\|}|\langle w, x\rangle + b| = \gamma|\langle w, x\rangle + b|$ and the rightmost expression is equal to γ if and only if $\langle w, x\rangle + b = \pm 1$ holds.

The gutter condition allows for the following characterization of affine linear classifiers with maximum margin via an optimization problem.

Theorem 14.8. *Let* $D = \{(x_i, y_i) \mid i = 1, \ldots, n\} \subseteq \mathbb{R}^d \times \{-1, 1\}$ *be a linearly separable dataset, in which not all labels coincide. Then* $h^*\colon \mathbb{R}^d \to \mathbb{R}$ *is a* *correct affine linear classifier with maximum margin if and only if there exists*

$$(w^*, b^*) \in \underset{(w,b)\in\mathcal{R}(D)}{\operatorname{argmax}} \frac{1}{\|w\|}$$

such that $h^* = h_{w^*,b^*}$ *holds.*

Proof. "$\Longrightarrow$" Let $h^* \in \mathcal{K}(D)$ be given with maximal margin. According to Proposition 14.6(i), there exists $(w^*, b^*) \in \mathcal{R}(D)$ with $h^* = h_{w^*,b^*}$. Now let $(w, b) \in \mathcal{R}(D)$ be arbitrary and let $h := h_{w,b} \in \mathcal{K}(D)$ be the corresponding classifier. Then, with Proposition 14.6(ii), we have

$$1/\|w\| = \gamma(h) \leqslant \gamma(h^*) = 1/\|w^*\|$$

which shows that (w^*, b^*) is a maximizer.

"$\Longleftarrow$" Let (w^*, b^*) be as in the theorem. Then $h^* := h_{w^*,b^*} \in \mathcal{K}(D)$ and for every other $h \in \mathcal{K}(D)$ we can choose $(w, b) \in \mathcal{R}(D)$ such that $h = h_{w,b}$. Analogous to the above, we get

$$\gamma(h) = 1/\|w\| \leqslant 1/\|w^*\| = \gamma(h^*)$$

by applying Proposition 14.6(ii). □

The above result says nothing yet about the existence or uniqueness of h^*. In the following theorem we will however cover both by reformulating the above optimization problem into a (strictly) convex problem.

Theorem 14.9. (on the associated convex optimization problem) *Let* $D = \{(x_i, y_i) \mid i = 1, \ldots, n\} \subseteq \mathbb{R}^d \times \{-1, 1\}$ *be a linearly separable dataset in which not all labels coincide. Then there exists a unique affine linear classifier* $h^*\colon \mathbb{R}^d \to \mathbb{R}$ *with maximum margin. This classifier is precisely given by the parameters* $\alpha \cdot (w^*, b^*)$ *with* $\alpha > 0$, *where*

$$(w^*, b^*) = \underset{\substack{(w,b)\in\mathbb{R}^{d+1} \\ \forall i:\, y_i(\langle w, x_i\rangle + b)\geqslant 1}}{\operatorname{argmin}} \|w\|^2$$

is the only solution of the given convex minimization problem.

Proof. ① We show the existence of (w^*, b^*). Since D is linearly separable, there exist $(w_L, b_L) \in \mathbb{R}^{d+1}$ and $\varepsilon > 0$ such that $y_i(\langle w_L, x_i\rangle + b_L) \geqslant \varepsilon$ holds for all $i = 1, \ldots, n$ and $w_L \neq 0$. It follows

$$y_i\big(\langle \tfrac{w_L}{\varepsilon}, x_i\rangle + \tfrac{b_L}{\varepsilon}\big) \geqslant 1$$

and consequently

$$\begin{aligned} M &:= \big\{(w,b) \in \mathbb{R}^{d+1} \,\big|\, \forall\, i=1,\dots,n\colon y_i(\langle w, x_i\rangle + b) \geqslant 1\big\} \\ &= \bigcap_{i=1}^{n} \big\{(w,b) \,\big|\, y_i(\langle w, x_i\rangle + b) \geqslant 1\big\} \end{aligned}$$

is a non-empty, convex and closed set, as it is the intersection of finitely many closed affine half-spaces (division by y_i leads either to $\langle w, x_i\rangle + b \geqslant 1$ or to $\langle w, x_i\rangle + b \leqslant -1$). We define $f\colon M \to \mathbb{R}$, $f(w,b) := \|w\|^2$ and obtain

$$\inf_{(w,b)\in M} f(w,b) \leqslant f(\tfrac{w_L}{\varepsilon}, \tfrac{b}{\varepsilon}) = \frac{\|w_L\|^2}{\varepsilon^2}.$$

We put $C := \|w_L\|/\varepsilon$ and observe that it is sufficient to minimize over $(w,b) \in M$ with $\|w\| \leqslant C$. For such (w,b) we have $y_i\langle w, x_i\rangle + y_i b = y_i(\langle w, x_i\rangle + b) \geqslant 1$, from which, with the Cauchy-Schwarz-Bunyakovsky inequality, we get

$$-y_i b \leqslant y_i\langle w, x_i\rangle - 1 \leqslant |\langle w, x_i\rangle| + 1 \leqslant \|w\|\|x_i\| + 1 \leqslant C\|x_i\| + 1 \leqslant R$$

for all i, where we define $R := 1 + C\max_{j=1,\dots,n}\|x_j\|$. By assumption, there are $i,j \in \{1,\dots,n\}$ with $y_i = 1$ and $y_j = -1$. With these, we get $-b \leqslant R$ and $b \leqslant R$, i.e., $|b| \leqslant R$, and it follows

$$\operatorname*{argmin}_{(w,b)\in M} f(w,b) \;=\; \operatorname*{argmin}_{\substack{(w,b)\in M \\ \|w\|\leqslant C,\, |b|\leqslant R}} f(w,b) \;\neq\; \emptyset,$$

because we are minimizing a continuous function over a compact set.

② As preparation for the uniqueness of (w^*, b^*), we first show that for every minimizer $(w^*, b^*) \in \operatorname{argmin}_{(w,b)\in M} f(w,b)$ there exist indices $i,j \in \{1,\dots,n\}$ with

$$y_i = 1,\;\; \langle w^*, x_i\rangle + b^* = 1,\;\; y_j = -1 \;\text{ and }\; \langle w^*, x_j\rangle + b^* = -1.$$

Assuming the contrary, there is a minimizer (w^*, b^*) such that, for all $i \in \{1,\dots,n\}$ with $y_i = 1$, the inequality $\langle w^*, x_i\rangle + b^* > 1$ holds. This means

$$m := \min_{\substack{i=1,\dots,n \\ y_i=1}} \langle w^*, x_i\rangle + b^* - 1 > 0.$$

We define $\hat{w} := \frac{w^*}{1+m/2}$ and $\hat{b} := \frac{b^*-m/2}{1+m/2}$ and obtain for i with $y_i = 1$ the estimate

$$\langle \hat{w}, x_i\rangle + \hat{b} = \frac{\langle w^*, x_i\rangle + b^* - m/2}{1+m/2} \geqslant \frac{1+m/2}{1+m/2} = 1,$$

while for i with $y_i = -1$ the following holds:

$$\langle \hat{w}, x_i \rangle + \hat{b} = \frac{\langle w^*, x_i \rangle + b^* - m/2}{1 + m/2} \leqslant \frac{-1 - m/2}{1 + m/2} = -1.$$

Combining the last two statements implies $(\hat{w}, \hat{b}) \in M$, but $\|\hat{w}\| < \|w^*\|$ contradicts the fact that $(w^*, b^*) \in M$ is a minimizer of f. The existence of j is shown analogously.

③ We now prove the uniqueness of (w^*, b^*). Suppose we have two minimizers (w_1^*, b_1^*) and (w_2^*, b_2^*). Since M is convex, $M_1 := \{w \in \mathbb{R}^d \mid (w, b) \in M\}$ is also convex and $f|_{M_1} = \|\cdot\|^2$ is strictly convex according to Example 17.15. Therefore, by Proposition 17.14, we have $w_1^* = w_2^* =: w^*$. Assume wl.o.g. $b_1^* < b_2^*$. Then we choose for (w^*, b_2^*) an index i as in part ② and obtain

$$y_i(\langle w^*, x_i \rangle + b_1^*) = \langle w^*, x_i \rangle + b_1^* < \langle w^*, x_i \rangle + b_2^* = 1$$

which contradicts the fact that (w^*, b_1^*) belongs to M.

④ We now need to bridge the gap from the parameters to the classifier. The previous parts of the proof show that, in the following chain of equations

$$\begin{aligned}(w^*, b^*) &= \underset{\substack{(w,b)\in\mathbb{R}^{d+1} \\ \forall\, i:\; y_i(\langle w, x_i\rangle + b) \geqslant 1}}{\operatorname{argmin}} \|w\|^2 = \underset{\substack{(w,b)\in\mathbb{R}^{d+1} \\ \min\limits_{i=1,\dots,n} y_i(\langle w, x_i\rangle + b) = 1}}{\operatorname{argmin}} \|w\|^2 \\ &= \underset{(w,b)\in\mathcal{R}(D)}{\operatorname{argmin}} \|w\| = \underset{(w,b)\in\mathcal{R}(D)}{\operatorname{argmax}} \frac{1}{\|w\|},\end{aligned}$$

apparently, the first equation holds. But also the second is valid, because part ② implies in particular that there is an i with $y_i(\langle w, x_i \rangle + b) = 1$ holds. For the third equation, we use the abbreviation $\mathcal{R}(D)$ from Proposition 14.6. We may then drop the square, as $(\,\cdot\,)^2 \colon (0, \infty) \to \mathbb{R}$ is strictly increasing. The last equation finally holds as the minimum $\|w^*\| > 0$ is strictly positive and since $(\,\cdot\,)^{-1} \colon (0, \infty) \to \mathbb{R}$ is strictly decreasing.

⑤ According to Theorem 14.8 we have $h_{w^*,b^*} = \operatorname{argmax}_{h\in\mathcal{K}(D)} \gamma(h)$. The classification of all (other) parameters that also yield the maximizer $h^* := h_{w^*,b^*}$ follows from Lemma 14.2. □

Let us mention that part ③ of the preceding proof only used the first of the statements proven in part ②, namely that at least one of the positively labeled points lies in the positive gutter $\mathcal{R}_+$. The second statement provides that there is always at least one negatively labeled point in the negative gutter $\mathcal{R}_-$ as well.

Definition 14.10. Let $D = \{(x_i, y_i) \mid i = 1, \dots, n\} \subseteq \mathbb{R}^d \times \{-1, 1\}$ be a linearly separable dataset such that not all labels coincide. Let $h^* \colon \mathbb{R}^d \to \mathbb{R}$ be the unique classifier with maximum margin $\gamma^* > 0$, according to Theorem 14.9. Let $\mathcal{H}$ and $\mathcal{R}_\pm$ denote the decision boundary and, respectively, the gutters of h^*. Then the data points x_i that satisfy $\text{dist}(\mathcal{H}, x_i) = \gamma^*$ are called the *support vectors* and h^* is called the *support vector machine* or short *SVM* associated with D.

The following picture illustrates the new notation using our initial example.

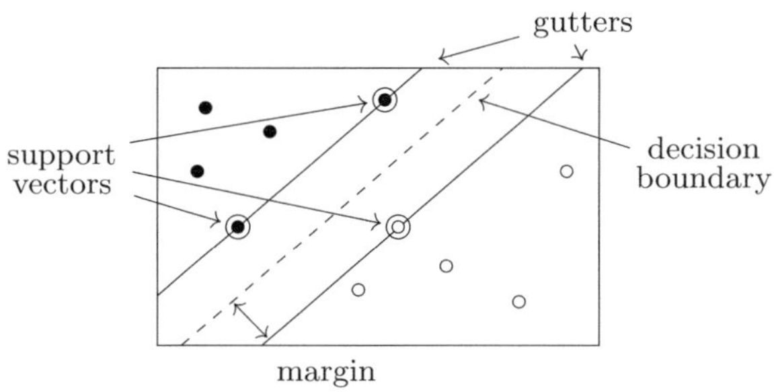

In view of Theorem 14.9, one could try to minimize the function $\|\cdot\|^2$ over the set specified in the theorem, cf. Exercise 14.4. Instead, we will now transform the maximization task of Theorem 14.9 into a quadratic optimization problem for which efficient solution methods exist. The central theorem is as follows.

Theorem 14.11. (on the associated quadratic optimization problem) *Let a dataset* $D = \{(x_i, y_i) \mid i = 1, \dots, n\} \subseteq \mathbb{R}^d \times \{-1, 1\}$ *be given in which not all labels coincide, and let* $y := (y_1, \dots, y_n) \in \mathbb{R}^n$ *denote the vector of labels. Then* D *is linearly separable if and only if there exists at least one solution to the* associated quadratic optimization problem

$$\lambda^* \in \underset{\substack{\lambda \in \mathbb{R}^n_{\geqslant 0} \\ \langle \lambda, y \rangle = 0}}{\operatorname{argmin}} \frac{1}{2} \sum_{i,j=1}^{n} \lambda_i \lambda_j y_i y_j \langle x_i, x_j \rangle - \sum_{i=1}^{n} \lambda_i .$$

Provided this is the case, every *solution* λ^* *is different from zero and leads to the support vector machine* $h^* = h_{w^*, b^*}$ *associated with* D *via the following steps:*

1. *Define* $w^* := \lambda_1^* y_1 x_1 + \dots + \lambda_n^* y_n x_n$.
2. *Choose* $i_0 \in \{1, \dots, n\}$ *such that* $\lambda_{i_0}^* \neq 0$.
3. *Define* $b^* := y_{i_0} - \langle w^*, x_{i_0} \rangle$.

Proof. In order to prove the theorem, we need a version of the so-called Karush-Kuhn-Tucker theorem that fits exactly to our situation. Detailed references can be found at the end of the chapter.

Theorem 14.12. (Karush-Kuhn-Tucker) *Let $f\colon \mathbb{R}^p \to \mathbb{R}$ be convex and continuously differentiable, and let $g_1, \dots, g_q,\ h_1, \dots, h_r\colon \mathbb{R}^p \to \mathbb{R}$ be affine linear. Let*

$$L\colon \mathbb{R}^p \times \mathbb{R}^q \times \mathbb{R} \to \mathbb{R},\ L(x, \theta, \mu) := f(x) - \sum_{i=1}^{q} \theta_i g_i(x) + \sum_{j=1}^{r} \mu_j h_j(x)$$

be the associated Lagrange function. *Then the following holds:*

(i) For every minimizer

$$x^* \in \operatorname*{argmin}_{\substack{x \in \mathbb{R}^p \\ g(x) \geqslant 0,\, h(x)=0}} f(x)$$

there exist Lagrange multipliers $\theta^* = (\theta_1^*, \dots, \theta_q^*) \in \mathbb{R}^q$ *as well as* $\mu^* = (\mu_1^*, \dots, \mu_r^*) \in \mathbb{R}^r$, *such that* (x^*, θ^*, μ^*) *satisfies the* KKT conditions

$$\begin{aligned} &\text{(KKT-1)} && \nabla_x L(x^*) = 0 \\ &\text{(KKT-2)} && h(x) = 0 \\ &\text{(KKT-3)} && \theta^* \geqslant 0,\ g(x^*) \geqslant 0,\ \langle \theta^*, g(x^*) \rangle = 0 \end{aligned}$$

where inequalities for vectors are to be understood componentwise.

(ii) For every triple $(x^*, \theta^*, \mu^*) \in \mathbb{R}^p \times \mathbb{R}^q \times \mathbb{R}^r$ that satisfies the KKT conditions, x^* is a minimizer of the above optimization problem. ◇

In the following, we will bring the convex problem from Theorem 14.9, as well as the quadratic problem from the current theorem, into the above form and use the KKT conditions for each of them. In fact, in the subsequent proof we will use each of the two parts of Theorem 14.12 twice.

We start with the existence statement.

① Let D be linearly separable and (w^*, b^*) be the unique solution of the convex optimization problem in Theorem 14.9. By adding a purely cosmetic factor of $1/2$, we obtain

$$(w^*, b^*) = \operatorname*{argmin}_{\substack{(w,b) \in \mathbb{R}^{d+1} \\ \forall i:\, y_i(\langle w, x_i \rangle + b) \geqslant 1}} \tfrac{1}{2}\|w\|^2, \tag{14.1}$$

which is of the form given in Theorem 14.12 if we put $p = d+1$, $q = n$, $r = 0$, as well as $f(w, b) = \frac{1}{2}\|w\|^2$ and $g_i(w, b) = y_i(\langle w, x_i \rangle + b) - 1$. The Lagrange function is (with λ instead of θ for good reasons!):

$$L(w, b, \lambda) = \tfrac{1}{2}\|w\|^2 - \sum_{i=1}^{n} \lambda_i [y_i(\langle w, x_i \rangle + b) - 1],$$

and, according to Theorem 14.12(i), there exists $\lambda^* \in \mathbb{R}^n$, such that (w^*, b^*, λ^*)

satisfies the KKT conditions. Now (KKT-1) yields

$$0 = \nabla_{(w,b)} L(w^*, b^*, \lambda^*) = \Big(\frac{\partial L}{\partial w}, \frac{\partial L}{\partial b}\Big)(w^*, b^*, \lambda^*)$$
$$= \Big(w^* - \sum_{i=1}^n \lambda_i^* y_i x_i, -\sum_{i=1}^n \lambda_i^* y_i\Big),$$

that is, $w^* = \lambda_1^* y_1 x_1 + \cdots + \lambda_n^* y_n x_n$ and $\langle \lambda^*, y\rangle = 0$. By (KKT-3) we have $\lambda^* \geqslant 0$, $(y_i(\langle w^*, x_i\rangle + b^*) - 1 \geqslant 0$ for all $i = 1, \ldots, n$, as well as $\big\langle \lambda^*, (y_i(\langle w^*, x_i\rangle + b^*) - 1)_{i=1,\ldots,n}\big\rangle = 0$.

② We now claim that

$$\lambda^* \in \operatorname*{argmin}_{\substack{\lambda \in \mathbb{R}^n_{\geqslant 0} \\ \langle \lambda, y\rangle = 0}} \tfrac{1}{2} \sum_{i,j=1}^n \lambda_i \lambda_j y_i y_j \langle x_i, x_j\rangle - \sum_{i=1}^n \lambda_i \tag{14.2}$$

holds, and we show this by using Theorem 14.12(ii). In fact, (14.2) is also of the form considered in Theorem 14.12, but this time $p = q = n$, $r = 1$, $f(\lambda)$ is the expression to be minimized in (14.2), $g_i(\lambda) = \lambda_i$ and $h_1(\lambda) = \langle \lambda, y\rangle$. We particularly point out that λ now takes the role of the variable and is no longer a Lagrange multiplier! The Lagrange function is

$$L(\lambda, \theta, \mu) = f(\lambda) - \sum_{i=1}^n \theta_i \lambda_i + \mu_1 \langle \lambda, y\rangle$$
$$= \tfrac{1}{2} \sum_{i,j=1}^n \lambda_i \lambda_j y_i y_j \langle x_i, x_j\rangle - \sum_{i=1}^n \lambda_i - \sum_{i=1}^n \theta_i \lambda_i + \mu_1 \sum_{i=1}^n \lambda_i y_i$$
$$= \tfrac{1}{2}\Big\langle \begin{bmatrix} \lambda_1 \\ \vdots \\ \lambda_n \end{bmatrix}, \underbrace{\begin{bmatrix} y_1 y_1 \langle x_1, x_1\rangle & \cdots & y_1 y_n \langle x_1, x_n\rangle \\ \vdots & & \vdots \\ y_n y_1 \langle x_n, x_1\rangle & \cdots & y_n y_n \langle x_n, x_n\rangle \end{bmatrix}}_{=:P} \begin{bmatrix} \lambda_1 \\ \vdots \\ \lambda_n \end{bmatrix}\Big\rangle + \Big\langle \underbrace{\begin{bmatrix} -1-\theta_1+\mu_1 y_1 \\ \vdots \\ -1-\theta_n+\mu_1 y_n \end{bmatrix}}_{=:q}, \begin{bmatrix} \lambda_1 \\ \vdots \\ \lambda_n \end{bmatrix}\Big\rangle.$$

If we look again at the last two lines in part ① of this proof, we see that λ^* fulfills the condition (KKT-2) and the last two conditions from (KKT-3) for the problem (14.2). It remains to show that $\theta^* \geqslant 0$ and $\mu^* \in \mathbb{R}$ exist, such that (KKT-1) holds. To see this, we calculate

$$\nabla_\lambda L(\lambda, \theta, \mu) = \nabla_\lambda \big(\tfrac{1}{2}\langle \lambda, P\lambda\rangle + \langle q, \lambda\rangle\big) = P\lambda + q$$

and write down the i-th entry of this vector equation, evaluated at $\lambda = \lambda^*$. Using the equation $w^* = \lambda_1^* y_1 x_1 + \cdots + \lambda_n^* y_n x_n$, which we have shown at the end of part ①, this gives us

$$(\nabla_\lambda L(\lambda^*, \theta, \mu))_i = \sum_{j=1}^n y_i y_j \langle x_i, x_j\rangle \lambda_j^* - 1 - \theta_i + \mu_1 y_i$$

$$
\begin{aligned}
&= y_i\Big\langle x_i, \sum_{j=1}^{n} \lambda_j^* y_j x_j \Big\rangle - 1 - \theta_i + \mu_1 y_i \\
&= y_i \langle x_i, w^* \rangle - 1 - \theta_i + \mu_1 y_i \\
&= y_i(\langle x_i, w^* \rangle + \mu_1) - 1 - \theta_i.
\end{aligned}
$$

If we choose $\mu_1^* := b^*$ and $\theta_i^* := y_i(\langle x_i, w^* \rangle + \mu_1^*) - 1$, we obtain $\theta^* \geqslant 0$ due to the (in-)equalities shown at the end of part ③, and $\nabla_\lambda L(\lambda^*, \theta^*, \mu^*) = 0$ by construction. According to Theorem 14.12(ii), λ^* solves the quadratic optimization problem as claimed.

Now we get to the "for all" statement.

③ Let λ^* be any minimizer of the quadratic problem in the theorem and thus also of the problem in (14.2). We have already brought the latter into the corresponding form in part ② and therefore we may infer that, due to Theorem 14.12(i), multipliers $\theta^* \geqslant 0$ and $\mu \geqslant 0$ exist, such that $(\lambda^*, \theta^*, \mu^*)$ fulfills the KKT conditions. We have also calculated the gradient of the Lagrange function already in part ②, and thus we get

$$
0 = (\nabla_\lambda L(\lambda^*, \theta^*, \mu^*))_i = y_i\Big\langle x_i, \sum_{j=1}^{n} \lambda_j^* y_j x_j \Big\rangle - 1 - \theta_i^* + \mu_1^* y_i
$$

for all $i = 1, \ldots, n$, due to (KKT-1). Moreover, we obtain $\langle \lambda^*, y \rangle = 0$ from (KKT-2), as well as $\theta^* \geqslant 0$, $\lambda^* \geqslant 0$ and $\langle \theta^*, \lambda^* \rangle = 0$ from (KKT-3).

④ We now define $w^* := \lambda_1^* y_1 x_1 + \cdots + \lambda_n^* y_n x_n$, $b^* := \mu_1^*$ and claim that (w^*, b^*, λ^*) satisfies the KKT conditions of the problem in (14.1). In part ① we have already explained how the latter can be brought into the form given in Theorem 14.12. Note that now again (w, b) is the variable and λ is the vector of Lagrange multipliers for the inequality conditions! We first check (KKT-1) and see that

$$
\begin{aligned}
\nabla_{(w,b)} L(w^*, b^*, \lambda^*) &= \Big(\frac{\partial L}{\partial w}, \frac{\partial L}{\partial b} \Big)(w^*, b^*, \lambda^*) \\
&= \Big(w^* - \sum_{i=1}^{n} \lambda_i^* y_i x_i, -\sum_{i=1}^{n} \lambda_i^* y_i \Big) = 0
\end{aligned}
$$

holds if we apply the definition of w^* and use the equation $\langle \lambda^*, y \rangle = 0$ from part ③. Since we have no equality constraint in (14.1), we omit (KKT-2). For (KKT-3) we have $\lambda^* \geqslant 0$ and

$$
\begin{aligned}
g_i(w^*, b^*) &= y_i(\langle w^*, x_i \rangle + b^*) - 1 \\
&\underset{\substack{\uparrow \\ \text{plug in} \\ w^* \text{ and } b^*}}{=} y_i\Big(\Big\langle \sum_{j=1}^{n} \lambda_j^* y_j x_j, x_i \Big\rangle + \mu_1^* \Big) - 1
\end{aligned}
$$

$$\underset{\substack{\uparrow\\ \text{add}\\ \text{zero}}}{=} \underbrace{y_i\Big\langle \sum_{j=1}^{n} \lambda_j^* y_j x_j, x_i \Big\rangle - 1 - \theta_i^* + y_i\mu_1^*}_{=0 \text{ by part ③}} + \theta_i^* \;=\; \theta_i^*$$

holds for $i = 1, \ldots, n$. It follows that $g(w^*, b^*) = \theta^* \geqslant 0$ and $\langle \lambda^*, g(w^*, b^*) \rangle = \langle \lambda^*, \theta^* \rangle = 0$, each by employing the (in-)equalities that we have shown in part ③. According to Theorem 14.12(ii), (w^*, b^*) is a minimizer of the problem in (14.1), which in particular classifies all data points correctly. Hence, D is linearly separable and according to Theorem 14.9, (w^*, b^*) is the only minimizer for (14.1) and h_{w^*,b^*} is the SVM for the dataset D. Since h_{w^*,b^*} classifies all data points correctly and not all data points have the same label, w^* cannot be the zero vector. With the equation $w^* = \lambda_1^* y_1 x_1 + \cdots + \lambda_n^* y_n x_n$ it follows that λ^* cannot be the zero vector, either.

⑤ It remains to verify that, given λ^*, steps 1–3 actually lead to (w^*, b^*): Indeed, above we saw $w^* = \lambda_1^* y_1 x_1 + \cdots + \lambda_n^* y_n x_n$ as in step 1. Because $\lambda^* \neq 0$, it is possible to choose i_0 with $\lambda_{i_0}^* \neq 0$ as required in step 2. Taking into account that $y_i \in \{-1, 1\}$, we have $1/y_i = y_i$ for any i and therefore

$$b^* \underset{\substack{\uparrow\\ \text{Dfn}\\ \text{of } b^*}}{=} \mu_1^* \underset{\substack{\uparrow\\ \text{1st Equa-}\\ \text{tion in ③}}}{=} \frac{1}{y_i}\Big(-y_i\Big\langle x_i, \sum_{j=1}^{n} \lambda_j^* y_j x_j \Big\rangle + 1 + \theta_i^*\Big) \underset{\substack{\uparrow\\ \text{Dfn}\\ \text{of } w^*}}{=} -\langle x_i, w^* \rangle + y_i + y_i\theta_i^*.$$

We claim that $\theta_i^* = 0$ holds if $\lambda_i^* \neq 0$. For this, we note that, according to part ③, all θ_i^* and all λ_i^* are greater than or equal to zero, while their scalar product $\langle \theta^*, \lambda^* \rangle$ vanishes. In the following sum

$$\sum_{i=1}^{n} \theta_i^* \lambda_i^* = 0$$

each individual term must therefore be zero. But then $\lambda_i^* \neq 0$ implies $\theta_i = 0$ as claimed. □

Corollary 14.13. *Let D be as in Theorem 14.11, let (w^*, b^*) be the solution of the convex optimization problem and let λ^* be a solution of the quadratic optimization problem. Then the following holds:*

(i) All x_i with $\lambda_i \neq 0$ are support vectors.

(ii) The sum $\lambda_1^* + \cdots + \lambda_n^* = \|w^*\|^2 = (1/\gamma^*)^2$ is independent of λ^* and equal to the squared reciprocal of the SVM's margin γ^*.

Proof. (i) According to Theorem 14.11, $b^* = y_i + \langle w^*, x_i \rangle$ holds for each $i \in \{1, \ldots, n\}$ with $\lambda_i \neq 0$. Rearranging gives $\langle w^*, x_i \rangle - b^* = y_i \in \{-1, 1\}$. Since h_{w^*,b^*} is the SVM for D, (w^*, b^*) satisfies the gutter condition and it follows with Remark 14.7 that $x_i \in \mathcal{R}_+ \cup \mathcal{R}_-$ is a support vector.

(ii) Projection of the equation $w^* = \lambda_1^* y_1 x_1 + \cdots + \lambda_n^* y_n x_n$ onto $\operatorname{span}\{w^*\}$ yields

$$\langle w^*, w^* \rangle = \Big\langle w^*, \sum_{i=1}^{n} \lambda_i^* y_i x_i \Big\rangle = \sum_{i=1}^{n} \lambda_i^* y_i \langle w^*, x_i \rangle,$$

from which, with $y_i(\langle w^*, x_i\rangle + b^*) = 1$ and $\langle \lambda^*, y \rangle = 0$, the identity

$$\|w^*\|^2 = \sum_{i=1}^{n} \lambda_i^* y_i \langle w^*, x_i \rangle = \sum_{i=1}^{n} \lambda_i^* (1 - y_i b^*) = \sum_{i=1}^{n} \lambda_i^* + b^* \langle \lambda^*, y \rangle = \sum_{i=1}^{n} \lambda_i^*$$

follows. The equation $\|w^*\| = 1/\gamma^*$ follows from Proposition 14.6. □

Remark 14.14. (i) Solutions to the quadratic optimization problem in Theorem 14.11 can be computed using ready-made packages for quadratic optimization. Typically, one needs to input the problem in the form

$$\underset{\substack{\lambda \in \mathbb{R}^n \\ G\lambda \leqslant h,\, A\lambda = b}}{\operatorname{argmin}} \ \tfrac{1}{2}\lambda^\intercal P \lambda + q^\intercal \lambda$$

and then provide matrices P, G, A; and (possibly 1-dimensional) vectors q, h, b. In our case, we have already determined P and q in the proof of Theorem 14.11 and can choose $G = -\operatorname{id}_{\mathbb{R}^n}$, $h = 0_{\mathbb{R}^n}$, $A = [y_1 \ \cdots \ y_n]$ and $b = 0_{\mathbb{R}^1}$, cf. Exercise 14.7. In this book we cannot cover the (important and interesting!) question of how such problems are solved numerically. Instead, we refer to the classic optimization literature, see, e.g., the references at the end of the chapter.

(ii) In general, the quadratic optimization problem in Theorem 14.11 has multiple solutions, see Example 14.3 and Exercise 14.7. The above Corollary 14.13(ii) states that the sum of the λ_i's is constant on solutions, i.e., the sum only depends on the dataset. The concrete value of the sum of the λ_i's can be arbitrarily large or small; consider, for example, a dataset with only two data points. Moreover, Corollary 14.13(i) is not an equivalence, and there can indeed be support vectors x_i with $\lambda_i = 0$, see Example 14.15 and Exercise 14.9.

Example 14.15. (i) We consider the following dataset

Data point	1	2	3	4	5	6	7	8	9	10
Abscissa	3.00	4.50	5.10	2.50	3.60	2.83	1.00	0.50	1.40	0.60
Ordinate	1.50	0.55	2.20	0.70	1.00	3.10	1.50	2.20	2.70	3.00
Label	+1	+1	+1	+1	+1	−1	−1	−1	−1	−1

belonging to the picture shown several times in this chapter. Solving the corresponding quadratic optimization problem with a suitable package gives rise to

$$\lambda^* = (\overset{1}{1.154}, 0.000, 0.000, 0.000, 0.000, \overset{6}{0.714}, \overset{7}{0.439}, 0.000, 0.000, 0.000),$$

which leads to

$$w^* = 1.154 \cdot \begin{bmatrix} 3.00 \\ 1.50 \end{bmatrix} - 0.714 \cdot \begin{bmatrix} 2.83 \\ 3.10 \end{bmatrix} - 0.439 \cdot \begin{bmatrix} 1.00 \\ 1.50 \end{bmatrix} = \begin{bmatrix} 1.00 \\ -0.75 \end{bmatrix}$$

and

$$b^* = 1 - \left\langle \begin{bmatrix} 1.00 \\ -0.75 \end{bmatrix}, \begin{bmatrix} 3.00 \\ 1.50 \end{bmatrix} \right\rangle = 1.87.$$

In the following picture

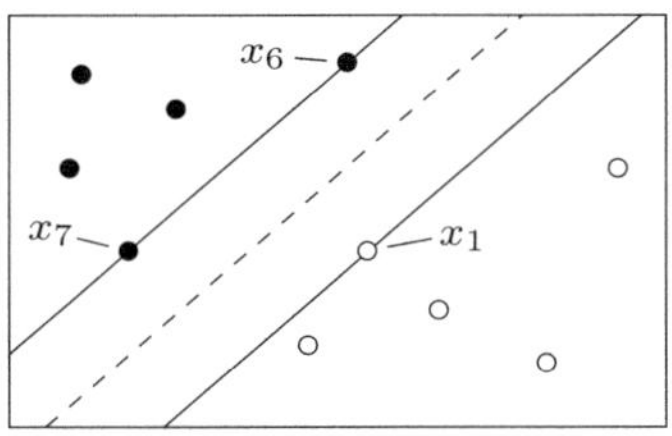

we see that, in this example, precisely those data points x_i for which $\lambda_i \neq 0$ are support vectors. In fact, λ^* is the unique solution of the quadratic optimization problem: Since only x_1, x_6 and x_7 are located in the gutters, at most λ_1, λ_6 and λ_7 can be non-zero. Since w^* is uniquely determined by the dataset, Theorem 14.11(i) and Corollary 14.13(ii) provide three linear equations for the three variables to be determined.

(ii) We now add an 11th data point to the dataset from (i) with coordinates $(1.915, 2.300)$ and label -1. This point lies exactly between x_6 and x_7 and thus in the gutter $\mathcal{R}_-$:

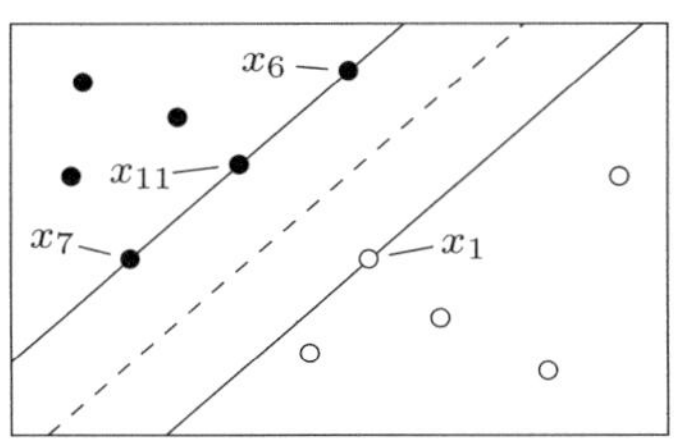

The SVM remains unchanged, but there are now four support vectors. If we solve the optimization problem for the extended dataset by computer, we obtain the solution

$$\mu^* = (\overset{1}{1.154}, 0.000, 0.000, 0.000, 0.000, \overset{6}{0.532}, \overset{7}{0.256}, 0.000, 0.000, 0.000, \overset{11}{0.364}),$$

in which now four entries are non-zero. On the other hand, if we extend the solution from (i) by a zero at the 11th position, we obtain with

$$\lambda^* = (\overset{1}{1.154}, 0.000, 0.000, 0.000, 0.000, \overset{6}{0.714}, \overset{7}{0.439}, 0.000, 0.000, 0.000, \overset{11}{0.000})$$

a second solution. In the latter, $\lambda_{11} = 0$ and x_{11} is nevertheless a support vector.

Example 14.16. (i) We stick to the data given in Example 14.15 and now consider the subsets

$$D_1 = \{(x_i, y_i) \mid i = 1, 6, 7\} \quad \text{and} \quad D_2 = \{(x_i, y_i) \mid i = 1, 6, 11\}.$$

By deleting zeros in Example 14.15(i) and by solving the optimization problem for D_2, we obtain the values λ_i^* for the respective datasets given in the following picture, which are unique with the argument from Example 14.15(i).

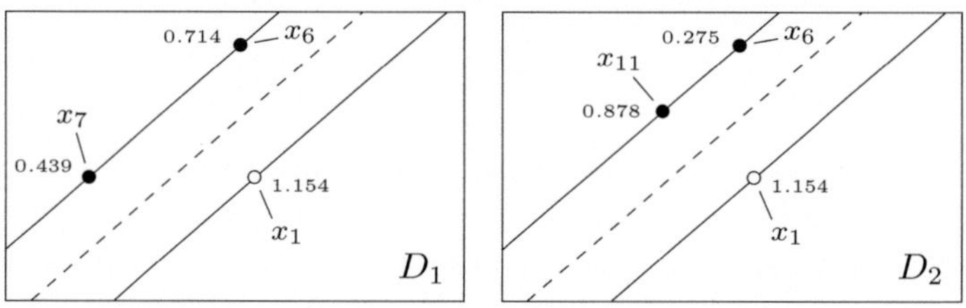

The values of λ_i^* can be interpreted quantitatively as follows: The sum of the λ_i^*'s, both on the left and on the right, is equal to 2.307. Since both datasets were created by omitting points from Example 14.15 and the SVM has not changed in the process, this is not a coincidence, but follows from Corollary 14.13. Furthermore, λ_1^* has the same value on the left and right, which is equal to $0.5 \cdot 2.307$, i.e., half of the sum of all entries, modulo rounding errors. This reflects that x_1 is alone on one side of the decision boundary. On the left, the values of λ_6^* and λ_7^* are not equal, but also not too dissimilar, which fits the fact that the points x_6 and x_7 are almost aligned symmetrically with respect to x_1. If on the left either one of x_7 or x_1 is omitted, then the SVM changes similarly in each case. On the right-hand side, this is quite different: There, λ_{11}^* is significantly larger than λ_6^* and, due to the location of the data points, we see that the SVM hardly changes if x_6 is omitted, but that it changes significantly if x_{11} is omitted.

(ii) Next, we consider two datasets D_1 and D_2, each consisting of only two points, one of which has label $+1$ and the other label -1. The distance between the two points is smaller in the dataset D_1 than in the dataset D_2. The solution vectors $\lambda_{D_i}^* = (\lambda_{D_i}, \lambda_{D_i})$ then each have two equal entries, and since the margin of the SVM for D_1 is smaller than the margin of the SVM for D_2, $\lambda_{D_1} > \lambda_{D_2}$ holds, according to Corollary 14.13(ii). If one of the two data points is now changed in the same way for both datasets (in the following picture, for example, the ordinate of the point labeled with $+1$ is increased by the same value),

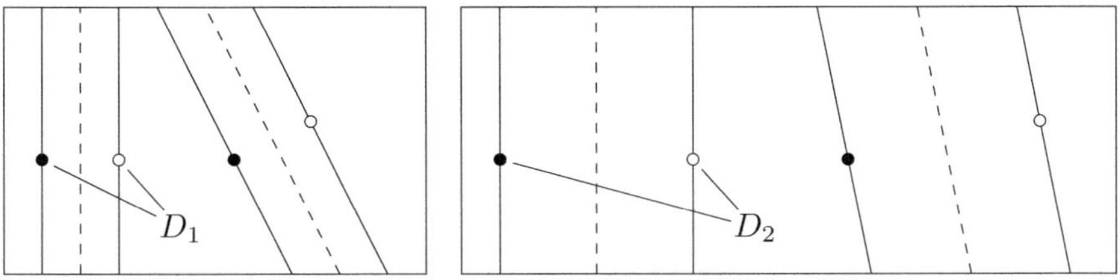

then the SVM for the dataset D_1, i.e., the one with the larger $\lambda^*_{D_1}$, changes more than for the dataset D_2 with the smaller $\lambda^*_{D_2}$.

Remark 14.17. (i) The examples suggest that the entries λ_i^* of each solution λ^* of the quadratic optimization problem can be interpreted as an indication of the "importance" of the data point x_i for the SVM. This can be understood qualitatively, in the sense that omitting x_i with $\lambda_i^* = 0$ leads to the same SVM, see Example 14.15. But "importance" can also be understood quantitatively: With a large λ_i^*, changing or omitting the data point x_i leads to a larger change in the SVM than with a small λ_i^*. Here, "large" and "small" can be understood relative to the sum of the λ_i^*, see Example 14.16(i), but also absolutely as in Example 14.16(ii).

(ii) Formally, we have proven that the omission of a non-support vector does not change the SVM (this follows from the second equation in part ④ of the proof of Theorem 14.9), and that the sum of the λ_i^*'s for a fixed dataset is always equal to the square of the reciprocal of the SVM's margin, see Corollary 14.13. The merely heuristic explanations in (i) should therefore be treated with caution, cf. Exercise 14.6.

In conclusion of this chapter, we point out that we have restricted ourselves to *hard* SVMs. In contrast, *soft* SVMs allow a few misclassifications. This is achieved by introducing so-called *slack variables* $\xi_1, \ldots, \xi_n \geqslant 0$ and accordingly replacing the constraint $y_i(\langle w, x_i\rangle + b) \geqslant 1$ with $y_i(\langle w, x_i\rangle + b) \geqslant 1 - \xi_i$. A violation of the original constraint by ξ_i is thus allowed, with the aim, of course, that the ξ_i are as small as possible. This can be achieved by

$$\underset{\substack{(w,b,\xi)\in\mathbb{R}^{d+1}\times\mathbb{R}^n_{\geqslant 0}\\ \forall i:\, y_i(\langle w,x_i\rangle+b)\geqslant 1-\xi_i}}{\operatorname{argmin}} \|w\|^2 + \tfrac{1}{n}\sum_{i=1}^{n}\xi_i,$$

where one adds the average of the ξ_i, as a summand, to the function that is to be minimized, as in this way violations of the original constraint get "punished". In this case, one will in general lose the uniqueness properties of the hard case, but one gains that datasets which are "almost" linearly separable, e.g., the following two examples

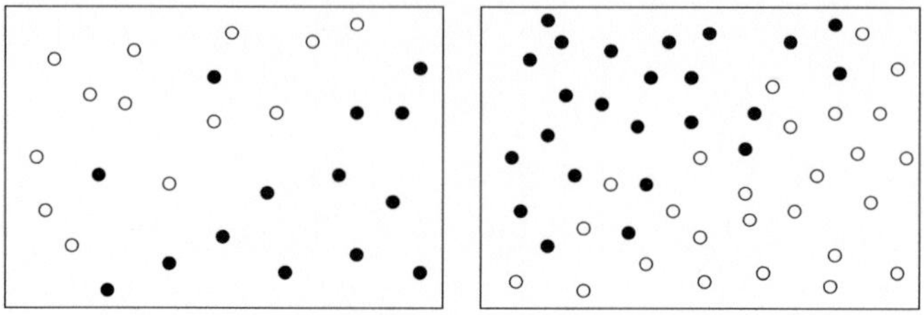

can now be treated. Notice that in the left picture only a few data points disturb the linear separability, while in the right picture the data points labeled with $+1$ transition along a hyperplane into the data labeled with -1. Notice further, that a logistic regressor is also suitable for overlapping data as in the pictures above, compare Chapter 2.4.

References

We first note that the minimization problem in Theorem 14.12 is precisely [Roc70, (P) on p. 273], but with a minus sign added to $f_1, \ldots, f_r$. Accordingly, in the Lagrange function [Roc70, middle of p. 280] a minus has to be added before the summands with indices $1, \ldots, r$, and in the (K)KT conditions [Roc70, (a)–(c) in Theorem 28.3] the inequalities for $f_1, \ldots, f_r$ have to be reversed. Theorem 14.12 is then [Roc70, Theorem 28.3]. Much of this chapter is based on the excellent video [AM12], but has been extended and furnished with complete proofs. The uniqueness part ② of the proof of Theorem 14.9 follows [Ber17, BC99]. Exercise 14.4 is due to [Nos16]. For more information on soft SVMs we refer to [SSBD14, Chapter 15.2]. A good guide to using the Python package `CVXOPT` can be found in [MC].

Exercises

Exercise 14.1. Given the following three datasets, draw the classifier that has maximum margin. Sketch the gutters and mark the support vectors.

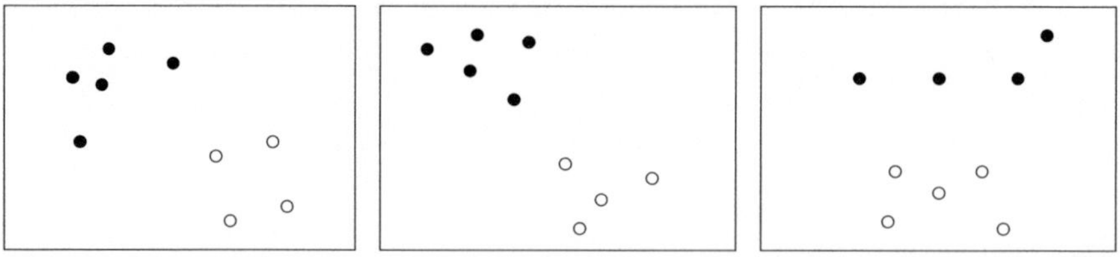

If one wants to choose a minimal set of support vectors that uniquely determine the SVM, how many support vectors are needed in the examples?

Exercise 14.2. Calculate the distance of the point $(1, 4, 0)$ from the plane given by the equation $3x + 4y = 4$ in $\mathbb{R}^3$, using Lemma 14.3. Does the formula from Lemma

14.3, perhaps with the additional condition $\|w\| = 1$, remind you of something you encountered in high school?

Exercise 14.3. Let $D = \{(x_i, y_i) \mid i = 1, \ldots, n\} \subseteq \mathbb{R}^d \times \{-1, 1\}$ be a dataset. Show that the normalization condition $\|w\| = 1$, instead of the gutter condition, leads to the optimization problem

$$\underset{\substack{(w,b) \in \mathbb{S}^1 \times \mathbb{R} \\ \forall i:\, y_i(\langle w, x_i \rangle + b) > 0}}{\operatorname{argmax}} \ \min_{i=1,\ldots,n} |\langle w, x_i \rangle + b|$$

where $\mathbb{S}^1 = \{w \in \mathbb{R}^d \mid \|w\| = 1\}$ denotes the d-dimensional unit sphere.

Exercise 14.4. We consider the dataset

$$\begin{aligned} D = \big\{&(x_1, y_1) := \left(\left[\begin{smallmatrix}1\\1\end{smallmatrix}\right], 1\right), (x_2, y_2) := \left(\left[\begin{smallmatrix}1\\3\end{smallmatrix}\right], 1\right), \\ &(x_3, y_3) := \left(\left[\begin{smallmatrix}2\\2\end{smallmatrix}\right], -1\right), (x_4, y_4) := \left(\left[\begin{smallmatrix}3\\2\end{smallmatrix}\right], -1\right)\big\}. \end{aligned}$$

(i) Write down the optimization task from Theorem 14.9, but leave out the square. Visualize the area $M \subseteq \mathbb{R}^3$, over which we minimize, by a 3D plot, e.g., with a suitable CAS.

(ii) Read off your plot $w^* = \operatorname{argmin}_{w \in M} \|w\|$ (= the point in M closest to the origin) and find an inequality constraint where equality holds when subbing w^* in. Determine b^* from this.

(iii) Now sketch the dataset, as well as decision boundary and gutters, of the classifier with maximum margin given by (w^*, b^*). Don't be upset if you notice that in this simple example the parameters (w^*, b^*) could have been easily guessed.

Exercise 14.5. We consider the dataset

$$D := \left\{(x_1, y_1) := \left(\left[\begin{smallmatrix}0\\0\end{smallmatrix}\right], -1\right), (x_2, y_2) := \left(\left[\begin{smallmatrix}1\\0\end{smallmatrix}\right], 1\right), (x_3, y_3) := \left(\left[\begin{smallmatrix}0\\1\end{smallmatrix}\right], 1\right)\right\}$$

consisting of three points in $\mathbb{R}^2$. Write down the quadratic optimization task from Theorem 14.11 and calculate a solution by hand.

Exercise 14.6. Test the heuristic from Remark 14.17 as follows.

(i) Consider in the situation of Example 14.15(ii) the dataset consisting only of x_1, x_6, x_7 and x_{11}. Then

$$\mu^* = (\overset{1}{1.154}, \overset{6}{0.532}, \overset{7}{0.256}, \overset{11}{0.364})$$

is a solution of the associated quadratic optimization problem. How much does the corresponding SVM change if one of the points x_6, x_7, or x_{11} is omitted?

(ii) Now consider the dataset D_2 from Example 14.16(i):

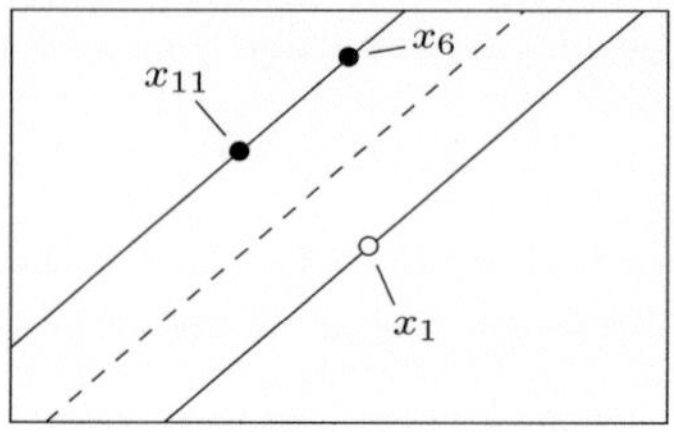

Can you add a fourth point such that (a) the quadratic optimization problem still has the unique solution $\lambda^* = (1.154, 0.275, 0.878)$ and (b) the removal of the point x_{11} hardly changes the SVM at all?

Exercise 14.7. Consider the dataset $D = \{(x_i, y_i) \mid i = 1, \ldots, 7\} \subseteq \mathbb{R}^2 \times \{-1, 1\}$, where the coordinates of the x_i and the values of the labels y_i are listed in the following table.

Data point	1	2	3	4	5	6	7
Abscissa	0.00	-1.0	2.5	1.0	-1.5	-3.0	1.0
Ordinate	3.0	-3.0	3.0	1.0	0.0	1.0	-1.0
Label	+1	-1	-1	-1	+1	+1	−1

(i) Write down the quadratic optimization problem from Theorem 14.11 and calculate a solution, for example using the Python package `CVXOPT`.

(ii) Based on the above, determine the affine linear classifier $h\colon \mathbb{R}^2 \to \mathbb{R}$ with maximum margin and compute the margin.

(iii) Sketch the dataset, the decision boundary, and the gutters. Mark the support vectors and indicate where the margin can be seen in the picture. Find at least one more solution in (i).

Exercise 14.8. Research how to calculate the support vector machine — for example, with Python — without getting your hands dirty, but by using a ready-made package instead. Determine a SVM for the dataset from Exercise 3.1 and classify the unlabeled picture given there as either "B" or "8".

Exercise 14.9. Show that the number γ, which appeared in the proof of Theorem 13.13, has the margin γ^* of the uniquely determined support vector machine as a lower bound. Employ this, to deduce an upper bound for the runtime of the perceptron algorithm. Finally, find a way to estimate the margin — at least roughly — *without* explicitly calculating the SVM.

Note: Keep in mind that we used the data points $\widehat{x}_i$, i.e., x_i extended by 1, in Theorem 13.13, while we worked without the extension by 1 in Chapter 14.

15

Kernel Method

Next, we consider datasets that are not linearly separable. In order to nevertheless obtain a classifier with the methods of the last two chapters, we will map our given dataset, which is not linearly separable, into a higher-dimensional (and sometimes even infinitely dimensional!) space to achieve linear separability for the image. It is particularly noteworthy that we are now doing, in a sense, the opposite of dimensionality *reduction*, cf. Chapters 11 and 7.1.

In this chapter, we use several concepts from Hilbert space theory; we will cite the orthogonal projection theorem and the completion theorem for pre-Hilbert spaces below without proof.

We now begin with an example to illustrate the idea of "embedding" not linearly separable data into a higher-dimensional space.

Example 15.1. Let $D = \{(x_i, y_i) \mid i = 1, \ldots, n\} \subseteq \mathbb{R} \times \{-1, 1\}$ be the 1-dimensional dataset shown in the following picture, which is clearly not linearly separable.

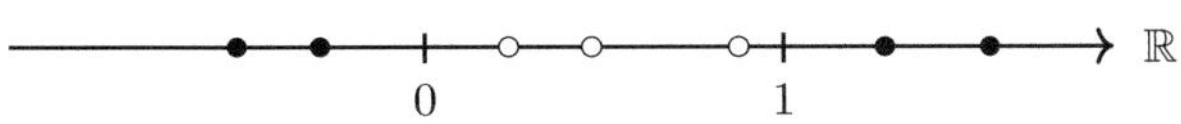

We map the feature parts of the data points via $\psi\colon \mathbb{R} \to \mathbb{R}^2$, $\psi(x) = (x, x^2)$ to $\mathbb{R}^2$ and assign to the value $\psi(x_i)$ of x_i under ψ the label y_i of its preimage. That is, we consider the 2-dimensional dataset $\hat{D} := \{(\psi(x_i), y_i) \mid i = 1, \ldots, n\} \subseteq \mathbb{R}^2 \times \{-1, 1\}$, which we call the *image dataset*. In the following picture, we see that $\hat{D}$ is linearly separable.

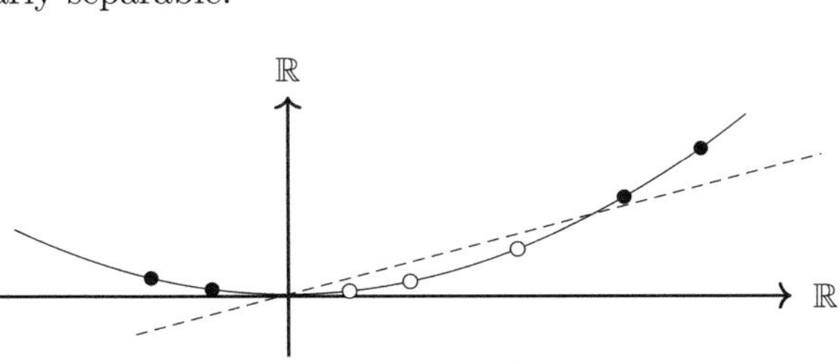

S. A. Wegner, *Mathematical Introduction to Data Science*,
https://doi.org/10.1007/978-3-662-69426-8_15

For $\hat{D}$, we can thus determine the support vector machine $h\colon \mathbb{R}^2 \to \mathbb{R}$ using the methods discussed in Chapter 14. The composition $h \circ \psi\colon \mathbb{R} \to \mathbb{R}$ then yields a classifier for the dataset D, which we call the *pullback classifier*. The latter is, of course, no longer affine linear.

In the above example, the choice of the mapping ψ may seem somewhat arbitrary. On the other hand, the first picture in Example 15.1 suggests that linear separability can be achieved with a quadratic function. Our choice of ψ is then the simplest candidate, and as it turned out, it leads to the desired result. Alternatively, one may adapt ψ even better to the data: Using, for example, $\psi(x) = (x, p(x))$ with $p(x) = -(x-0)(x-1)$ we get that all points of D with label $+1$ get mapped to the upper half-plane, while all points with label -1 get mapped to the lower half-plane. Putting more thought into the definition of ψ thus pays off, as we get the simple classifier $h = \operatorname{sgn}(\langle [\begin{smallmatrix}0\\1\end{smallmatrix}], \cdot\rangle + 0)$ without further computation. The underlying approach, namely constructing ψ employing a polynomial whose linear factors correspond to the data points, works for any dataset in $\mathbb{R}$, see Exercise 15.1 for another example.

If one is looking for a function ψ with properties as above for a given dataset, we cannot expect to always get off as easily as in Example 15.1, but that complicated functions with high-dimensional codomains will be needed. We firstly define what kind of codomains we want to allow in general. Since we want to find an affine linear classifier for $\hat{D}$, we definitely need a scalar product on the codomain, which however by no means has to be the standard scalar product of $\mathbb{R}^d$.

Definition 15.2. Let H be a (possibly infinite-dimensional, but always real) vector space. A map $\langle\cdot\,,\cdot\rangle\colon H \times H \to \mathbb{R}$ is called a *scalar product* on H, if for all $u, v, w \in H$ and $\alpha \in \mathbb{R}$ the following holds:

(SP1) $\langle u, u\rangle \geqslant 0$, (positivity)

(SP2) $\langle u, u\rangle = 0 \implies u = 0$, (definiteness)

(SP3) $\langle u, v\rangle = \langle v, u\rangle$, (symmetry)

(SP4) $\langle \alpha u + v, w\rangle = \alpha\langle u, w\rangle + \langle v, w\rangle$. (linearity in the 1st argument)

Obviously, (SP3) and (SP4) imply that $\langle\cdot\,,\cdot\rangle$ is also linear in the 2nd argument, i.e., $\langle\cdot\,,\cdot\rangle$ is bilinear. From this it follows $\langle \alpha u, \alpha u\rangle = \alpha^2\langle u, u\rangle$ and consequently in (SP2) the reverse implication holds as well.

We now show the following fundamental inequality, and we point out that we do not need all properties of a scalar product for its proof. This will be of importance in Theorem 15.16.

Lemma 15.3. *Let H be a vector space and let $\langle\cdot\,,\cdot\rangle\colon H \times H \to \mathbb{R}$ be a mapping with the properties (SP1), (SP3) and (SP4). Then for all $u, v \in H$, the* Cauchy-

Schwarz-Bunyakovsky inequality

$$\langle u, v\rangle^2 \leqslant \langle u, u\rangle\langle v, v\rangle.$$

holds. If $\langle\cdot,\cdot\rangle$ also satisfies (SP2), then $\langle u, v\rangle^2 = \langle u, u\rangle\langle v, v\rangle$ holds if and only if u and v are linearly dependent.

Proof. For $\alpha \in \mathbb{R}$ we have

$$0 \underset{\text{(SP1)}}{\overset{}{\leqslant}} \langle u + \alpha v, u + \alpha v\rangle \underset{\text{(SP3) and (SP4)}}{=} \langle u, u\rangle + 2\alpha\langle u, v\rangle + \alpha^2\langle v, v\rangle$$

and we now consider two cases.

① If $\langle u, u\rangle = \langle v, v\rangle = 0$, then we divide in the above inequality by 2 and sub in $\alpha = \pm 1$ to get on the one hand $0 \leqslant \langle u, v\rangle$ and on the other hand $0 \leqslant -\langle u, v\rangle$. Both together is however only possible if $\langle u, v\rangle = 0$ holds.

② Now let $\langle v, v\rangle \neq 0$. Then we sub $\alpha = -\langle u, v\rangle / \langle v, v\rangle$ in the initial inequality and obtain

$$0 \leqslant \langle u, u\rangle - 2\frac{\langle u, v\rangle}{\langle v, v\rangle}\langle u, v\rangle + \frac{\langle u, v\rangle^2}{\langle v, v\rangle^2}\langle v, v\rangle = \langle u, u\rangle - \frac{\langle u, v\rangle^2}{\langle v, v\rangle}.$$

Addition of $-\langle u, v\rangle^2 / \langle v, v\rangle$ and multiplication with $\langle v, v\rangle$, which is positive due to (SP2), gives rise to the desired inequality.

It remains to prove the characterization of equality: If, for example, $u = \alpha v$, then by subbing in we see that $\langle u, v\rangle^2 = \langle \alpha v, v\rangle^2 = \alpha^2\langle v, v\rangle\langle v, v\rangle = \langle u, u\rangle\langle v, v\rangle$. For the other direction, we note that in case ① only $u = v = 0$ is possible. For case ② we note that we have only estimated in one single step of our computation. If equality holds there, then $u + \alpha v = 0$ follows with (SP2) and $\alpha = -\langle u, v\rangle^2 / \langle v, v\rangle$. □

Lemma 15.4. *Let $\langle\cdot,\cdot\rangle\colon H \times H \to \mathbb{R}$ be a scalar product. Then $\|\cdot\|\colon H \to \mathbb{R}$, $\|u\| := \sqrt{\langle u, u\rangle}$ is a* norm *on H, i.e., for $u, v \in H$ and $\alpha \in \mathbb{R}$ the following hold:*

(N1) $\|u\| \geqslant 0$ and $\|u\| = 0 \iff u = 0$, (positive definiteness)

(N2) $\|\alpha u\| = |\alpha|\|u\|$, (homogeneity)

(N3) $\|u + v\| \leqslant \|u\| + \|v\|$. (triangle inequality)

Proof. The properties (N1) and (N2) follow immediately from the definition of $\|\cdot\|$ and from (SP1)–(SP4). Property (N3) can be seen by

$$\begin{aligned}\|u + v\|^2 &= \langle u + v, u + v\rangle = \|u\|^2 + 2\langle u, v\rangle + \|v\|^2 \\ &\leqslant \|u\|^2 + 2\|u\|\|v\| + \|v\|^2 = (\|u\| + \|v\|)^2,\end{aligned}$$

where the estimate is just the Cauchy-Schwarz-Bunyakovsky inequality from Lemma 15.3. □

Definition 15.5. Let H be a vector space and let $\langle\cdot,\cdot\rangle$ be a scalar product on H. Then we call $(H,\langle\cdot,\cdot\rangle)$ a *pre-Hilbert space*. If $(H,\|\cdot\|)$, where $\|\cdot\|=\sqrt{\langle\cdot,\cdot\rangle}$ denotes the *induced norm*, is complete in the sense that every $\|\cdot\|$-Cauchy sequence in H converges, then we call $(H,\langle\cdot,\cdot\rangle)$ a *Hilbert space*.

The study of Hilbert spaces is a part of functional analysis which indeed provides a rich theory for spaces of this type. We restrict ourselves here and present only those aspects relevant for our purposes in this chapter. We start with two examples.

Example 15.6. (i) The Euclidean space $(\mathbb{R}^d,\langle\cdot,\cdot\rangle)$ with the standard scalar product is a Hilbert space. Indeed, every finite-dimensional Hilbert space $(H,\langle\cdot,\cdot\rangle)$ is isometrically isomorphic to $(\mathbb{R}^d,\langle\cdot,\cdot\rangle)$ with $d=\dim H$: Choosing an orthonormal basis $\mathcal{U}=\{u_1,\dots,u_d\}$ in H and defining $A\colon H\to\mathbb{R}^d$ as a linear extension of $Au_i=\mathrm{e}_i$ provides an isomorphism of vector spaces and it follows that $\langle Au,Av\rangle=\langle u,v\rangle$ holds by subbing in the $\mathcal{U}$-expansions of u and v.

(ii) The space of square summable real sequences

$$\ell^2=\Big\{(u_i)_{i\in\mathbb{N}}\subseteq\mathbb{R}\;\Big|\;\sum_{i=1}^{\infty}|u_i|^2<\infty\Big\}\;\text{ with }\;\big\langle(u_i)_{i\in\mathbb{N}},(v_i)_{i\in\mathbb{N}}\big\rangle:=\sum_{i=1}^{\infty}u_iv_i$$

is an infinite-dimensional Hilbert space: The scalar product is well-defined by applying the CSB inequality to the series truncated at d. With the help of (N3) it follows that ℓ^2 is a subspace of the vector space $\mathbb{R}^{\mathbb{N}}$ of all mappings from $\mathbb{N}$ to $\mathbb{R}$. To get the completeness, take into account that an ℓ^2-Cauchy sequence is in particular coordinate-wise Cauchy, thus coordinatewise convergent thanks to the completeness of $\mathbb{R}$. This gives a candidate for the limit and then, using the triangle inequality, one derives that (a) this candidate actually belongs to ℓ^2, and (b) the given Cauchy sequence converges in norm to the candidate.

After this excursion into functional analysis, it will be no surprise for the reader that we want to consider ψ's that map our data into a Hilbert space. If the Hilbert space is finite-dimensional we can calculate the support vector machine by solving the quadratic optimization problem from Theorem 14.11. If H is infinite-dimensional, the following proposition shows that we can always restrict ourselves to a finite-dimensional subspace containing the feature parts of the image dataset. This is indeed very intuitive. If, for example, a two-dimensional dataset is given, such that all points belong to a one-dimensional subspace, the following picture

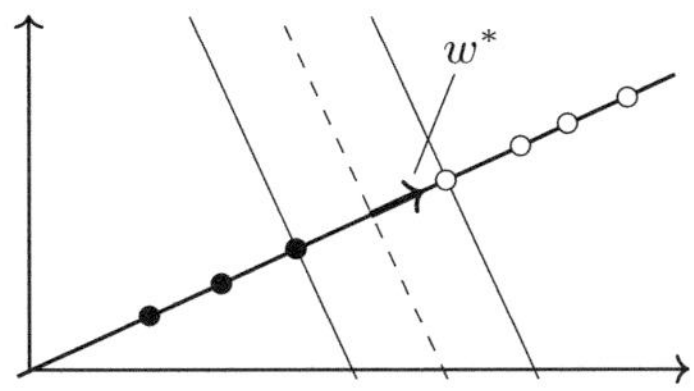

shows that the vector w^*, which defines to the SVM, must also belong to that very subspace.

For abstract and possibly infinite-dimensional Hilbert spaces H, we first define linear separability in the same way as in Definition 13.1: $D \subseteq H \times \{-1,1\}$ is *linearly separable* if there exists a correct classifier of the form $h = \operatorname{sgn}(\langle w, \cdot\rangle + b)$ for D, where now of course $(w,b) \in H \times \mathbb{R}$.

Proposition 15.7. *Let $(H, \langle\cdot,\cdot\rangle)$ be a Hilbert space. Let $D = \{(x_i, y_i) \mid i = 1,\dots,n\} \subseteq H \times \{-1,1\}$ be a linearly separable dataset in which not all labels coincide. Then there exists precisely one affine linear classifier with maximum margin $h^*\colon\ H \to \mathbb{R}$ and it is given by $h^* = \operatorname{sgn}(\langle w^*, \cdot\rangle + b^*)$ with*

$$(w^*, b^*) \quad = \quad \underset{\substack{(w,b)\in \operatorname{span} D_1 \times \mathbb{R} \\ \forall i:\, y_i(\langle w, x_i\rangle + b) \geqslant 1}}{\operatorname{argmin}} \quad \|w\|^2 \quad = \quad \underset{\substack{(w,b)\in H \times \mathbb{R} \\ \forall i:\, y_i(\langle w, x_i\rangle + b) \geqslant 1}}{\operatorname{argmin}} \quad \|w\|^2,$$

where $D_1 := \pi_1(D) = \{x_1,\dots,x_n\}$, that is, $\operatorname{span} D_1 \subseteq H$ *is the finite-dimensional subspace spanned by the feature parts of the data points.*

Proof. For the proof, we need the following generalization of the projection theorem known from linear algebra. We refer to the end of the chapter for references and additional remarks.

Theorem 15.8. (Orthogonal projection theorem) *Let $(H, \langle\cdot,\cdot\rangle)$ be a Hilbert space and let $\|\cdot\|$ be the norm induced by the scalar product. Let $U \subseteq H$ be a subspace which is closed with respect to this norm. Then every element $w \in H$ can be written as a sum $w = u + v$ such that $u \in U$ and $v \in U^\perp$ are uniquely determined by w. The linear projection $\pi_U\colon H \to H$, $w \mapsto u$, satisfies $\|\pi_U w\| \leqslant \|w\|$ for all $w \in H$.* ◇

We now use the abbreviation $U := \operatorname{span} D_1$. The space $(U, \langle\cdot,\cdot\rangle)$ is then a finite-dimensional Hilbert space and, in particular, a closed subspace of H. We denote by $\pi_U\colon H \to H$ the orthogonal projection as mentioned in Theorem 15.8.

① Since $D \subseteq H \times \{-1,1\}$ is linearly separable, by definition there exists $(w,b) \in H \times \mathbb{R}$ such that $y_i(\langle w, x_i\rangle + b) > 0$ for all i. We claim that $(\pi_U w, b) \in U \times \mathbb{R}$ provides a correct classifier, too. Writing $w = \pi_U w + v \in U + U^\perp$ as in

Theorem 15.8, it follows for all i

$$\begin{aligned} y_i(\langle \pi_U w, x_i\rangle + b) &= y_i(\langle w - v, x_i\rangle + b) \\ &= y_i(\langle w, x_i\rangle - \langle v, x_i\rangle + b) \\ &= y_i(\langle w, x_i\rangle + b), \end{aligned}$$

since $v \in U^\perp$ and $x_i \in U$. This means in particular that $D \subseteq U \times \{-1, 1\}$ is linearly separable.

② With a suitable d, the finite-dimensional space $(U, \langle\cdot,\cdot\rangle)$ is isometrically isomorphic to $\mathbb{R}^d$, where $\mathbb{R}^d$ is equipped with the standard scalar product. According to Theorem 14.9, there exists a unique minimizer

$$(w^*, b^*) = \operatorname*{argmin}_{\substack{(w,b)\in U\times\mathbb{R} \\ \forall i:\, y_i(\langle w,x_i\rangle+b)\geqslant 1}} \|w\|^2.$$

Now let $(w, b) \in H \times \mathbb{R}$ be arbitrary, such that $y_i(\langle w, x_i\rangle + b) \geqslant 1$ for all i. With the same calculation as in part ①, one can see that $(\pi_U w, b) \in U \times \mathbb{R}$ also satisfies $y_1(\langle \pi_U w, x_i\rangle + b) \geqslant 1$ for all i. Hence we have $\|w^*\|^2 \leqslant \|\pi_U w\|^2$, by our selection of (w^*, b^*). Now we write again $w = \pi_U w + v \in U + U^\perp$. With Theorem 15.8 we get the estimate $\|w\|^2 \geqslant \|\pi_U w\|^2$. Both together show

$$(w^*, b^*) \in \operatorname*{argmin}_{\substack{(w,b)\in H\times\mathbb{R} \\ \forall i:\, y_i(\langle w,x_i\rangle+b)\geqslant 1}} \|w\|^2.$$

③ Next, we claim, regardless of the already proven existence, that at most one minimizer for the task specified above can exist — even in a possibly infinite-dimensional space. To see this, we go through the parts ② and ③ in the proof of Theorem 14.9 again: In part ② we indeed never used that the underlying space is finite-dimensional. In part ③ we used Proposition 17.14. There we consider functions defined on convex subsets of $\mathbb{R}^d$. But one can see that the implication

$$f\colon A \to \mathbb{R} \text{ strictly convex} \implies \exists \text{ at most one minimizer}$$

can be formulated and proven in the same way if A is a subset of any vector space. It remains to convince ourselves that the norm square $\|\cdot\|^2\colon H \to \mathbb{R}$ in our abstract Hilbert space is strictly convex. But here too, one can see that Example 17.15 does not use that the dimension of the underlying space is finite.

④ From what we have done so far, the three equalities in the proposition and the existence of the classifier h^* follow. For the uniqueness of h^*, we recycle our arguments from Chapter 13 and 14. An inspection of the proofs of Lemmas 13.3 and 14.2 shows that these are also correct in infinite dimensions:

Here, one only needs to be careful in Definition 13.2, and define hyperplanes by $\mathcal{H} = X + x_0$ with $x_0 \in H$ and a closed subspace $X \subseteq H$ with $\dim(H/X) = 1$. All further arguments were based on the orthogonal projection theorem from linear algebra, which we, in the current setting, may replace with Theorem 15.8. □

Remark 15.9. If we have mapped a dataset $D = \{(x_i, y_i) \mid i = 1 \ldots, n\}$ via a mapping ψ into an infinite-dimensional Hilbert space H such that $\hat{D} = \{(\psi(x_i), y_i) \mid i = 1 \ldots, n\} \subseteq H \times \{-1, 1\}$ is linearly separable, then Proposition 15.7 guarantees on the one hand that an SVM for $\hat{D}$ exists and on the other hand that its parameters (w^*, b^*) can be found by solving a finite-dimensional optimization problem. That the parameter w^* belongs to D_1 means explicitly that there are coefficients $\alpha_1, \ldots, \alpha_n \in \mathbb{R}$ with

$$w^* = \sum_{i=1}^{n} \alpha_i \psi(x_i).$$

The (first component of the) solution to the a priori ∞-dimensional optimization problem can thus be represented as a linear combination from the (finite) image dataset. Results of the type of Proposition 15.7 are therefore referred to as *representer theorems*.

The last missing piece in our puzzle is the domain of ψ. In Example 15.1, our dataset was a subset of $\mathbb{R}$ and it seemed natural there to define ψ on the whole real line. We may easily generalize to datasets in $\mathbb{R}^d$, and thus look for maps $\psi\colon \mathbb{R}^d \to (H, \langle\cdot,\cdot\rangle)$. However, our goal was precisely to shift the search for affine linear classifiers to the space $(H, \langle\cdot,\cdot\rangle)$ via ψ. This in particular means that the dataset itself no longer needs to lie in a space where we can talk about linear separability! It is therefore completely sufficient to specify a superset X of the data points themselves, which will then become the domain of our classifier.

In the following remark, we summarize our setup again before then formulating a central observation, which is often called the *kernel trick*.

Remark 15.10. (Kernel trick) Let X be a set and let a non-empty dataset $D = \{(x_i, y_i) \mid i = 1, \ldots, n\} \subseteq X \times \{-1, 1\}$ be given. Let $\psi\colon X \to (H, \langle\cdot,\cdot\rangle)$ be a map from X into a Hilbert space and let $\hat{D} = \{(\psi(x_i), y_i) \mid i = 1, \ldots, n\} \subseteq X \times \{-1, 1\}$ be the image dataset. Assuming $\hat{D}$ is linearly separable, Proposition 15.7 and Theorem 14.11 state that the SVM for $\hat{D}$ can be found by computing

$$\lambda^* \in \operatorname*{argmin}_{\substack{\lambda \in \mathbb{R}^n_{\geqslant 0} \\ \langle \lambda, y \rangle = 0}} \tfrac{1}{2} \sum_{i,j=1}^{n} \lambda_i \lambda_j y_i y_j \langle \psi(x_i), \psi(x_j) \rangle - \sum_{i=1}^{n} \lambda_i.$$

We must however anticipate that ψ will be non-linear and possibly complicated to calculate. Furthermore, we recall once again that $\langle\cdot,\cdot\rangle$ is not necessarily the standard scalar product, but any scalar product in a possibly infinite-dimensional space.

The crucial observation now is that we do not actually need to know the $\psi(x_i)$ themselves, or even the Hilbert space with its scalar product, to find λ^*: To find λ^*, it is completely sufficient if we know the *numbers* $\langle\psi(x_i),\psi(x_j)\rangle$ for the finitely many data points $x_1,\dots,x_n$! So, if we had a function $k\colon X\times X\to\mathbb{R}$ with

$$\forall\, x,y\in X\colon k(x,y)=\langle\psi(x),\psi(y)\rangle, \tag{15.1}$$

we could, based solely on this function, first calculate λ^*, and then the SVM.

Since the above contains a certain mental twist, we point out again that on the left-hand side of (15.1) we sub points from our set X into k and then get the scalar product of their values under ψ "directly", while on the right-hand side we would first have to explicitly calculate the vectors $\psi(x)$ and $\psi(y)$ to only then determine their scalar product.

The observation in Remark 15.10 leads to the following concept.

Definition 15.11. Let X be a non-empty set. A function $k\colon X\times X\to\mathbb{R}$ is called a *kernel function*, if there exists a Hilbert space $(H,\langle\cdot,\cdot\rangle)$ and a mapping $\psi\colon X\to H$ such that $k(x,y)=\langle\psi(x),\psi(y)\rangle$ holds for all $x,y\in X$.

We now note two typical examples in which the kernel property can be checked by explicitly specifying the function ψ.

Example 15.12. (i) Let $\emptyset\neq X\subseteq\mathbb{R}^d$. The function $k\colon X\times X\to\mathbb{R}$, $k(x,y)=(1+\langle x,y\rangle)^m$ with $m\in\mathbb{N}$ is called the *polynomial kernel of degree m*. We check Definition 15.11 only in the special case $d=3$, $m=2$. Here we get

$$\begin{aligned}
k(x,y) &= \big(1+\sum_{i=1}^{3}x_iy_i\big)^2 = 1+2\sum_{i=1}^{3}x_iy_i+\big(\sum_{i=1}^{3}x_iy_i\big)^2\\
&= 1+2x_1y_1+2x_2y_2+2x_3y_3+x_1^2y_1^2+x_2^2y_2^2+x_3^2y_3^2\\
&\quad +2x_1x_2y_2y_1+2x_1x_3y_3y_1+2x_2x_3y_2y_3\\
&= \left\langle \begin{bmatrix}1\\ \sqrt{2}x_1\\ \sqrt{2}x_2\\ \sqrt{2}x_3\\ x_1^2\\ x_2^2\\ x_3^2\\ \sqrt{2}x_1x_2\\ \sqrt{2}x_1x_3\\ \sqrt{2}x_2x_3\end{bmatrix}, \begin{bmatrix}1\\ \sqrt{2}y_1\\ \sqrt{2}y_2\\ \sqrt{2}y_3\\ y_1^2\\ y_2^2\\ y_3^2\\ \sqrt{2}y_1y_2\\ \sqrt{2}y_1y_3\\ \sqrt{2}y_2y_3\end{bmatrix}\right\rangle = \langle\psi(x),\psi(y)\rangle
\end{aligned}$$

with $\psi\colon X\to\mathbb{R}^{10}$, where $\mathbb{R}^{10}$ is equipped with the standard scalar product. For arbitrary dimension d and arbitrary degree m, one calculates $(1+\langle x,y\rangle)^m$ using the multinomial theorem, to obtain $\psi\colon X \to \mathbb{R}^q$ with $q = \binom{d+m}{m} = \frac{(d+m)!}{m!d!}$, see Exercise 15.3. To illustrate the kernel trick again using this example, we note that q increases quickly with d and m, and that the evaluation of ψ requires many computational steps. Therefore, calculating $\psi(x_i)$ and $\psi(x_j)$ first and then their q-dimensional scalar product would be expensive. But in our setup, we just need to calculate $k(x_i, x_j) = (1 + \langle x_i, y_i \rangle)^m$, which only requires one d-dimensional standard scalar product, adding a one, and then the formation of one m-th power.

(ii) Let $\emptyset \neq X \subseteq \mathbb{R}^d$. The function $k\colon X \times X \to \mathbb{R}$, $k(x,y) = \mathrm{e}^{-\frac{\|x-y\|^2}{2\sigma^2}}$ with $\sigma > 0$ is called *RBF kernel*, where RBF stands for *radial basis function*. We check Definition 15.11 again only in the special case $\sigma = 1$, $d = 2$, in which we get

$$\begin{aligned} k(x,y) &= \mathrm{e}^{-\|x-y\|^2} = \mathrm{e}^{-\langle x-y, x-y\rangle} = \mathrm{e}^{\langle x,y\rangle}\,\mathrm{e}^{-\|x\|^2/2}\,\mathrm{e}^{-\|y\|^2/2} \\ &= \sum_{j=0}^{\infty} \frac{\langle x,y\rangle^j}{j!}\,\mathrm{e}^{-\|x\|^2/2}\,\mathrm{e}^{-\|y\|^2/2} = \sum_{j=0}^{\infty} \frac{(x_1y_1 + x_2y_2)^j}{j!}\,\mathrm{e}^{-\|x\|^2/2}\,\mathrm{e}^{-\|y\|^2/2} \\ &= \sum_{j=0}^{\infty}\sum_{i=0}^{j} \tfrac{1}{j!}\tbinom{j}{i}(x_1y_1)^i(x_2y_2)^{j-i}\,\mathrm{e}^{-\|x\|^2/2}\,\mathrm{e}^{-\|y\|^2/2} \\ &= \sum_{j=0}^{\infty}\sum_{i=0}^{j} \Big(\mathrm{e}^{-\|x\|^2/2}\,\frac{x_1^i x_2^{j-i}}{(i!(j-i)!)^{1/2}}\Big)\Big(\mathrm{e}^{-\|y\|^2/2}\,\frac{y_1^i y_2^{j-i}}{(i!(j-i)!)^{1/2}}\Big) = \langle \psi(x), \psi(y)\rangle \end{aligned}$$

with

$$\psi\colon X \to \ell^2,\ \psi(x) = \mathrm{e}^{\|x\|^2/2}\big(\underbrace{\overset{i=0}{1}}_{j=0}, \underbrace{\overset{i=0}{x_2^1}, \overset{i=1}{x_1^1}}_{j=1}, \underbrace{\overset{i=0}{\tfrac{x_1^2}{2}}, \overset{i=1}{\tfrac{x_1x_2}{2}}, \overset{i=2}{\tfrac{x_2^2}{2}}}_{j=2}, \ldots\big).$$

That ψ is well-defined follows from $\|\psi(x)\|^2 = \mathrm{e}^{\|x-y\|^2/2} < \infty$, which holds according to the above calculation. For arbitrary d, the multinomial theorem can be applied again, and for arbitrary $\sigma > 0$, corresponding constants will show up in the computation. Analogous to (i), $k(x_i, x_j)$ can essentially be determined by a norm calculation and one evaluation of the exponential function — notice that, as the codomain of ψ is infinite-dimensional, it would even be impossible to first calculate $\psi(x_i)$ and $\psi(x_j)$ in a numerical way.

We now formulate the method that arises from the kernel trick as a theorem.

Theorem 15.13. (Kernel trick theorem) *Let $X \subseteq \mathbb{R}^d$ and $D = \{(x_i, y_i) \mid i = 1, \ldots, n\} \subseteq X \times \{-1, 1\}$ be a dataset such that not all labels coincide. Let $k\colon X \times X \to \mathbb{R}$ be a kernel function and let*

$$\lambda^* \in \operatorname*{argmin}_{\substack{\lambda \in \mathbb{R}^n_{\geqslant 0} \\ \langle \lambda, y\rangle = 0}} \frac{1}{2}\sum_{i,j=1}^{n} \lambda_i\lambda_j y_i y_j k(x_i, x_j) - \sum_{i=1}^{n} \lambda_i$$

be any solution of the associated optimization problem. Then $\lambda^* \neq 0$ *and, for each* $i_0 \in \{1, \ldots, n\}$ *with* $\lambda^*_{i_0} \neq 0$*, the following function is a correct classifier for* D*:*

$$h_k \colon X \to \mathbb{R},\ h_k(x) = \operatorname{sgn}\Big(\sum_{i=1}^{n} \lambda_i^* k(x_i, x) + y_{i_0} - \sum_{i=1}^{n} \lambda_i^* y_i k(x_i, x_{i_0}) \Big).$$

The latter is independent of the choice of λ^* *and* i_0*. Moreover, it is possible to calculate the classifier's values solely via the kernel function.*

Proof. By Definition 15.11, we choose $\psi \colon X \to (H, \langle \cdot, \cdot \rangle)$ such that $k(x, y) = \langle \psi(x), \psi(y) \rangle$ holds for all $x, y \in X$. We denote by $\hat{D}$ the image dataset, which lives in the finite-dimensional subspace

$$U := \operatorname{span}\{\psi(x_i) \mid i = 1, \ldots, n\} \subseteq H,$$

which is isometrically isomorphic to $\mathbb{R}^d$ with $d = \dim U$. According to Theorem 14.11, the existence of λ^*, which we have assumed above, implies that $\hat{D}$ is linearly separable. In combination with Proposition 15.7, Theorem 14.11 further implies that

$$w^* = \sum_{i=1}^{n} \lambda_i^* y_i \psi(x_i) \ \text{ and } \ b^* = y_{i_0} - \langle w^*, x_{i_0} \rangle$$

provide the uniquely determined affine linear classifier $h^* \colon H \to \mathbb{R}$ for $\hat{D}$. Here, w^* is independent of the choice of λ^*, and b^* is independent of the choice of i_0, as long as $\lambda^*_{i_0} \neq 0$ holds. The composition $h^* \circ \psi \colon X \to \mathbb{R}$ provides by construction a correct classifier for the original dataset D. By subbing in, we obtain

$$\begin{aligned} h^*(\psi(x)) &= \operatorname{sgn}(\langle w^*, x \rangle + b^*) \\ &= \operatorname{sgn}\Big(\sum_{i=1}^{n} \lambda_i^* k(x_i, x) + y_{i_0} - \sum_{i=1}^{n} \lambda_i^* y_i k(x_i, x_{i_0}) \Big), \end{aligned}$$

which is exactly the expression given in the theorem for $h_k(x)$. □

Definition 15.14. In the situation of Theorem 15.13, we call $h_k \colon X \to \mathbb{R}$ the *classifier induced by* k.

Remark 15.15. (i) The proof of Theorem 15.13 has shown $h_k = h^* \circ \psi$, i.e., the classifier induced by k coincides with the pullback classifier, compare p. 210. It is noteworthy that ψ is not uniquely determined by k, cf. Exercise 15.6, but that the induced classifier depends only on k.

(ii) In nice cases, h_k, h and ψ may be illustrated as follows.

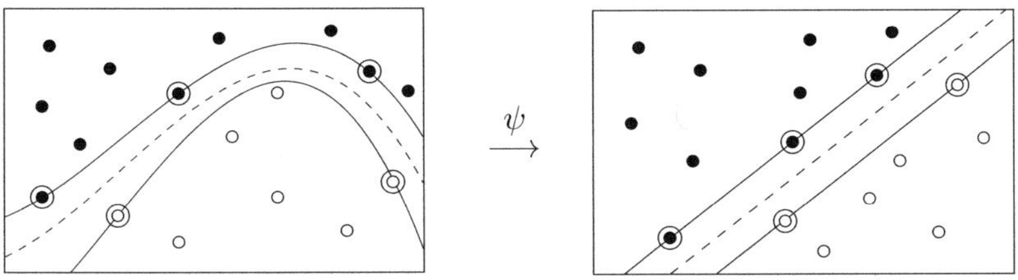

$h_k = h \circ \psi\colon X \to \mathbb{R}$ $\qquad$ $h\colon H \to \mathbb{R}$

In the right picture, the support vectors of h are marked by circles, while circles in the left picture mark *preimages of support vectors*. Note that the latter do not all have the same distances from the decision boundary. There may even be data points whose images are not support vectors, but which are closer to the decision boundary than all pre-images of support vectors.

(iii) The picture above raises the question of what properties the function $\psi\colon X \to H$ should actually have. In our Examples 15.1 and 15.12, the mappings ψ were necessarily non-linear, but always injective and continuous. In Example 15.1, one can easily see that the function ψ is open onto its image. On the other hand, the kernel trick does not requires a topology on X and imposes only a very weak injectivity condition: If the image dataset $\hat{D} \subseteq H \times \{-1, 1\}$ is linearly separable, then certainly

$$\forall\, i, j \in \{1, \dots, n\}\colon y_i \neq y_j \implies \psi(x_i) \neq \psi(x_j),$$

but it is very well possible that ψ sends data points with the same label to the same point, cf. Exercise 15.4. On the other hand, with regard to the induced classifier, it is of course not disadvantageous if k is chosen such that there is a corresponding map $\psi\colon X \to H$ which is injective and continuous. We refer to Remark 15.18 for more details.

As the last topic of this chapter, we want to discuss how kernel functions can be found. We have already seen one method in Example 15.12, where we established the kernel property by specifying a function $\psi\colon X \to H$. In the following theorem, we present another method, which provides a criterion for the kernel property — and which can be checked without explicit knowledge of ψ and H.

Condition (ii) in the following theorem is sometimes referred to as *Mercer's condition*, compare the remarks and references at the end of the chapter.

Theorem 15.16. (Mercer's condition) *Let X be a non-empty set and $k\colon X \times X \to \mathbb{R}$ be a function. Then the following are equivalent.*

(i) k is a kernel function.

(ii) For all $x_1, \ldots, x_m \in X$*, the following* Gram matrix *is symmetric and positive semi-definite:*

$$G := \begin{bmatrix} k(x_1,x_1) & \cdots & k(x_1,x_m) \\ \vdots & & \vdots \\ k(x_m,x_1) & \cdots & k(x_m,x_m) \end{bmatrix}.$$

Proof. (i)$\Longrightarrow$(ii): Let $x_1, \ldots, x_m \in X$ be given and let G be defined as above. By assumption, there exists $\psi\colon X \to (H, \langle\cdot,\cdot\rangle)$ such that $k(x,y) = \langle\psi(x), \psi(y)\rangle$ holds for all $x, y \in X$. Since the scalar product is symmetric, this implies that G is symmetric. Now let $\xi \in \mathbb{R}^m$ be given. Then the calculation

$$\begin{aligned}\langle\xi, G\xi\rangle &= \Big\langle \begin{bmatrix}\xi_1\\ \vdots \\ \xi_n\end{bmatrix}, \begin{bmatrix} k(x_1,x_1) & \cdots & k(x_1,x_m) \\ \vdots & & \vdots \\ k(x_m,x_1) & \cdots & k(x_m,x_m) \end{bmatrix}\begin{bmatrix}\xi_1\\ \vdots \\ \xi_n\end{bmatrix}\Big\rangle \\ &= \sum_{i,j=1}^{m} \xi_i\xi_j k(x_i,x_j) = \sum_{i,j=1}^{m} \xi_i\xi_j \langle\psi(x_i), \psi(x_j)\rangle \\ &= \langle \sum_{i=1}^{m} \xi_i\psi(x_i), \sum_{j=1}^{m} \xi_j\psi(x_j)\rangle = \big\|\sum_{i=1}^{m} \xi_i\psi(x_i)\big\|^2 \geqslant 0\end{aligned}$$

implies that G is positive semi-definite.

(ii) $\Longrightarrow$ (i): We define $\psi\colon X \to \mathbb{R}^X = \{f\colon X \to \mathbb{R}\}$ via $\psi(x) := k(\cdot, x)$ and, for illustration, we consider the example of the one-dimensional RBF kernel. In this special case, ψ sends each point $x \in \mathbb{R}$ to a Gaussian function that has x as its mean:

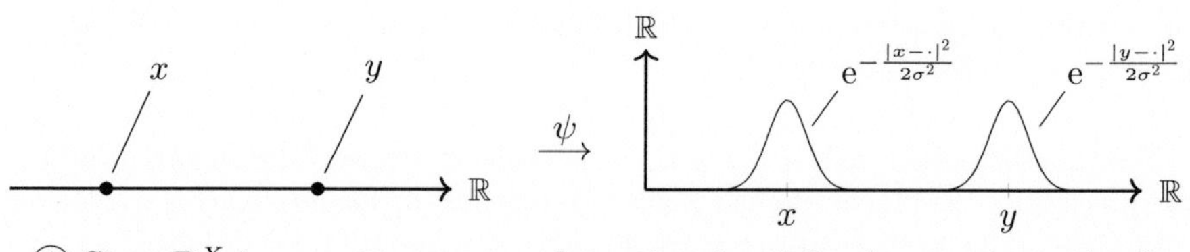

① Since $\mathbb{R}^X$ is a vector space under pointwise defined operations, the first idea would be to define on this space a suitable scalar product, such that the identity $k(x,y) = \langle\psi(x), \psi(y)\rangle$ holds. However, since $\mathbb{R}^X$ is very large, it is better to do this only on a subspace. The smallest candidate is obviously $H := \operatorname{span}\operatorname{ran}\psi$ and, by definition, every $f \in H$ can be written as

$$f = \sum_{i=1}^{n} \alpha_i k(\cdot, x_i)$$

with $n \in \mathbb{N}$, $\alpha_i \in \mathbb{R}$, $x_i \in X$.

② If f is as above and $g = \sum_{j=1}^{m} \beta_j k(\cdot, y_j)$ is another element of H, then we define

$$\langle f, g\rangle := \sum_{i=1}^{n}\sum_{j=1}^{m} \alpha_i\beta_j k(x_i, y_j).$$

As the representations of f and g are not unique, we first have to show that the right-hand side above is independent of the choices of $n, m, \alpha_i, \beta_j, x_i, y_j$. In fact, we have

$$\sum_{i=1}^{n}\sum_{j=1}^{m} \alpha_i\beta_j k(x_i, y_j) = \sum_{j=1}^{m} \beta_j \sum_{i=1}^{n} \alpha_i k(x_i, y_j) = \sum_{j=1}^{m} \beta_j f(y_j),$$

from whence it follows that $\langle f, g\rangle$ is independent of the choice of m, α_i, x_i. That the definition is independent of the choices for n, β_j, y_j, too, can be seen analogously. Thus, $\langle\cdot\,,\cdot\rangle\colon H \to \mathbb{R}$ is well-defined, and the scalar product properties now have to be verified. Here, (SP3) follows from the assumption that the matrices G are symmetric and (SP4) holds by definition. For f as above, the Gram matrix being positive semi-definite implies

$$\langle f, f\rangle = \sum_{i,j=1}^{m} \alpha_i\alpha_j k(x_i, x_j) = \langle \alpha, G\alpha\rangle \geqslant 0,$$

which shows that $\langle\cdot\,,\cdot\rangle$ satisfies condition (SP1). It remains to show (SP2), i.e., the implication $\langle f, f\rangle = 0 \implies f = 0$. For this, we observe that for f as above, the identity

$$\begin{aligned}\langle k(\cdot\,, x), f\rangle &= \Big\langle k(\cdot\,, x), \sum_{i=1}^{m} \alpha_i k(\cdot\,, x_i)\Big\rangle \\ &= \sum_{i=1}^{m} \alpha_i \langle k(\cdot\,, x), k(\cdot\,, x_i)\rangle \\ &\underset{\substack{\uparrow\\ \text{Dfn of SP}}}{=} \sum_{i=1}^{m} \alpha_i k(x, x_i) = f(x)\end{aligned}$$

holds. With the Cauchy-Schwarz-Bunyakovsky inequality, which by Lemma 15.3 holds even if (SP2) was not satiafied, we get

$$f(x)^2 = \langle k(\cdot\,, x), f\rangle^2 \leqslant \langle k(\cdot\,, x), k(\cdot\,, x)\rangle \cdot \langle f, f\rangle = k(x, x)\langle f, f\rangle$$

for any $x \in X$. Hence, if $\langle f, f\rangle = 0$ holds, then it follows $f(x) = 0$ for all $x \in X$, and this means that $f \in H \subseteq \mathbb{R}^X$ is the zero vector.

③ So far, we have at least a pre-Hilbert space $(H, \langle\cdot\,,\cdot\rangle)$ and a mapping $\psi\colon X \to H$, which fulfills the identity $k(x, y) = \langle \psi(x), \psi(y)\rangle$ for all $x \in X$. To get a Hilbert space, only the completeness is missing and for that we need the following theorem, which is part of the standard repertoire of functional analysis and can be thought of as a generalization of the transition from $\mathbb{Q}$ to $\mathbb{R}$.

Theorem 15.17. (Completion theorem) *Let* $(H, \langle\,\cdot\,,\,\cdot\,\rangle_H)$ *be a pre-Hilbert space. Then there exists a uniquely determined Hilbert space* $(\hat{H}, \langle\,\cdot\,,\,\cdot\,\rangle_{\hat{H}})$, *such that* $H \subseteq \hat{H}$, $\langle x, y\rangle_H = \langle x, y\rangle_{\hat{H}}$ *for all* $x, y \in H$, *and* $\overline{H}^{\|\cdot\|_{\hat{H}}} = \hat{H}$ *hold. The space* $\hat{H}$ *is called the* completion *of* H. ◇

If our space H from ① is not complete, Theorem 15.17 allows us to replace the space with its completion and to consider ψ as a mapping into the possibly enlarged space. The identity $k(x, y) = \langle\psi(x), \psi(y)\rangle$ remains unaffected. □

We point out that in Theorem 15.16(ii), the points $x_1, \ldots, x_m$ are not the data points of a given dataset, but that *any* points from X have to be tested here. Moreover, their cardinality $m \in \mathbb{N}$ is not fixed and thus symmetry and positive semi-definiteness of the Gram matrix have to be checked for infinitely many matrices of arbitrary dimension.

We have now seen two methods by which one can find kernel functions or test whether a given function is a kernel function. Beyond this, the kernel property has extremely good permanence properties. For example,

(a) restrictions of kernels are again kernels,
(b) linear combinations of kernels are again kernels,
(c) products of kernels are again kernels,
(d) pointwise limits of kernels are again kernels,

where we treat (a), (b) and (d) in Exercise 15.7 and refer to the references at the end of the chapter for (c).

Based on well-known examples and inheritance properties, there exist lists of well-known kernel functions that can be found in suitable books. On the other hand, one interpretation of the kernel trick is that it is actually not necessary to know whether a function k really has the kernel property: Assuming one has found a solution λ^* to the optimization problem in Theorem 15.13, one can implement the induced classifier h_k and check by simply subbing in data points whether h_k classifies correctly—if k was a kernel or not.

We conclude this chapter by highlighting that a much more comprehensive theory of kernel functions exists, in comparison to what we could present here in this chapter.

Remark 15.18. Let X be a non-empty set and $(H, \langle\,\cdot\,,\,\cdot\,\rangle)$ be a Hilbert space, whose elements are functions from X to $\mathbb{R}$. A function $k\colon X \times X \to \mathbb{R}$ is called a *reproducing kernel*, if

$$\forall\, f \in H,\ x \in X\colon \langle k(\,\cdot\,, x), f\rangle$$

holds, i.e., if every function in H can be "reproduced" by the scalar product and k. The space H is then called a *reproducing kernel Hilbert space* or *RKHS*. The specialization to reproducing kernels allows a richer theory than that of kernels in the sense of Definition 15.11. For details we refer the reader to the references mentioned below.

References

Theorem 15.8 (orthogonal projection) can be found in [Rud87, Theorem 4.11] or [Con90, Theorem I.2.7] and Theorem 15.17 (completion) in [Rud91, Exercise 12.1] or [Con90, Proposition I.1.9]. We note that within this chapter we used Theorem 15.8 only in the special case of a finite-dimensional subspace $U \subseteq X$ or a finite-dimensional quotient X/U. In this case, one could establish the projection onto U, or onto the algebraic complement of U, by choosing a finite basis as in linear algebra.

The main references for this chapter are [SSBD14, AM12, SC08]. Detailed calculations for Example 15.12 can be found in [Sha08, Chapter 4.3]. *Mercer's condition* is named after Mercer's theorem from 1909, see [SC08, p. 159]. There are much more general versions of the representer theorem, see [SC08, Chapter 4.5].

We also point out that some authors refer to the mapping ψ as a *feature map* and to the Hilbert space H as a *feature space*. Furthermore, one says that "the kernel function k implements the scalar product with respect to ψ". This in particular makes sense in view of Remark 15.10, according to which, one only needs to know k to solve the quadratic optimization problem (but not H and ψ).

Exercises

Exercise 15.1. We consider the following 1-dimensional dataset where the first entry indicates the feature and the second entry indicates the label:

$$D := \{(-3.5, 1), (0.5, 1), (0.75, 1), (-2.5, -1), (-1, -1), (5, -1)\} \subseteq \mathbb{R} \times \{-1, 1\},$$

(i) Sketch the dataset D.

(ii) Find a map $\psi\colon \mathbb{R} \to \mathbb{R}^2$ such that the image dataset $\hat{D} := \{(\psi(x), y) \mid (x, y) \in D\} \subseteq \mathbb{R}^2 \times \{-1, 1\}$ is linearly separable and sketch $\hat{D}$.

(iii) Determine a classifier of the form $h = \operatorname{sgn}(\langle w, \cdot\rangle + b)$ for $\hat{D}$.

(iv) Indicate which $x \in \mathbb{R}$ are classified by the pullback classifier $h \circ \psi$ with ± 1.

Exercise 15.2. Consider a dataset of the following form:

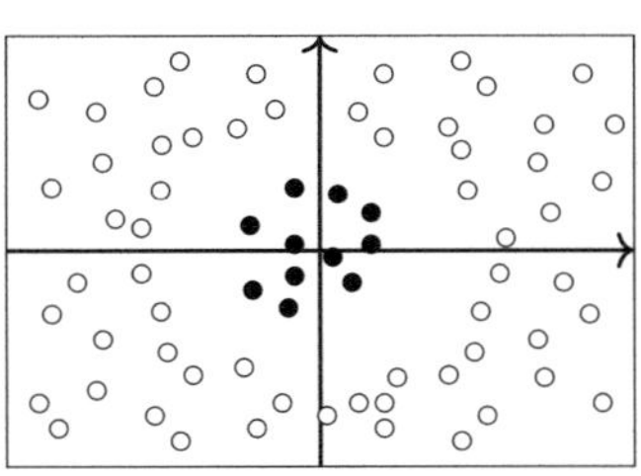

Find a function $\psi\colon \mathbb{R}^2 \to \mathbb{R}^3$, which can map the dataset injectively and continuously onto a linearly separable dataset.

Exercise 15.3. Show that the function $k\colon \mathbb{R} \times \mathbb{R} \to \mathbb{R}$, $k(x, y) = (1 + \langle x, y\rangle)^m$ with $m \in \mathbb{N}$ is a kernel function. Proceed as in Example 15.12(i) and use the multinomial theorem to calculate the m-th power.

Exercise 15.4. Let $\emptyset \neq X \subseteq \mathbb{R}$ and $D = \{(x_i, y_i) \mid i = 1, \ldots, n\} \subseteq X \times \{-1, 1\}$ be a dataset such that $x_i = x_j \implies y_i = y_j$ (recall Remark 1.2).

(i) Show that for every function $\psi\colon X \to \mathbb{R}^n$ with $\psi(x_i) = \mathrm{e}_i$ the image dataset $\hat{D} := \{(\psi(x_i), y_i) \mid i = 1, \ldots, n\} \subseteq \mathbb{R}^n \times \{-1, 1\}$ is linearly separable, by guessing the parameters of a classifier $h = \operatorname{sgn}(\langle w, \cdot\rangle + b)$ for $\hat{D}$.

(ii) Find a way to get by with $\mathbb{R}$ instead of $\mathbb{R}^n$ above.

(iii) Let $\psi\colon X \to H \subseteq \mathbb{R}^X$, $\psi(x) = \mathrm{e}^{-\frac{(x-\cdot)^2}{2\sigma^2}}$ be as in the proof of Theorem 15.16. Show that $\hat{D} := \{(\psi(x_i), y_i) \mid i = 1, \ldots, n\} \subseteq H \times \{-1, 1\}$ is linearly separable, provided that we choose a suitable σ.

(iv) Which of the three "embedding methods" do you find better and why?

Exercise 15.5. We consider the dataset from Example 15.1, which is given by the following table.

Data point	1	2	3	4	5	6	7
Value	−0.9	−0.3	0.2	0.4	0.9	1.3	1.6
Label	−1	−1	+1	+1	+1	−1	−1

Solve the optimization problem from Theorem 15.13, e.g., with the Python package `CVXOPT`, using

(i) the kernel function induced by ψ, as given in Example 15.1,

(ii) polynomial and RBF kernels, as given in Example 15.12,

(iii) for other functions $k\colon \mathbb{R}^2 \to \mathbb{R}$.

Plot the resulting classifiers as functions $h\colon \mathbb{R} \to \mathbb{R}$ and add the pairs (value, label) to the plot in order to see whether all points are correctly classified or not.

Exercise 15.6. Show that for a kernel function $k\colon X \times X \to \mathbb{R}$, neither the Hilbert space $(H, \langle\cdot, \cdot\rangle)$ nor the function $\psi\colon X \to H$ in Definition 15.11 are unique. Consider for this $k\colon \mathbb{R} \times \mathbb{R} \to \mathbb{R}$, $k(x, y) := xy$, as well as $\psi_i\colon \mathbb{R} \to \mathbb{R}^i$, defined by $\psi_1(x) = x$ and $\psi_2(x) = \frac{1}{\sqrt{2}}(x, x)$.

Exercise 15.7. Show that the kernel property is stable under restriction, under linear combinations and under pointwise limits.

Hint: For restrictions and linear combinations, one can guess the corresponding Hilbert spaces H and mappings ψ. That the pointwise limit k of a sequence of kernels $(k_n)_{n\in\mathbb{N}}$ is again a kernel can be shown with the help of Theorem 15.16.

16

Neural Networks

In this chapter, we deal with artificial neural networks. We have already encountered their central building block, the artificial neuron, in earlier chapters, without however having used this name. Before we give the formal definition, we point out to the reader that parts of this chapter require knowledge from functional analysis, measure theory, and Fourier transform: Two versions of the Stone-Weierstrass theorem, the Hahn-Banach theorem, as well as the Riesz representation theorem are therefore given below without proof. References can be found at the end of the chapter. The aforementioned prerequisites only apply to the second half of Subchapter 16.1 on the expressivity of neural networks. The reader not familiar with all of them is encouraged to nevertheless notice the main results 16.20, 16.21, 16.22 and 16.25 even if they, at least for the first read, decide to skip the proofs.

We begin with the definition of an artificial neuron.

Definition 16.1. A function $n\colon \mathbb{R}^d \to \mathbb{R}$ of the form

$$n(x) = \sigma(\langle w, x\rangle + b) = \sigma\Big(\sum_{i=1}^{d} w_i x_i + b\Big)$$

is called an *(artificial) neuron* with *weight* $w = (w_1, \dots, w_d) \in \mathbb{R}^d$, *bias* $b \in \mathbb{R}$ and *activation function* $\sigma\colon \mathbb{R} \to \mathbb{R}$. Neurons are often illustrated by pictures like the following.

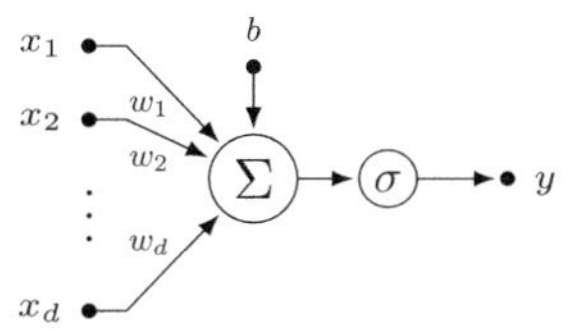

Artificial neurons are inspired by the biological neurons of the human brain. Given a biological neuron, one says that it *fires* if the incoming signals exceed

S. A. Wegner, *Mathematical Introduction to Data Science*,
https://doi.org/10.1007/978-3-662-69426-8_16

a certain threshold. This analogy fits particularly well if we think of $\sigma = \mathbb{1}_{[0,\infty)}$ being the Heaviside function which, along with other typical activation functions, is shown below.

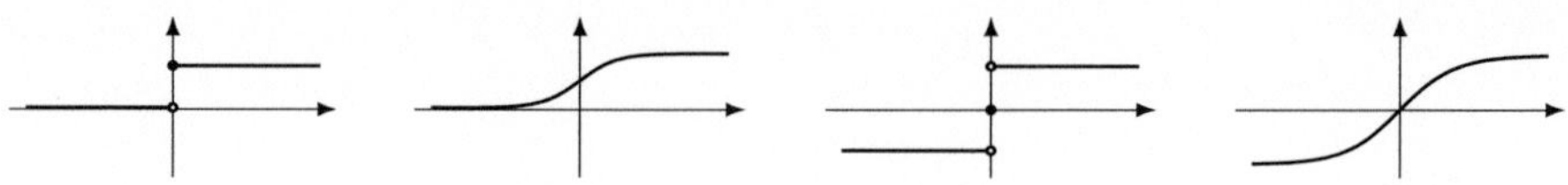

Heaviside	sigmoid	signum	tanh
$\mathbb{1}_{[0,\infty)}$	$\mathrm{sig}(t) = \frac{1}{1+\mathrm{e}^{-t}}$		$\frac{\mathrm{e}^t - \mathrm{e}^{-t}}{\mathrm{e}^t + \mathrm{e}^{-t}}$

A Heaviside-activated neuron outputs 1 if $\langle w, x\rangle \geqslant -b$, otherwise it outputs 0. A smooth variant of the latter is given by the sigmoid function. In fact, the logistic regressors that we considered in Chapter 2 are examples of sigmoid-activated neurons. The perceptron, and thus also every support vector machine, see Chapters 13 and 14, is on the other hand an example of a signum-activated neuron. A smooth version of the signum function is, for example, the hyperbolic tangent.

In addition to the aforementioned activation functions, in this chapter we will particularly consider the *Rectified Linear Unit (ReLU)* and its bounded version (see below), which we will denote with $\overline{\mathrm{ReLU}}$.

ReLU	$\overline{\mathrm{ReLU}}$
$\max(0, t)$	$\min(\mathrm{ReLU}, 1)$

In contrast to the previous chapters, in which we used individual neurons as predictors, we will now combine them into *neural networks*, i.e., functions of the type in Definition 16.1 are composed or added up in a suitable way to obtain a classifier or regressor. This idea is also inspired by nature, since in the human brain neurons can connect through *synapses*. In an artificial neural network, each neuron has its own weights and its own bias, but all neurons in a network usually have the same activation function.

In the following Subchapter 16.1, we will first focus on the question of which functions a neural network can represent or approximate in a sense yet to be specified. This is often referred to as the *expressivity* of a network or an entire class of networks.

16.1 Expressivity

In the above-mentioned question of which functions can be represented or approximated by neural networks of a certain type, it is useful to distinguish between classification problems and regression problems. We start with the following discrete results, which are more suitable for classification problems.

Proposition 16.2. *With $\sigma = \mathbb{1}_{[0,\infty)}$, the logical functions $\wedge$, $\vee$, $\neg$ can each be represented by a neuron.*

Proof. Indeed, we have

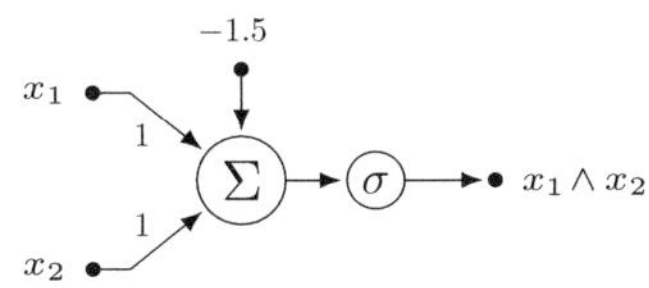

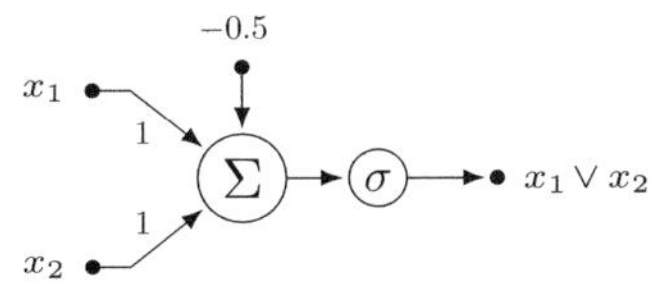

and

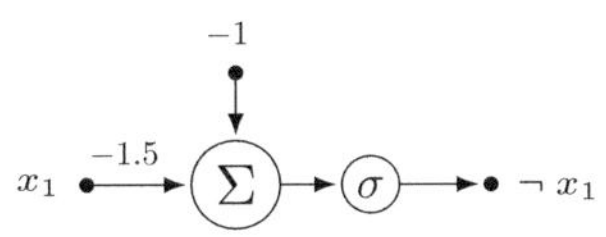

as claimed. □

Remark 16.3. In contrast to Proposition 16.2, the exclusive or function XOR: $\{0,1\}^2 \to \{0,1\}$ cannot be represented by a single neuron, see Exercise 16.1. Via the formula

$$\mathrm{XOR}(x_1, x_2) = (x_1 \vee x_2) \wedge \neg(x_1 \wedge x_2)$$

we can, however, assemble neurons from Proposition 16.2

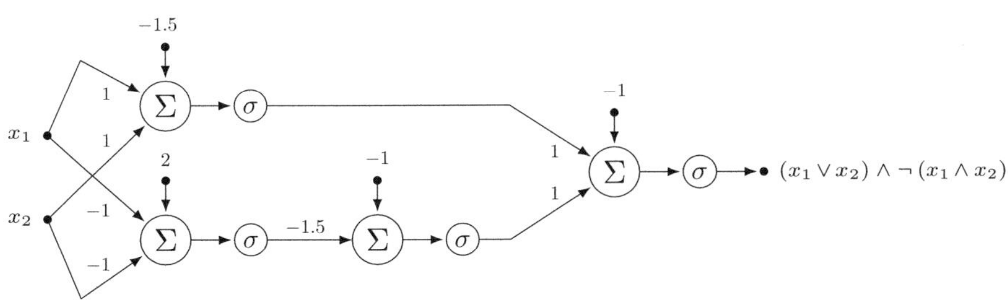

and obtain XOR using four neurons. In Exercise 16.1, we will show that a representation with only three neurons is also possible.

From the above, it directly follows that an arbitrary function $f\colon \{0,1\}^d \to \{0,1\}$, e.g., via its conjunctive normal form

$$N(z_1,\dots,z_d) = \bigwedge_{f(x_i)=0} y_{i1} \vee \cdots \vee y_{id}, \quad \text{where } y_{ik} = \begin{cases} \neg\, z_k, & \text{if } x_{ik} = 1, \\ z_k, & \text{otherwise,} \end{cases}$$

can be represented by a Heaviside-activated neural network. A concrete example can be found in Exercise 16.2. As we have already observed in our perceptron chapter, see Remark 13.14, it is sometimes advantageous to use the values $\{-1,1\}$ instead of $\{0,1\}$ for both the features and the labels, and to replace the Heaviside function with the signum function accordingly. If we do this, we obtain the following, very elegant, representation of the above-discussed "boolean functions".

Theorem 16.4. *Every function $f\colon \{-1,1\}^d \to \{-1,1\}$ can be represented by a neural network of the form*

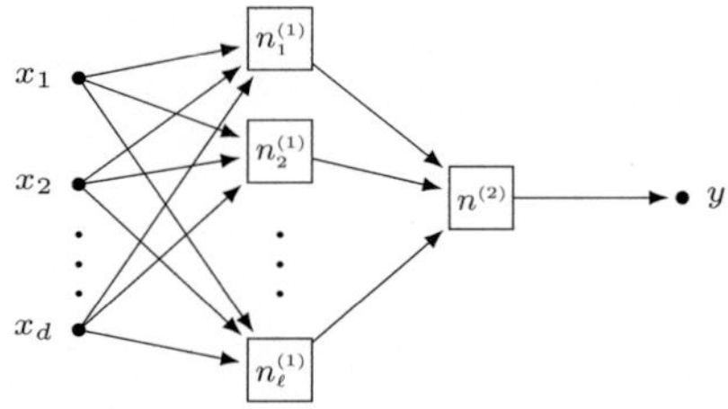

where the $n_i^{(k)}$ are signum-activated neurons.

Proof. Let $w_i \in \{-1,1\}^d$, $i = 1,\dots,\ell$, be those elements for which $f(w_i) = 1$ holds. Then we obtain

$$f = n^{(2)} \circ \begin{bmatrix} n_1^{(1)} \\ \vdots \\ n_\ell^{(1)} \end{bmatrix}$$

with $n^{(2)} = \operatorname{sgn}(\langle \mathbb{1}, \cdot\rangle + \ell - 1.5)$ and $n_i^{(1)} = \operatorname{sgn}(\langle w_i, \cdot\rangle - d + 1)$ for $i = 1,\dots,\ell$. In fact, we have for $x = w_i$:

$$n_i^{(1)}(x) = \operatorname{sgn}(\langle w_i, w_i\rangle - d + 1) = \operatorname{sgn}\big(\sum_{j=1}^{d} w_{ij}^2 - d + 1\big) = 1.$$

For $x = (x_1,\dots,x_d) \neq w_i$ there is at least one j_0 such that $x_{j_0} \neq w_{ij_0}$, and therefore $x_{j_0} \cdot w_{ij_0} = -1$. For such $x \in \{-1,1\}^d$ it follows

$$n_i^{(1)}(x) = \operatorname{sgn}\big(w_{ij_0}x_{j_0} + \sum_{\substack{j=1 \\ j\neq j_0}}^{d} w_{ij}x_j - d + 1\big) = -1,$$

since we can estimate the argument of the signum function from above by -1. For any $x \in \{-1,1\}^d$, the so-called first layer of the neural network thus yields a vector $(n_1^{(1)}(x), \ldots, n_\ell^{(1)}(x)) \in \{-1,1\}^\ell$, whose i-th entry has value 1 if and only if $x = w_i$ holds. It follows

$$\left\langle \begin{bmatrix} 1 \\ \vdots \\ 1 \end{bmatrix}, \begin{bmatrix} n_1^{(1)}(x) \\ \vdots \\ n_\ell^{(1)}(x) \end{bmatrix} \right\rangle = \sum_{i=1}^{\ell} n_i^{(1)}(x) = \begin{cases} -\ell + 2, & \text{if } \exists\, i \in \{1, \ldots, \ell\} \colon x = w_i, \\ -\ell, & \text{otherwise.} \end{cases}$$

Since the w_i are by construction precisely those elements of $\{-1,1\}^d$ for which $f(w_i) = 1$, we obtain

$$n^{(2)}\left(\begin{bmatrix} n_1^{(1)}(x) \\ \vdots \\ n_\ell^{(1)}(x) \end{bmatrix} \right) = \operatorname{sgn}\left(\left\langle \begin{bmatrix} 1 \\ \vdots \\ 1 \end{bmatrix}, \begin{bmatrix} n_1^{(1)}(x) \\ \vdots \\ n_\ell^{(1)}(x) \end{bmatrix} \right\rangle + \ell - 1.5 \right) = f(x)$$

for every $x \in \{-1,1\}^d$, as claimed. □

We note that for binary classifiers, it makes sense to use a neural network with *architecture* as in Theorem 16.4, i.e., a *hidden layer* is followed by an *output layer* with only one neuron, which ensures that the network becomes $\{0,1\}$-valued when working with Heaviside activation. Alternatively, with signum activation, for all neurons, or only for $n^{(2)}$, a $(0,1)$-valued network is obtained. In the latter case, the output can be interpreted as a probability, cf. Exercise 16.5. We will later also discuss classifiers in the context of multidimensional labels, see Theorem 16.34.

Now we study expressivity in an analytic context. We want to use neural networks as regressors and, in order to do so, we firstly consider so-called *shallow neural networks.* Their architecture will be very similar to that considered in Theorem 16.4, but we have to change something about the *output neuron* as the latter restricts heavily what kind of functions can be approximated by the network. So instead of feeding the outputs of the hidden layer into a single neuron $n^{(2)}$, we will allow linear combination of their values via an *output matrix*.

Definition 16.5. A function $N \colon \mathbb{R}^d \to \mathbb{R}^m$ of the form

$$N = \begin{bmatrix} a_{11} & \cdots & a_{1\ell} \\ \vdots & & \vdots \\ a_{m1} & \cdots & a_{md} \end{bmatrix} \begin{bmatrix} n_1 \\ \vdots \\ n_\ell \end{bmatrix}$$

with neurons $n_1, \ldots, n_\ell$ and a matrix $(a_{ij}) \in \mathbb{R}^{\ell \times m}$ is called a *shallow, fully connected, feedforward network of width ℓ, with linear output and activation $\sigma \colon \mathbb{R} \to \mathbb{R}$.* Such a network can be illustrated as follows:

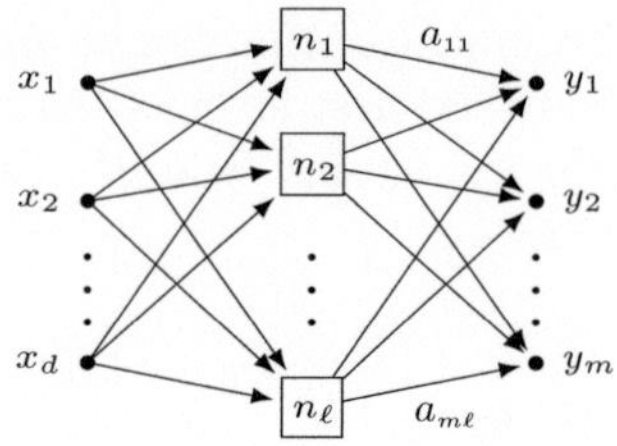

We denote by $\mathcal{S}^{\sigma,\ell}(\mathbb{R}^d, \mathbb{R}^m)$ the set of all neural networks of the above form and put

$$\mathcal{S}^{\sigma}(\mathbb{R}^d, \mathbb{R}^m) := \bigcup_{\ell=1}^{\infty} \mathcal{S}^{\sigma,\ell}(\mathbb{R}^d, \mathbb{R}^m).$$

In short, the elements of $\mathcal{S}^{\sigma}(\mathbb{R}^d, \mathbb{R}^m)$ are referred to as *shallow networks*. We put $\mathcal{S}^{\sigma,\ell}(\mathbb{R}^d) := \mathcal{S}^{\sigma,\ell}(\mathbb{R}^d, \mathbb{R})$ and $\mathcal{S}^{\sigma}(\mathbb{R}^d) := \mathcal{S}^{\sigma}(\mathbb{R}^d, \mathbb{R})$. If we read the application of the activation function coordinatewise, a shallow network $N \in \mathcal{S}^{\sigma}(\mathbb{R}^d, \mathbb{R}^m)$ can be represented by the formula

$$N(x) = A\,\sigma(Wx + b)$$

with $W \in \mathbb{R}^{\ell \times d}$, $b \in \mathbb{R}^{\ell}$ and $A \in \mathbb{R}^{\ell \times m}$. Here, the i-th row of W contains the weights of the neuron n_i.

We now want to investigate for which activation functions σ, every continuous function $f\colon \mathbb{R}^d \to \mathbb{R}^m$ can be approximated by a shallow neural network, in the sense of uniform convergence on compact sets. We note that this is on the one hand a very natural concept of convergence that appears already in foundational real analysis courses; for example, the series expansions of sin, cos and exp converge exactly in this sense. On the other hand, datasets often arise naturally as subsets of some compact set, for instance because, for features representing physical measurements, an upper and lower bound is known a priori. We then obtain uniform convergence on this compact set and thus a very strong approximation statement.

Definition 16.6. A subset $\mathcal{S} \subseteq \mathrm{C}(\mathbb{R}^d, \mathbb{R}^m)$ of the space of continuous functions from $\mathbb{R}^d$ to $\mathbb{R}^m$ is called *dense*, or more precisely *uniformly dense on compact sets*, if for any norm $\|\cdot\|_{\mathbb{R}^m}$ on $\mathbb{R}^m$ we have

$$\forall\, F \in \mathrm{C}(\mathbb{R}^d, \mathbb{R}^m),\ \Omega \subseteq \mathbb{R}^d \text{ compact},\ \varepsilon > 0\ \exists\, S \in \mathcal{S}:$$
$$\|F - S\|_{\Omega,\infty} := \sup_{x \in \Omega} \|F(x) - S(x)\|_{\mathbb{R}^m} < \varepsilon.$$

If $\sigma\colon \mathbb{R} \to \mathbb{R}$ is continuous, then obviously $\mathcal{S}^{\sigma}(\mathbb{R}^d, \mathbb{R}^m) \subseteq \mathrm{C}(\mathbb{R}^d, \mathbb{R}^m)$, and we can ask whether this is a dense subspace in the sense of Definition 16.6. We give two examples.

Example 16.7. (i) Let $P \in \mathbb{R}[X]$ be a polynomial. If we write out $\mathcal{S}^P(\mathbb{R})$ explicitly, we immediately see that this space is finite-dimensional. For compact $\Omega \subseteq \mathbb{R}$, $\mathcal{S}^P(\Omega) \subseteq (\mathrm{C}(\Omega), \|\cdot\|_{\Omega,\infty})$ is thus closed and not dense.

(ii) Let $\exp\colon \mathbb{R} \to \mathbb{R}$ be the exponential function. Then $\mathcal{S}^{\exp}(\mathbb{R}) \subseteq \mathrm{C}(\mathbb{R})$ is uniformly dense on compact sets. This follows by applying the Stone-Weierstrass Theorem 16.10, see Exercise 16.3.

The next proposition allows us to reduce to the scalar-valued case in the remainder of this chapter.

Proposition 16.8. *Let $\sigma\colon \mathbb{R} \to \mathbb{R}$ be continuous. If $\mathcal{S}^\sigma(\mathbb{R}^d) \subseteq \mathrm{C}(\mathbb{R}^d)$ is uniformly dense on compact sets, then $\mathrm{S}^\sigma(\mathbb{R}^d, \mathbb{R}^m) \subseteq \mathrm{C}(\mathbb{R}^d, \mathbb{R}^m)$ is uniformly dense on compact sets, too.*

Proof. Given $F \in \mathrm{C}(\mathbb{R}^d, \mathbb{R}^m)$, we approximate each coordinate function F_i, according to the assumption, with some $N_i \in \mathcal{S}^\sigma(\mathbb{R}^d)$. Now we stack the N_i as shown in the following picture for $m = 2$. The lines not drawn below correspond to zeros in the output matrix A.

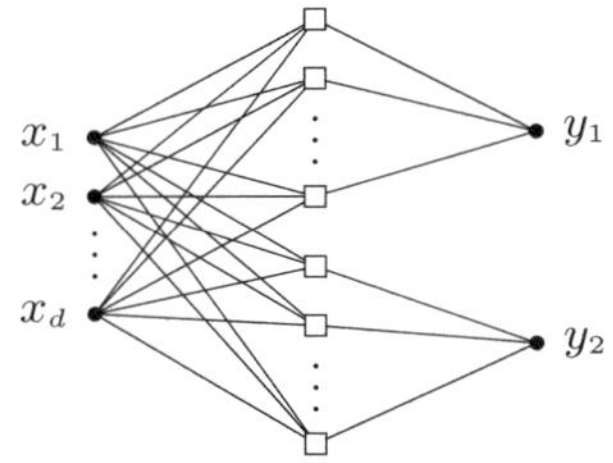

In this way, we obtain a shallow network N that approximates F in the desired sense. □

Next, we want to reduce to one-dimensional domains as well. This requires some preparation.

Lemma 16.9. *Let $\mathcal{P} := \{\langle\,\cdot\,, a\rangle^r\colon \mathbb{R}^d \to \mathbb{R} \mid r \in \mathbb{N}_0,\, a \in \mathbb{R}^d\}$. Then* $\operatorname{span}\mathcal{P} \subseteq \mathrm{C}(\mathbb{R}^d)$ *is uniformly dense on compact sets.*

Proof. We need the following version of the Stone-Weierstrass theorem. At the end of the chapter, we will provide detailed references.

Theorem 16.10. (Stone-Weierstrass, version 1) *Let $\Omega \subseteq \mathbb{R}^d$ be compact and let $A \subseteq \mathrm{C}(\Omega)$ be a subset such that the following holds:*

(i) $A \subseteq \mathrm{C}(\Omega)$ is a subalgebra,

(ii) $\mathbb{1}_\Omega \in A$,

(iii) $A \subseteq \mathrm{C}(\Omega)$ separates points, i.e., $\forall\, x \neq y \in \Omega\; \exists\, g \in A\colon g(x) \neq g(y)$.

Then $A \subseteq \mathrm{C}(\Omega)$ is dense with respect to the supremum norm. ◇

We fix a compact set $\Omega \subseteq \mathbb{R}^d$. Since the elements of $\mathcal{P}$ are uniquely determined by the parameters a and r, we can consider $A := \operatorname{span} \mathcal{P} \subseteq \mathrm{C}(\Omega)$ as a subset, and we now have to check the conditions in Theorem 16.10. Since A is by definition a subspace, only the closure under pointwise multiplication needs to be shown to establish (i). For this, let $a, b \in \mathbb{R}^d$ and $r, s \in \mathbb{N}_0$. We choose pairwise different $\beta_0, \ldots, \beta_{r+s} \in \mathbb{R}$ and put

$$c = \begin{bmatrix} c_0 \\ \vdots \\ c_{r+s} \end{bmatrix} := \binom{r+s}{s}^{-1} \underbrace{\begin{bmatrix} \beta_0^0 & \cdots & \beta_{r+s}^0 \\ \vdots & & \vdots \\ \beta_0^{r+s} & \cdots & \beta_{r+s}^{r+s} \end{bmatrix}}_{=:B}^{-1} \begin{bmatrix} 0 \\ \vdots \\ 0 \\ 1 \\ 0 \\ \vdots \\ 0 \end{bmatrix},$$

where B is invertible as a transposed Vandermonde matrix. Denoting by δ_{ij} the Kronecker symbol, i.e., $\delta_{ij} = 1$ for $i = j$ and $\delta_{ij} = 0$ for $i \neq j$, we obtain for $0 \leqslant i \leqslant r + s$ first

$$\sum_{j=0}^{r+s} \beta_j^i c_j = (Bc)_i = \binom{r+s}{s}^{-1} \delta_{is} = \binom{r+s}{i}^{-1} \delta_{is},$$

and thus

$$\delta_{is} = \binom{r+s}{i} \sum_{j=0}^{r+s} \beta_j^i c_j.$$

For $x \in \mathbb{R}^d$ we now calculate

$$\begin{aligned}
\langle x, a\rangle^r \langle x, b\rangle^s &= \sum_{i=0}^{r+s} \delta_{is} \langle x, a\rangle^{r+s-i} \langle x, b\rangle^i \\
&= \sum_{i=0}^{r+s} \Big[\binom{r+s}{i} \sum_{j=0}^{r+s} \beta_j^i c_j\Big] \langle x, a\rangle^{r+s-i} \langle x, b\rangle^i \\
&= \sum_{j=0}^{r+s} c_j \Big[\sum_{i=0}^{r+s} \binom{r+s}{i} \beta_j^i c_j \langle x, a\rangle^{r+s-i} \langle x, b\rangle^i\Big] \\
&= \sum_{j=0}^{r+s} c_j \big[\langle x, a\rangle + \beta_j \langle x, b\rangle\big]^{r+s} \\
&= \sum_{j=0}^{r+s} c_j \langle x, a + \beta_j b\rangle^{r+s},
\end{aligned}$$

from whence it follows that $\langle \cdot, a\rangle^r \langle \cdot, b\rangle^s \in A$ holds. The subspace A is therefore indeed a subalgebra. By choosing $r = 0$, it further follows that $\mathbb{1}_\Omega \in A$ and for $x \neq y$ in Ω, for example, $\langle \cdot, x - y\rangle$ maps these points to different values, as otherwise $\|x - y\|^2 = \langle x - y, x - y\rangle = 0$ would hold. Hence, A separates points.

Given $F \in \mathrm{C}(\mathbb{R}^d)$ and $\varepsilon > 0$, we consider $f := F|_\Omega$, and by Theorem 16.10 there exists a function $g \in A$ with $\|F - g\|_{\Omega,\infty} < \varepsilon$. It is enough now to read g as a function on $\mathbb{R}^d$. □

Lemma 16.11. *Let $\mathcal{F} \subseteq \mathrm{C}(\mathbb{R})$ be uniformly dense on compact sets. Then $\mathcal{G} := \operatorname{span}\{f(\langle w, \cdot\rangle) \mid w \in \mathbb{R}^d,\ f \in \mathcal{F}\} \subseteq \mathrm{C}(\mathbb{R}^d)$ is uniformly dense on compact sets.*

Proof. Let $F \in \mathrm{C}(\mathbb{R}^d)$, $\Omega \subseteq \mathbb{R}^d$ be compact, and let $\varepsilon > 0$ be given. Let $\mathcal{P}$ be defined as in Lemma 16.9. Then there exist $w_1, \dots, w_m \in \mathbb{R}^d$, $r_1, \dots, r_m \in \mathbb{N} \cup \{0\}$ and $\alpha_1, \dots, \alpha_m \in \mathbb{R}$, such that

$$\Big\|F - \sum_{j=1}^{m} \alpha_j \langle w_j, \cdot\rangle^{r_j}\Big\|_{\Omega,\infty} < \varepsilon/2$$

holds. For fixed $1 \leqslant j \leqslant k$ we define $K_j := \{\langle w_j, x\rangle \mid x \in \Omega\}$. Since $K_j \subseteq \mathbb{R}$ is compact, we can choose $f_{j1}, \dots, f_{j\ell} \in \mathcal{F}$ and $\alpha_{j1}, \dots, \alpha_{j\ell} \in \mathbb{R}$ with

$$\Big\|(\cdot)^{r_j} - \sum_{k=1}^{\ell} \alpha_{jk} f_{jk}\Big\|_{K_j,\infty} < \frac{\varepsilon}{2\|\alpha\|_1}.$$

By construction we obtain

$$\Big\|\langle w_j, \cdot\rangle^{r_j} - \sum_{k=1}^{\ell} \alpha_{jk} f_{jk}(\langle w_j, \cdot\rangle)\Big\|_{\Omega,\infty} < \frac{\varepsilon}{2\|\alpha\|_1},$$

where the linear combination of the $f_{jk}(\langle w_j, \cdot\rangle)$ belongs to $\mathcal{G}$ by definition. Consequently, we have

$$g := \sum_{j=1}^{m} \alpha_j \sum_{k=1}^{\ell} \alpha_{jk} f_{jk}(\langle w_j, \cdot\rangle) \in \mathcal{G}$$

and we obtain the estimate

$$\begin{aligned}
\|F - g\|_{\Omega,\infty} &\leqslant \Big\|F - \sum_{j=1}^{m} \alpha_j \langle w_j, \cdot\rangle^{r_j}\Big\|_{\Omega,\infty} + \Big\|\sum_{j=1}^{m} \alpha_j \langle w_j, \cdot\rangle^{r_j} - g\Big\|_{\Omega,\infty} \\
&< \varepsilon/2 + \sum_{j=1}^{m} |\alpha_j| \Big\|\langle w_j, \cdot\rangle^{r_j} - \sum_{k=1}^{\ell} \alpha_{jk} f_{jk}(\langle w_j, \cdot\rangle)\Big\|_{\Omega,\infty} \\
&< \varepsilon/2 + (\varepsilon/2) \sum_{j=1}^{m} |\alpha_j|/\|\alpha\|_1 \;=\; \varepsilon
\end{aligned}$$

as desired. □

With the two Lemmas 16.9 and 16.11, we are now able to prove the following reduction result for the domain.

Theorem 16.12. (Chui-Li-Lin-Pinkus) *Let $\sigma\colon \mathbb{R} \to \mathbb{R}$ be continuous and assume that the shallow networks $\mathcal{S}^\sigma(\mathbb{R}) \subseteq \mathrm{C}(\mathbb{R})$ are uniformly dense on compact sets. Then $\mathcal{S}^\sigma(\mathbb{R}^d) \subseteq \mathrm{C}(\mathbb{R}^d)$ is uniformly dense on compact sets as well.*

Proof. By definition, the elements $f \in \mathcal{S}^\sigma(\mathbb{R})$ are of the form $f(x) = \sigma(vx + b)$ with $v, b \in \mathbb{R}$. Thus we obtain

$$\begin{aligned}
\mathcal{S}^\sigma(\mathbb{R}^d) &= \Big\{ \sum_{i=1}^{\ell} a_i\, \sigma(\langle w_i, \cdot \rangle + b_i) \;\Big|\; a_i \in \mathbb{R},\ w_i \in \mathbb{R}^d,\ b_i \in \mathbb{R},\ \ell \geqslant 1 \Big\} \\
&= \operatorname{span}\big\{ \sigma(\langle w, \cdot \rangle + b) \;\big|\; w \in \mathbb{R}^d,\ b \in \mathbb{R} \big\} \\
&= \operatorname{span}\big\{ \sigma(v \langle w, \cdot \rangle + b) \;\big|\; v \in \mathbb{R},\ w \in \mathbb{R}^d,\ b \in \mathbb{R} \big\} \\
&\underset{\substack{\uparrow \\ f = \sigma(v\,\cdot + b) \\ v, b \in \mathbb{R}}}{=} \operatorname{span}\big\{ f(\langle w, \cdot \rangle) \mid w \in \mathbb{R}^d,\ f \in \mathcal{S}^\sigma(\mathbb{R}) \big\}.
\end{aligned}$$

The last set is equal to $\mathcal{G}$ as considered in Lemma 16.11 and, by applying Lemma 16.11, we see that it is dense in $\mathrm{C}(\mathbb{R}^d)$. □

Based on Proposition 16.8 and Theorem 16.12, we may restrict ourselves, without loss of generality, to the case $d = m = 1$, when asking whether the shallow networks $\mathcal{S}^\sigma(\mathbb{R}^d, \mathbb{R}^m) \subseteq \mathrm{C}(\mathbb{R}^d, \mathbb{R}^m)$ are uniformly dense on compact sets in the space the continuous functions. In order to answer this question, we have to introduce another concept.

Definition 16.13. For compact $\Omega \subseteq \mathbb{R}$, we denote by $\mathrm{M}(\Omega)$ the space of complex Borel measures and call a measurable function $\sigma\colon \mathbb{R} \to \mathbb{R}$ *discriminatory* if, for every compact set $\Omega \subseteq \mathbb{R}$ and for $\mu \in \mathrm{M}(\Omega)$, the following implication holds:

$$\Big(\forall\, w, b \in \mathbb{R}\colon \int_\Omega \sigma(wx + b)\, \mathrm{d}\mu(x) = 0 \Big) \implies \mu = 0.$$

The next two lemmas are the final ingredients for the subsequent approximation theorems with continuous activation functions.

Lemma 16.14. *If the Fourier transform $\hat{\mu}(s) = \int_{\mathbb{R}^d} \mathrm{e}^{-ist}\, \mathrm{d}\mu(t)$ of a measure $\mu \in \mathrm{M}(\mathbb{R})$ is equal to zero for every $s \in \mathbb{R}$, then already the measure is zero, i.e., $\mu = 0$.*

Proof. ① For $\varphi \in \mathrm{L}^1(\mathbb{R}, \mathbb{C})$ we consider the "normal" Fourier transform

$$[\mathcal{F}\varphi](\xi) \equiv \hat{\varphi}(\xi) := \int_{\mathbb{R}} \mathrm{e}^{-ix\xi}\, \varphi(x)\, \mathrm{d}x,$$

for which, according to the Riemann-Lebesgue lemma, $\hat{f} \in \mathrm{C}_0(\mathbb{R}, \mathbb{C})$ holds. We consider the set $A := \{\hat{f} \mid f \in \mathrm{L}^1(\mathbb{R}, \mathbb{C})\} \subseteq \mathrm{C}_0(\mathbb{R}, \mathbb{C})$ and show that it is dense in $\mathrm{C}_0(\mathbb{R}, \mathbb{C})$ by using the following version of the Stone-Weierstrass theorem.

Theorem 16.15. (Stone-Weierstrass, version 2) *Let A be a subset of the space $\mathrm{C}_0(\mathbb{R}, \mathbb{C}) := \{f \in \mathrm{C}(\mathbb{R}, \mathbb{C}) \mid \lim_{|x|\to\infty} f(x) = 0\}$ of continuous functions that vanish at infinity. Assume that $A \subseteq \mathrm{C}_0(\mathbb{R}, \mathbb{C})$*

(i) is a subalgebra,

(ii) separates points, i.e., $\forall\, x \neq y \in \mathbb{R}\; \exists\, g \in A\colon g(x) \neq g(y)$,

(iii) vanishes nowhere, i.e., $\forall\, x \in \mathbb{R}\; \exists\, g \in A\colon g(x) \neq 0$.

(iv) is closed under complex conjugation, i.e., $\forall\, g \in A\colon \overline{g} \in A$.

Then $A \subseteq \mathrm{C}_0(\mathbb{R}, \mathbb{C})$ is dense with respect to the supremum norm. ◇

Since the Fourier transform is linear, it follows from the easily verifiable and well-known identity $\mathcal{F}(f_1 * f_2) = \mathcal{F}f_1 \cdot \mathcal{F}f_2$ that A is a subalgebra. Fourier transformation of $f(x) := \mathrm{e}^{-x}\, \mathbb{1}_{[0,\infty)}(x)$ yields

$$\hat{f}(\xi) = \int_0^\infty \mathrm{e}^{-ix\xi}\, \mathrm{e}^{-x}\, \mathrm{d}x = \mathrm{e}^{-(i\xi+1)x}\, \frac{1}{i\xi+1}\Big|_0^\infty = \frac{1}{i\xi+1},$$

where we see that $\hat{f}(\xi) \neq \hat{f}(\zeta)$ holds for all $\xi \neq \zeta$, and $\hat{f}(\xi) \neq 0$ for every $\xi \in \mathbb{R}$. Thus, A vanishes nowhere and separates points. With another well-known identity, namely $[\mathcal{F}\overline{f}](\xi) = \overline{[\mathcal{F}f]}(-\xi)$ for $\xi \in \mathbb{R}$, we obtain that A is closed under complex conjugation.

② We now denote by $\mathrm{C}_0(\mathbb{R}, \mathbb{C})'$ the dual space of $\mathrm{C}_0(\mathbb{R}, \mathbb{C})$, i.e., the space of all linear and continuous functionals $\varphi\colon \mathrm{C}_0(\mathbb{R}, \mathbb{C}) \to \mathbb{C}$, equipped with the operator norm. We denote by $\mathrm{M}(\Omega)$ the space of complex, regular Borel measures whose total variation norm is finite. To proceed in the proof we need the following result from functional analysis:

Theorem 16.16. (Riesz representation theorem, version 1) *The map $T\colon \mathrm{M}(\mathbb{R}) \to \mathrm{C}_0(\mathbb{R}, \mathbb{C})'$, $[T\mu](f) = \int_{\mathbb{R}} f\, \mathrm{d}\mu$, is an isomorphism of Banach spaces.* ◇

To prove Lemma 16.14, let now $\mu \in \mathrm{M}(\mathbb{R})$, with $\hat{\mu} = 0$, be given. Let T be the isomorphism from Theorem 16.16. Then $T\mu \in \mathrm{C}_0(\mathbb{R}, \mathbb{C})'$ and we claim that this functional vanishes on A. To this end, let $f \in \mathrm{L}^1(\mathbb{R}, \mathbb{C})$. By Fubini's theorem (for complex measures!) we have

$$\begin{aligned}[T\mu](\hat{f}) &= \int_{\mathbb{R}} \hat{f}(\xi)\, \mathrm{d}\mu(\xi) = \int_{\mathbb{R}} \int_{\mathbb{R}} \mathrm{e}^{-ix\xi}\, f(x)\, \mathrm{d}x\, \mathrm{d}\mu(\xi) \\ &= \int_{\mathbb{R}} f(x) \int_{\mathbb{R}} \mathrm{e}^{-ix\xi}\, \mathrm{d}\mu(\xi)\, \mathrm{d}x = \int_{\mathbb{R}} f(x)\hat{\mu}(\xi)\, \mathrm{d}x = 0.\end{aligned}$$

According to part ①, $A \subseteq C_0(\mathbb{R}, \mathbb{C})$ is dense, and it thus follows $[T\mu](f) = 0$ for every $f \in C_0(\mathbb{R}, \mathbb{C})$. But then $T\mu = 0$, and since T is an isomorphism, we arrive at $\mu = 0$. □

Lemma 16.17. *Let $\sigma\colon \mathbb{R} \to \mathbb{R}$ be continuous and discriminatory. Then $\mathcal{S}^\sigma(\mathbb{R}) \subseteq C(\mathbb{R})$ is uniformly dense on compact sets.*

Proof. For the proof we need the following consequence of the Hahn-Banach theorem, which we note without proof.

Theorem 16.18. (Corollary to the Hahn-Banach theorem) *Let X be a normed space, let $Y \subseteq X$ be a closed subspace and let $x \in X \backslash Y$. Then there exists a linear and continuous map $\varphi\colon X \to \mathbb{R}$ with $\varphi|_Y = 0$ and $\varphi(x) \neq 0$.* ◇

Assume that $\mathcal{S}^\sigma(\mathbb{R}) \subseteq C(\mathbb{R})$ is not uniformly dense on compact sets. Then there exists a compact set $\Omega \subseteq \mathbb{R}$, such that $\overline{\mathcal{S}^\sigma(\Omega)} \subset C(\Omega)$ is a closed proper subspace. According to Theorem 16.18, we choose a functional $0 \neq \varphi \in C(\Omega)'$, that vanishes on $\mathcal{S}^\sigma(\Omega)$. Now we apply the following slightly different variant of the Riesz representation theorem.

Theorem 16.19. (Riesz representation theorem, version 2) *Let $\Omega \subseteq \mathbb{R}$ be compact. Then the map $T\colon M(\Omega) \to C(\Omega, \mathbb{C})'$, $[T\mu](f) = \int_\Omega f \,\mathrm{d}\mu$, is an isomorphism of Banach spaces.* ◇

We choose $\mu \in M(\Omega)$ with $T\mu = \varphi$ according to Theorem 16.19. That is, it holds

$$\forall\, f \in C(\Omega)\colon \varphi(f) = \int_\Omega f(x) \,\mathrm{d}\mu(x).$$

We specialize $f := \sigma(w \cdot + b) \in \mathcal{S}^\sigma(\Omega)$ and obtain

$$\forall\, w, b \in \mathbb{R}\colon \int_\Omega \sigma(wx + b) \,\mathrm{d}\mu(x) = \varphi(f) = 0.$$

Since σ is discriminatory, we get $\mu = 0$ and thus $\varphi = 0$. Contradiction. □

After many preparations, we are now ready to prove our first approximation theorem. Note that this in particular covers the "S-shaped" activation functions tanh, sigmoid and ReLU from the beginning of the chapter, but of course can be applied far beyond these three examples.

Theorem 16.20. (Cybenko) *Let $\sigma\colon \mathbb{R} \to \mathbb{R}$ be continuous and assume that f has finite limits $\ell := \lim_{t\to-\infty} \sigma(t) \neq \lim_{t\to+\infty} \sigma(t) =: r$ at $\pm\infty$. Then $\mathcal{S}^\sigma(\mathbb{R}^d, \mathbb{R}^m) \subseteq C(\mathbb{R}^d, \mathbb{R}^m)$ is uniformly dense on compact sets. More precisely, the following holds for every norm $\|\cdot\|_{\mathbb{R}^m}$ on $\mathbb{R}^m$:*

$$\forall\ F \in C(\mathbb{R}^d, \mathbb{R}^m),\ \Omega \subseteq \mathbb{R}^d \text{ compact},\ \varepsilon > 0\ \exists\, N \in \mathcal{S}^\sigma(\mathbb{R}^d, \mathbb{R}^m)\colon$$

$$\|F - N\|_{\Omega,\infty} = \sup_{x\in\Omega} \|F(x) - N(x)\|_{\mathbb{R}^m} < \varepsilon.$$

Proof. ① We first show that we may assume $\ell = 0$ without loss of generality. To this end, we claim that for any discriminatory, non-constant function $\sigma\colon \mathbb{R} \to \mathbb{R}$, and arbitrary $\ell \in \mathbb{R}$, $\sigma + \ell$ is discriminatory, too. Let $\Omega \subseteq \mathbb{R}$ be compact and let $\mu \in \mathrm{M}(\Omega)$ be such that

$$\forall\, w \in \mathbb{R}^n,\, b \in \mathbb{R}\colon \int_\Omega \sigma(\langle w, x\rangle + b) + \ell \,\mathrm{d}\mu(x) = 0$$

holds. We specialize in the above condition $w = 0$, and use the fact that the integrand then becomes constant. It follows

$$\forall\, b \in \mathbb{R}\colon (\sigma(b) + \ell) \cdot \mu(\Omega) = 0.$$

Because σ is not constant by assumption, we find $b \in \mathbb{R}$ such that $\sigma(b) + \ell \neq 0$. From the above, it follows $\mu(\Omega) = 0$. (Notice that this does not yet imply that $\mu = 0$, as the measure $\mu \in \mathrm{M}(\Omega)$ may not be positive!) If we use the very first condition again, we get

$$\begin{aligned} \forall\, w,\, b \in \mathbb{R}\colon 0 &= \int_\Omega \sigma(wx + b) + \ell \,\mathrm{d}\mu(x) \\ &= \int_\Omega \sigma(wx + b) \,\mathrm{d}\mu(x) + \ell \cdot \mu(\Omega) \\ &= \int_\Omega \sigma(wx + b) \,\mathrm{d}\mu(x). \end{aligned}$$

Since σ is discriminatory, this now implies $\mu = 0$, as desired.

② Let σ be as in the theorem; by part ① we assume without loss of generality that $\ell = 0$ holds. For compact $\Omega \subseteq \mathbb{R}$ and $\mu \in \mathrm{M}(\Omega)$, assume that

$$\forall\, w \in \mathbb{R}^d,\, b \in \mathbb{R}\colon \int_\Omega \sigma(wx + b) \,\mathrm{d}\mu(x) = 0$$

holds. We claim that $\mu(\Omega \cap I_{w,b}) = 0$ holds for all $w, b \in \mathbb{R}$, where $I_{w,b} := \{x \in \mathbb{R} \mid wx + b > 0\}$ is equal to the interval $(-b/w, \infty)$ for $w > 0$, while we get the interval $(-\infty, -b/w)$ if $w < 0$. For $w = 0$, $I_{w,b}$ is either empty and there is nothing to show, or $I_{w,b} = \mathbb{R}$ and $\mu(\Omega) = 0$ follows immediately with arguments as in ①. For $w, b, a \in \mathbb{R}$ we calculate

$$\lim_{j\to\infty} \sigma\big(j(wx + b) + a\big) = \left\{ \begin{matrix} r, & \text{if } wx + b > 0, \\ \sigma(a), & \text{if } wx + b = 0, \\ 0, & \text{otherwise} \end{matrix} \right\} = r\mathbb{1}_{I_{w,b}} + \sigma(a)\mathbb{1}_{\{-b/w\}}$$

and see that by assumption, for each $j \in \mathbb{N}$,

$$\int_\Omega \sigma(j(wx + b) + a)\,\mathrm{d}\mu(x) = \int_\Omega \sigma(jwx + jb + a)\,\mathrm{d}\mu(x) = 0$$

holds. For fixed $w, b \in \mathbb{R}$, we apply the dominated convergence theorem and obtain

$$\forall\, a \in \mathbb{R}\colon\; r\mu(I_{w,b} \cap \Omega) + \sigma(a)\mu(\{-b/w\} \cap \Omega) = 0.$$

As the first summand is independent of a, and σ is not constant, $\mu(\{-b/w\} \cap \Omega) = 0$ must hold, and we get

$$\forall\, w,\, b \in \mathbb{R}\colon\; \mu(I_{w,b} \cap \Omega) = 0.$$

③ Now we will finally show that $\mu = 0$ holds. For fixed $w \in \mathbb{R}$ we choose a compact interval $J \supseteq \{wx \mid x \in \Omega\}$ and consider

$$\varphi_w\colon\; \mathrm{L}^\infty(J) \to \mathbb{C},\; \varphi_w(f) := \int_\Omega f(wx)\,\mathrm{d}\mu(x).$$

For real $a < b$ with $(a, b] \subseteq \Omega$ we calculate

$$\begin{aligned} \varphi_w(\mathbb{1}_{(a,b]}) &= \int_\Omega \mathbb{1}_{(a,b]}(wx)\,\mathrm{d}\mu(x) \\ &= \int_\Omega \mathbb{1}_{(a,\infty)}(wx)\,\mathrm{d}\mu(x) - \int_\Omega \mathbb{1}_{(b,\infty)}(wx)\,\mathrm{d}\mu(x) \\ &= \mu(\Omega \cap I_{w,-a}) - \mu(\Omega \cap I_{w,-b}) \;=\; 0. \end{aligned}$$

Since the simple functions are dense in $\mathrm{L}^\infty(J)$, it follows that $\varphi_w \equiv 0$. We now consider on the one hand $f \in \mathrm{L}^\infty(J)$, defined by $f(t) = \mathrm{e}^{-it}$, and on the other hand $\mu \in \mathrm{M}(\mathbb{R})$, by extending $\mu \in \mathrm{M}(\Omega)$ by zero. Calculating the Fourier transform $\hat{\mu} \in \mathrm{C_b}(\mathbb{R})$ yields

$$\hat{\mu}(w) = \tfrac{1}{\sqrt{2\pi}} \int_\Omega \mathrm{e}^{-iwx}\,\mathrm{d}\mu(x) = \tfrac{1}{\sqrt{2\pi}}\varphi_w(f) = 0.$$

With Lemma 16.14 it follows $\mu = 0$ and we have proven that σ is discriminatory.

④ Lemma 16.17 finally implies that $\mathcal{S}^\sigma(\mathbb{R}) \subseteq \mathrm{C}(\mathbb{R})$ is uniformly dense on compacts, and Proposition 16.8 and Theorem 16.12 show that the statement of the theorem follows. □

As a corollary of the above theorem and our preparatory work, we obtain an analogous approximation result for the non-truncated ReLU function.

Corollary 16.21. *The subspace $\mathcal{S}^{\mathrm{ReLU}}(\mathbb{R}^d, \mathbb{R}^m) \subseteq \mathrm{C}(\mathbb{R}^d, \mathbb{R}^m)$ is uniformly dense on compact sets.*

Proof. It suffices to show that ReLU is discriminatory. Let $\Omega \subseteq \mathbb{R}$ be compact and let $\mu \in \mathrm{M}(\Omega)$ be such that $\int_\Omega \mathrm{ReLU}(wx+b)\,\mathrm{d}\mu(x) = 0$ holds for all $w, b \in \mathbb{R}$. For arbitrary $w, b \in \mathbb{R}$, we have by definition

$$\overline{\mathrm{ReLU}}(wx+b) = \begin{cases} 0, & \text{if } wx+b<0, \\ wx+b, & \text{if } 0 \leqslant wx+b<1, \\ 1, & \text{if } 1 \leqslant wx+b, \end{cases}$$

which by case analysis shows that

$$\begin{aligned} \overline{\mathrm{ReLU}}(wx+b) &= \max(0, wx+b) - \max(0, wx+b-1) \\ &= \mathrm{ReLU}(wx+b) - \mathrm{ReLU}(wx+b-1) \end{aligned}$$

holds for all $x \in \mathbb{R}$. Applying the assumption of the vanishing integrals twice, we obtain $\int_\Omega \overline{\mathrm{ReLU}}(wx+b)\,\mathrm{d}\mu(x) = 0$, and this holds for arbitrary $w, b \in \mathbb{R}$. Since the proof of Theorem 16.20 has in particular shown that $\overline{\mathrm{ReLU}}$ is discriminatory, it follows that $\mu = 0$. □

Next, we turn to discontinuous activation functions. To address the fact that for such a function σ the inclusion $\mathcal{S}^\sigma(\mathbb{R}^d, \mathbb{R}^m) \subseteq \mathrm{C}(\mathbb{R}^d, \mathbb{R}^m)$ no longer holds, we replace the supremum over Ω in Definition 16.6 with the essential supremum, i.e., we consider

$$\|f\|_{\Omega,\,\mathrm{ess\,sup}} = \operatorname{ess\,sup}_{x\in\Omega} \|f(x)\|_{\mathbb{R}^m} := \inf_{\substack{N\in\Sigma \\ \lambda(N)=0}} \|f|_{\Omega\setminus N}\|_\infty,$$

where λ is the Lebesgue measure on Ω and Σ is the σ-algebra of Borel sets of Ω. With this adjustment, we obtain the following theorem.

Theorem 16.22. (Leshno-Lin-Pinkus-Schocken) *Let $t_0 \in \mathbb{R}$ and let $\sigma \colon \mathbb{R} \to \mathbb{R}$ be continuous on $[-s, t_0) \cup (t_0, s]$ for some $s \in \mathbb{R}$, satisfying also that the limits $\lim_{t\nearrow t_0} \sigma(t) \neq \lim_{t\searrow t_0} \sigma(t)$ exist in $\mathbb{R}$. Then we have:*

$$\begin{aligned} \forall\ F \in \mathrm{C}(\mathbb{R}^d, \mathbb{R}^m),\ \Omega \subseteq \mathbb{R}^d \text{ compact},\ \varepsilon > 0\ \exists\, N \in \mathcal{S}^\sigma(\mathbb{R}^d, \mathbb{R}^m)\colon& \\ \|F - N\|_{\Omega,\,\mathrm{ess\,sup}} < \varepsilon.& \end{aligned}$$

Proof. According to Proposition 16.8 and Theorem 16.12, it is sufficient to consider the case $d = m = 1$. Here we have

$$\mathcal{S}^\sigma(\mathbb{R}) = \operatorname{span}\big\{\sigma(w\,\cdot + b) \mid w, b \in \mathbb{R}\big\}.$$

For $w = \pm 1$ we choose b such that $\sigma(b) \neq 0$. If we then take $1/\sigma(b)$ as a coefficient, we see that $\mathbb{1}_\mathbb{R} \in \mathcal{S}^\sigma(\mathbb{R})$. For $w = 1$ and suitable $a \in \mathbb{R}$, it holds

$$\ell := \lim_{t \nearrow 0} a\sigma(t - t_0) < \lim_{t \searrow 0} a\sigma(t - t_0) =: r,$$

and consequently $\tau \in \mathcal{S}^\sigma(\mathbb{R})$, defined as $\tau(t) := \frac{1}{r-\ell}(a\sigma(t-t_0) - \ell \mathbb{1}_{\mathbb{R}})$, satisfies

$$\lim_{t \nearrow 0} \tau(t) = 0 \quad \text{and} \quad \lim_{t \searrow 0} \tau(t) = 1.$$

We now consider $\mathcal{S}^\tau(\mathbb{R}) \subseteq \mathcal{S}^\sigma(\mathbb{R})$ and get, for $c \in \mathbb{R}$ and $a < b$ in $\mathbb{R}$,

$$\|\tau(w \cdot + c) - \mathbb{1}_{[c,\infty)}\|_{[a,b],\infty} = \sup_{t \in [a,b]} |\tau(wt + c) - \mathbb{1}_{[c,\infty)}| \xrightarrow{w \to 0} 0.$$

From this it follows that $\lim_{w \to 0} \tau(w \cdot + c) = \mathbb{1}_{[c,\infty)}$ converges uniformly on compact subsets of $\mathbb{R}$. Thus, we have

$$\begin{aligned}\mathbb{1}_{[c,\infty)} \in \overline{\mathcal{S}^\sigma(\mathbb{R})} := \big\{ F \colon \mathbb{R} \to \mathbb{R} \;\big|\; &\forall\, \Omega \subseteq \mathbb{R} \text{ compact}, \varepsilon > 0 \\ &\exists\, N \in \mathcal{S}^\sigma(\mathbb{R}) \colon \|F - N\|_{\Omega, \mathrm{ess\,sup}} < \varepsilon \big\}\end{aligned}$$

for any $c \in \mathbb{R}$. Since $\overline{\mathcal{S}^\sigma(\mathbb{R})}$ is a vector space, it contains all step functions (with finitely many steps!). From real analysis, it is well-known that any continuous function can be uniformly approximated by step functions on a fixed compact interval. This proves the statement of the theorem. □

Theorem 16.22 covers in particular shallow neural networks with Heaviside or signum activation. In fact, in these special cases, the first part of the proof can even be omitted, or at least significantly shortened.

In practice, it turns out that the approximation with a shallow neural network requires a lot of neurons, while in comparison, one can get by with significantly fewer neurons by just arranging them differently. We refer to Exercise 16.8 for an example that underpins the aformentioned claim and continue here with the formal definition of what is behind the term "deep learning".

Definition 16.23. Let $\sigma \colon \mathbb{R} \to \mathbb{R}$. A function $N \colon \mathbb{R}^d \to \mathbb{R}^m$ of the form

$$N = A \circ n^{(t)} \circ \cdots \circ n^{(1)} \quad \text{with} \quad n^{(k)} = \begin{bmatrix} n_1^{(k)} \\ \vdots \\ n_\ell^{(k)} \end{bmatrix},$$

with σ-activated neurons $n_1^{(1)}, \ldots, n_\ell^{(1)} \colon \mathbb{R}^d \to \mathbb{R}$ and $n_1^{(k)}, \ldots, n_\ell^{(k)} \colon \mathbb{R}^\ell \to \mathbb{R}$ for $k = 2, \ldots, t$ and $A \in \mathbb{R}^{\ell \times m}$, is called a *deep, fully connected, forward-propagating neural network of width ℓ and depth $t \geqslant 2$ with linear output.* Such a network can be illustrated by the following picture.

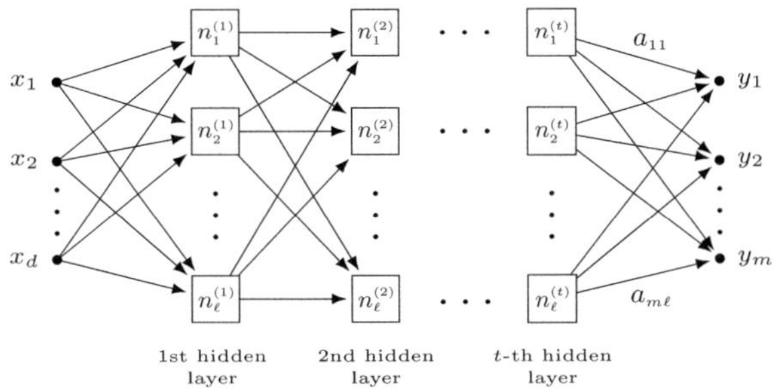

We denote with $\mathcal{D}^{\sigma,\ell,t}(\mathbb{R}^d,\mathbb{R}^m)$ the set of neural networks of the above type, which we briefly refer to as *deep networks*. Further, we define

$$\mathcal{D}^{\sigma,\ell}(\mathbb{R}^d,\mathbb{R}^m) = \bigcup_{t=1}^{\infty} \mathcal{D}^{\sigma,\ell,t}(\mathbb{R}^d,\mathbb{R}^m),$$

that is, the set of arbitrarily deep neural networks with fixed width ℓ. We put $\mathcal{D}^{\sigma,\ell,t}(\mathbb{R}^d) := \mathcal{D}^{\sigma,\ell,t}(\mathbb{R}^d,\mathbb{R})$ and $\mathcal{D}^{\sigma,\ell}(\mathbb{R}^d) := \mathcal{D}^{\sigma,\ell}(\mathbb{R}^d,\mathbb{R})$.

The next theorem shows that, under certain conditions, a trade-off between width and depth is possible. We first deal with the scalar-valued case.

Theorem 16.24. (Hanin) *Let $F \in \mathcal{S}^{\mathrm{ReLU}}(\mathbb{R}^d)$ be a shallow network with width ℓ and let $\Omega \subseteq \mathbb{R}^d_{\geqslant 0}$ compact. Then there is a deep network $N \in \mathcal{D}^{\mathrm{ReLU},d+3,\ell}(\mathbb{R}^d)$ with $F(x) = N(x)$ for all $x \in \Omega$.*

Proof. The shallow network $F\colon \mathbb{R}^d \to \mathbb{R}$ can be illustrated by the following picture

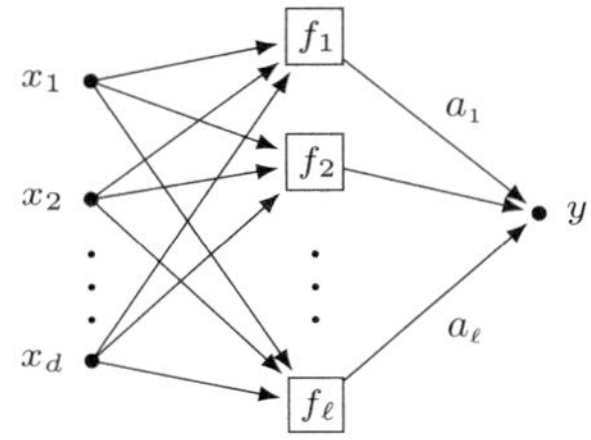

and is formally represented by the formula

$$F = [a_1 \cdots a_\ell] \circ \begin{bmatrix} f_1 \\ \vdots \\ f_\ell \end{bmatrix} = \sum_{i=1}^{\ell} a_i f_i,$$

where $a_i \in \mathbb{R}$ and the $f_i\colon \mathbb{R}^d \to \mathbb{R}$ are σ-activated neurons. Since Ω is compact and F is continuous, there exists $c_0 \geqslant 0$ with $F + c_0 \geqslant 0$ on Ω. We construct the deep network N according to the following picture, where missing arrows correspond to a zero in the weight vector, respectively, in the output matrix.

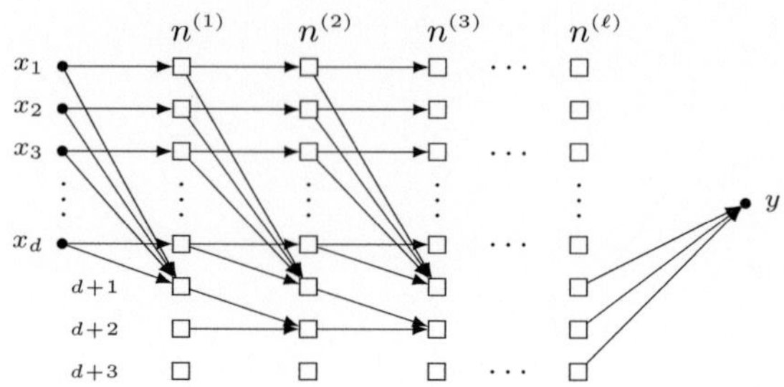

The neurons in rows 1 to d duplicate the inputs. More precisely, we put $n_j^{(k)} := \operatorname{ReLU}(\langle \mathrm{e}_j, \cdot\rangle) + 0)$ for $k = 1, \dots, \ell$ and $j = 1, \dots, d$, where e_j is the j-th unit vector in $\mathbb{R}^d$, respectively in $\mathbb{R}^{d+3}$. Consequently, we obtain $n_j^{(k)}(x) = x_j$ for each $x = (x_1, \dots, x_d) \in \Omega$.

The $(d+1)$-th row is now supposed to recreate the output of the given shallow network F. To achieve this, we put $n_{d+1}^{(1)} := f_1$. For $k = 2, \dots, \ell$, the function $n_{d+1}^{(k)}$ depends on $(d+3)$-many variables. So if w_k is the weight vector and b_k is the bias of f_k, we define

$$n_{d+1}^{(k)} := \operatorname{ReLU}\Big(\Big\langle \begin{bmatrix} w_k \\ 0 \\ 0 \\ 0 \end{bmatrix}, \cdot \Big\rangle + b_k\Big)$$

for $k \geqslant 2$ and obtain the desired result. In the $(d+2)$-th row, we define

$$n_{d+2}^{(1)} = \operatorname{ReLU}(\langle 0, \cdot\rangle + c_0) \ \text{ and } \ n_{d+2}^{(k)} = \operatorname{ReLU}(\langle \left[\begin{smallmatrix} a_{k-1} \\ 1 \end{smallmatrix}\right], \cdot\rangle + 0) \ \text{ for } k \geqslant 2.$$

Successively subbing in shows

$$\begin{aligned} n_{d+2}^{(1)} &= \operatorname{ReLU}(\langle 0, \text{-}\rangle + c_0) = c_0 \\ n_{d+2}^{(2)} \circ n^{(1)} &= \operatorname{ReLU}(\langle \left[\begin{smallmatrix} a_1 \\ 1 \end{smallmatrix}\right], n^{(1)}\rangle) = \operatorname{ReLU}(a_1 f_1 + c_0) = a_1 f_1 + c_0 \\ &\;\;\vdots \\ n_{d+2}^{(\ell)} \circ n^{(\ell-1)} \circ \dots \circ n^{(1)} &= \operatorname{ReLU}(\langle \left[\begin{smallmatrix} a_{\ell-1} \\ 1 \end{smallmatrix}\right], n^{(\ell-1)} \circ \dots \circ n^{(1)}\rangle) = \sum_{i=1}^{\ell-1} a_i f_i + c_0 \end{aligned}$$

on Ω. We are still missing the ℓ-th summand and we also still need to get rid of the extra c_0-summand. To fix both, we define the neurons of the $(d+3)$-th row by

$$\begin{aligned} n_{d+3}^{(k)} &:= \operatorname{ReLU}(\langle 0, \cdot\rangle + 0) \ \text{ for } k = 1, \dots, \ell-1, \\ n_{d+3}^{(\ell)} &:= \operatorname{ReLU}(\langle 0, \cdot\rangle + c_0) = c_0. \end{aligned}$$

For the last neuron in the $(d+1)$-th row, we obtain $n_{d+1}^{(\ell)} \circ n^{(\ell-1)} \circ \dots \circ n^{(1)}(x) =$

$f_\ell(x)$ for $x \in \Omega$ by construction. We now choose $B = [\,0 \cdots 0 \;\; a_\ell \;\; 1 \;\; -1\,]$ as output matrix and obtain

$$N = B \circ n^{(\ell)} \circ \cdots \circ n^{(1)} = a_\ell f_\ell + \sum_{i=1}^{\ell-1} a_i f_i + c_0 - c_0 = F$$

on Ω. □

In the above proof, we defined the deep network N very lavishly; in fact, $(n \cdot \ell)$-many neurons only copy the inputs, and $(\ell - 1)$-many more define the zero function. This suggests on the one hand that we should be able to achieve the same with much less depth, while keeping the same width. On the other hand, for a density result we need arbitrarily many neurons, i.e., we must allow arbitrary depth given that we want the width to be fixed.

Corollary 16.25. *Let $\Omega \subseteq \mathbb{R}^n$ be compact and $\sigma = \mathrm{ReLU}$. Then the deep networks $\mathcal{D}^{\sigma,n+3m}(\Omega, \mathbb{R}^m) \subseteq \mathrm{C}(\Omega, \mathbb{R}^m)$ are dense.*

Proof. Let $F \in \mathrm{C}(\mathbb{R}^d, \mathbb{R}^m)$, $\Omega \subseteq \mathbb{R}^d$ be compact and $\varepsilon > 0$.

① We claim that we may assume $\Omega \subseteq \mathbb{R}^d_{\geqslant 0}$ without loss of generality. In order to see this, we choose $x_0 \in \mathbb{R}^d$ such that $\Omega_0 := x_0 + \Omega \subseteq \mathbb{R}^d_{\geqslant 0}$ and put $F_0 := F(\,\cdot - x_0)$. Now let $N_0 \in \mathcal{D}^{\sigma,n+3m}(\mathbb{R}^d, \mathbb{R}^m)$ be such that $\|F_0 - N_0\|_{\Omega_0,\infty} < \varepsilon$. With $N := N_0(\,\cdot + x_0)$ we have $\|F - N\|_{\Omega,\infty} < \varepsilon$ and it remains to show that N is a deep neural network. We will do this by modifying the biases in the first hidden layer of N_0 as follows. For $x \in \mathbb{R}^d$ we have

$$N(x) = N_0(x + x_0) = [A \circ n^{(t)} \circ \cdots \circ n^{(1)}](x - x_0) = [A \circ n^{(t)} \circ \cdots \circ \tilde{n}^{(1)}](x),$$

where the $n^{(i)}$'s are the layers of N and $\tilde{n}_i^{(1)}(x) = \mathrm{ReLU}(\langle w_i^{(1)}, x\rangle + \tilde{b}_i)$ with $\tilde{b}_i^{(1)} = \langle w_i^{(1)}, x_0\rangle + b_i^{(1)}$. This establishes the claim and allows us to assume $\Omega \subseteq \mathbb{R}^d_{\geqslant 0}$ for the remainder of this proof.

② As our second step, we observe that we can make Theorem 16.24 $\mathbb{R}^m$-valued by starting as in the proof of Theorem 16.24 and then adding three more rows to the network for each additional coordinate function. In this sense, we obtain

$$\mathcal{S}^{\mathrm{ReLU},\ell}(\mathbb{R}^d, \mathbb{R}^m) \subseteq \mathcal{D}^{\mathrm{ReLU},d+3m,\ell}(\mathbb{R}^d, \mathbb{R}^m)$$

for each $\ell \in \mathbb{N}$, and taking the union over ℓ yields

$$\mathcal{S}^{\mathrm{ReLU}}(\mathbb{R}^d, \mathbb{R}^m) \subseteq \mathcal{D}^{\mathrm{ReLU},d+3m}(\mathbb{R}^d, \mathbb{R}^m) \subseteq \mathrm{C}(\mathbb{R}^d, \mathbb{R}^m).$$

By Corollary 16.21 we find $N \in \mathcal{S}^{\mathrm{ReLU}}(\mathbb{R}^d, \mathbb{R}^m)$ with $\|F - N\|_{\Omega,\infty} < \varepsilon$, and thus are done due to the above inclusion. □

The following simple example shows, that we cannot compensate for arbitrarily small widths with greater depth.

Example 16.26. We consider a neural network with width $\ell = 1$, depth $t \geqslant 2$ and ReLU activation as follows:

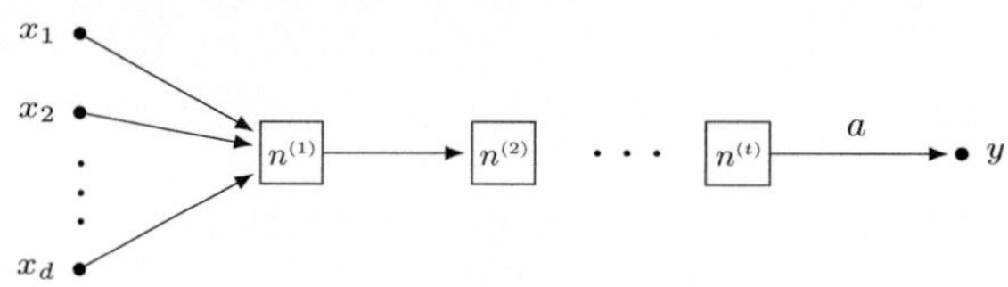

If we write down the corresponding function, we see that it is of the form $N = a\,\mathrm{ReLU}(\cdots)$ and consequently, depending on the sign of a, the latter will be, on the whole space $\mathbb{R}^d$, either nonnegative or nonpositive.

There are results on expressivity, sometimes under additional assumptions on the functions to be approximated, that provide asymptotic bounds for the width and depth necessary to get density. References can be found at the end of the chapter.

16.2 Backpropagation

Chapter 16.1 provides pure existence statements and reveals nothing about how to choose weights and biases if a neural network is to be designed as a classifier or regressor for a given dataset. In the following, we deal with this question, and first note that we have already discussed something similar in the very first chapter on affine linear regression, see the explicit formula in Theorem 2.9, and especially the reference to the gradient descent method at the end of Subchapter 2.2. Afterwards, in the context of logistic regression, we have even used (individual) neurons as predictors/approximants. The procedure with neural networks is essentially the same: We define a cost function, which calculates the deviations of the neural network's predictions from the labels on the given dataset, and then we minimize over the parameters of the network. The latter procedure is often referred to as the "training" of a neural network. We consider the following example

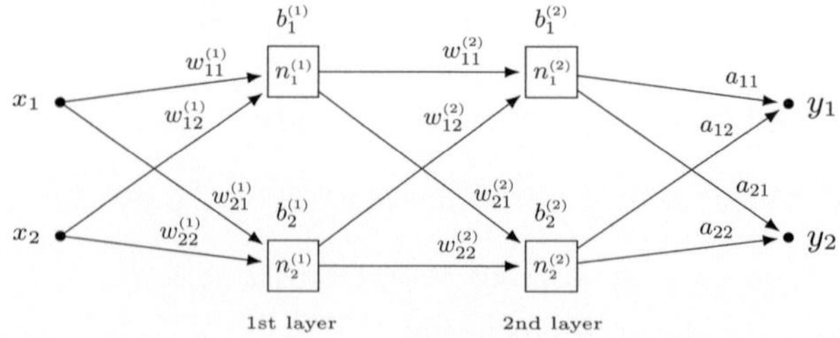

of a network as in Definition 16.23, with a differentiable activation function $\sigma\colon \mathbb{R} \to \mathbb{R}$, weights and biases as indicated and linear output $A = \left[\begin{smallmatrix} a_{11} & a_{12} \\ a_{21} & a_{22} \end{smallmatrix}\right]$. In the following, we introduce the terminology common for neural networks, but point out that most of it is just new names for well-known mathematical concepts.

Let a dataset

$$D = \big\{(x^{(p)}, y^{(p)}) \;\big|\; p = 1, \ldots, n\big\} \subseteq \mathbb{R}^2 \times \mathbb{R}^2,$$

with $x^{(p)} = (x_1^{(p)}, x_2^{(p)})$ and $y^{(i)} = (y_1^{(p)}, y_2^{(p)})$, be given, whose elements we will below use as training data. We refer to the network depicted on p. 245 as $N \in \mathcal{D}^{\sigma,2,2}(\mathbb{R}^2, \mathbb{R}^2)$ and define the *cost function*

$$C\colon \mathbb{R}^{16} \to \mathbb{R}, \quad C(w_{ij}^{(k)}, b_i^{(k)}, a_{ij}) := \tfrac{1}{2n} \sum_{p=1}^{n} \|N(x^{(p)}) - y^{(p)}\|^2.$$

Let us emphasize that the $x^{(p)}$'s and $y^{(p)}$'s are constant, while C is a function of the weights, biases and matrix entries, which occur in our neural network N.

The reason for the omission of the variables in the formula above is obvious: Even our very small neural network would lead to a lengthy expression including 16 real parameters/variables if we wanted to write everything out explicitly. Below we will often further abbreviate and, e.g., write "$C(w, b, a) = \cdots$" or even omit the arguments completely for the sake of readability, and thus simply write "$C = \cdots$" instead of "$C(w_{ij}^{(k)}, b_i^{(k)}, a_{ij}) = \cdots$".

We adopt now the heuristic that our network N approximates the dataset well if C is small, hence we seek a minimizer

$$(w^*, b^*, a^*) \in \operatorname*{argmin}_{(w,b,a) \in \mathbb{R}^{16}} C(w, b, a)$$

and we use the gradient descent method, which we will discuss in more detail in Chapter 17. We initialize the weights, biases, and matrix entries arbitrarily and update them by taking a step in the direction of $-\nabla C$.

Algorithm 16.27. *The following pseudocode outlines the gradient descent method for the cost function C given above. Note that the dataset D is already built into that cost function.*

1: **function** GRADIENT DESCENT $(C, \gamma, \varepsilon, T)$
2: $(w, b, a) \leftarrow$ arbitrary point in $\mathbb{R}^{16}$
3: **for** $\tau \leftarrow 1$ to T **do**
4: $w \leftarrow w - \gamma \cdot \frac{\partial C}{\partial w}(w, b, a)$
5: $b \leftarrow b - \gamma \cdot \frac{\partial C}{\partial b}(w, b, a)$
6: $a \leftarrow a - \gamma \cdot \frac{\partial C}{\partial a}(w, b, a)$
7: **if** $C(w, b, a) \leqslant \varepsilon$ **then** break
8: **return** (w, b, a)

The parameter $\gamma > 0$, which will be referred to as *step size* in Chapter 17, is often called the *learning rate* in the context of neural networks. It controls how far we go in the direction of $-\nabla C$, and can be constant as above, but may alternatively be changed in each iteration, compare Example 17.6 and the subsequent discussion. In Algorithm 16.27, w, b and a depend of course on the iteration τ. To avoid confusion with the indices i, j and k (that we omitted above), we refrain from explicitly indicating the dependence on τ. In addition to the termination criterion in line 7, which stops the algorithm once a given accuracy $\varepsilon > 0$ is reached, the maximum number of iterations to be performed is limited by $T \in \mathbb{N}$.

In Chapter 17 on the gradient descent method, we will prove the convergence of the recursively defined sequence of points (w, b, a) under mild conditions, especially for convex cost functions. Under the conditions of this chapter, however, C may not be convex. The results from Chapter 17 therefore suggest that our approach is promising — we will, however, not prove the convergence in the current chapter.

What we will prove in the following, however, are recursion formulas for calculating the derivatives of C, which are known under the name *backpropagation*. The following observations will turn out helpful for this.

Remark 16.28. (i) Firstly, we note that C is differentiable; this is ensured by (a) the taking squares of the Euclidean norms; (b) the special form of deep neural networks; and (c) our assumption that the activation function σ is differentiable. Furthermore, we have

$$\begin{aligned} \frac{\partial C}{\partial w_{ij}^{(k)}} &= \frac{\partial}{\partial w_{ij}^{(k)}} \Big(\frac{1}{2n} \sum_{p=1}^{n} \|N(x^{(p)}) - y^{(p)}\|^2 \Big) \\ &= \frac{1}{n} \sum_{p=1}^{n} \frac{\partial}{\partial w_{ij}^{(k)}} \underbrace{\big(\tfrac{1}{2}\|N(x^{(p)}) - y^{(p)}\|^2\big)}_{=:C_p} \end{aligned}$$

and corresponding formulas also apply to the partial derivatives with respect to $b^{(k)}$ and a_{ij}. Therefore, it is sufficient if we determine the derivatives of all

C_p. To increase readability, we will omit the index p and from now on assume that the dataset $D = \{(x, y) = ((x_1, x_2), (y_1, y_2))\} \in \mathbb{R}^2 \times \mathbb{R}^2$ contains only one point. The cost function then simplifies to:

$$C = \tfrac{1}{2}\|N(x) - y\|^2.$$

(ii) If weights and biases are given, then $N(x)$ can be calculated by going layer-by-layer from left to right. This is called *forward propagation*. We will show that

$$\frac{\partial C}{\partial w_{ij}^{(k)}}, \quad \frac{\partial C}{\partial b_i^{(k)}} \quad \text{and} \quad \frac{\partial C}{\partial a_{ij}}$$

can be calculated recursively, i.e., layer-by-layer from right to left. This is called *backpropagation*.

(iii) To write the latter efficiently, we define

$$z^{(k)} := \begin{bmatrix} z_1^{(k)} \\ z_2^{(k)} \end{bmatrix} \quad \text{with } z_i^{(k)} = \left\langle \begin{bmatrix} w_{i1}^{(k)} \\ w_{i2}^{(k)} \end{bmatrix}, \begin{bmatrix} n_1^{(k-1)} \\ n_2^{(k-1)} \end{bmatrix} \right\rangle + b_i^{(k)}$$

for $k = 1, 2$, $i = 1, 2$ where $n_i^{(0)} = x_i$. The $z_i^{(k)}$'s thus each express what we calculate *inside* a neuron, *before* we apply the activation function.

We first deal with the output matrix and the 2nd layer.

Proposition 16.29. *Let $\sigma\colon \mathbb{R} \to \mathbb{R}$ be differentiable, $N \in \mathcal{D}^{\sigma,2,2}(\mathbb{R}^2, \mathbb{R}^2)$ be a neural network, $D = \{(x, y)\} \subseteq \mathbb{R}^2 \times \mathbb{R}^2$ be a dataset with a single data point and $C\colon \mathbb{R}^{16} \to \mathbb{R}$, $C(w, b, a) = \|N(x) - y\|^2$. Then we have*

$$\frac{\partial C}{\partial a_{ij}} = (N_i - y_i) \cdot n_j^{(2)} \quad \textit{as well as} \quad \frac{\partial C}{\partial w_{ij}^{(2)}} = \delta_i^{(2)} \cdot n_j^{(1)} \quad \textit{and} \quad \frac{\partial C}{\partial b_i^{(2)}} = \delta_i^{(2)}$$

with $\delta_i^{(2)} = \sum\limits_{\mu=1}^{2} a_{\mu i}(N_\mu - y_\mu) \cdot \sigma'(z_i^{(2)})$ *and* $N = \left[\begin{smallmatrix} N_1 \\ N_2 \end{smallmatrix}\right]$.

Proof. We have

$$N = \begin{bmatrix} N_1 \\ N_2 \end{bmatrix} = \begin{bmatrix} a_{11} & a_{12} \\ a_{21} & a_{22} \end{bmatrix} \begin{bmatrix} n_1^{(2)} \\ n_2^{(2)} \end{bmatrix} = \begin{bmatrix} a_{11}n_1^{(2)} + a_{12}n_2^{(2)} \\ a_{21}n_1^{(2)} + a_{22}n_2^{(2)} \end{bmatrix}$$

and applying the chain rule

$$\begin{aligned}
\frac{\partial C}{\partial a_{ij}} &= \frac{\partial}{\partial a_{ij}}\left(\tfrac{1}{2}\|N(x) - y\|^2\right) = \frac{1}{2}\frac{\partial}{\partial a_{ij}}\left\| \begin{bmatrix} a_{11}n_1^{(2)} + a_{12}n_2^{(2)} - y_1 \\ a_{21}n_1^{(2)} + a_{22}n_2^{(2)} - y_2 \end{bmatrix} \right\|^2 \\
&= \frac{1}{2}\frac{\partial}{\partial a_{ij}}\left[(a_{11}n_1^{(2)} + a_{12}n_2^{(2)} - y_1)^2 + (a_{21}n_1^{(2)} + a_{22}n_2^{(2)} - y_2)^2\right] \\
&= (a_{i1}n_1^{(2)} + a_{i2}n_2^{(2)} - y_i) \cdot n_j^{(2)} = (N_i - y_i) \cdot n_j^{(2)}
\end{aligned}$$

we get the far right derivative in the proposition. To prove the first two derivatives, we use the auxiliary functions $z_i^{(2)}$ from Remark 16.28(iii), with which we can write the neurons of the 2nd layer as

$$\begin{bmatrix} n_1^{(2)} \\ n_2^{(2)} \end{bmatrix} = \begin{bmatrix} \sigma\left(\left\langle \begin{bmatrix} w_{11}^{(2)} \\ w_{12}^{(2)} \end{bmatrix}, \begin{bmatrix} n_1^{(1)} \\ n_2^{(1)} \end{bmatrix} \right\rangle + b_1^{(2)}\right) \\ \sigma\left(\left\langle \begin{bmatrix} w_{21}^{(2)} \\ w_{22}^{(2)} \end{bmatrix}, \begin{bmatrix} n_1^{(1)} \\ n_2^{(1)} \end{bmatrix} \right\rangle + b_2^{(2)}\right) \end{bmatrix} = \begin{bmatrix} \sigma(z_1^{(2)}) \\ \sigma(z_2^{(2)}) \end{bmatrix}.$$

We start deriving as above, but now we have to take into account that both terms in the square bracket of equation (16.1) depend on $z_i^{(2)}$. We therefore get:

$$\begin{aligned} \frac{\partial C}{\partial z_i^{(2)}} &= \frac{1}{2} \frac{\partial}{\partial z_i^{(2)}} \left[(a_{11} n_1^{(2)} + a_{12} n_2^{(2)} - y_1)^2 + (a_{21} n_1^{(2)} + a_{22} n_2^{(2)} - y_2)^2 \right] \qquad (16.1) \\ &= \sum_{\mu=1}^{2} (a_{\mu 1} n_1^{(2)} + a_{\mu 2} n_2^{(2)} - y_\mu) \cdot \frac{\partial}{\partial z_i^{(2)}} (a_{\mu 1} n_1^{(2)} + a_{\mu 2} n_2^{(2)} - y_\mu) \\ &= \sum_{\mu=1}^{2} (N_\mu - y_\mu) \cdot a_{\mu i} \sigma'(z_i^{(2)}) = \sum_{\mu=1}^{2} a_{\mu i} (N_\mu - y_\mu) \cdot \sigma'(z_i^{(2)}) = \delta_i^{(2)}. \end{aligned}$$

By the chain rule, it follows that

$$\frac{\partial C}{\partial w_{ij}^{(2)}} = \frac{\partial C}{\partial z_i^{(2)}} \cdot \frac{\partial z_i^{(2)}}{\partial w_{ij}^{(2)}} = \delta_i^{(2)} \cdot n_j^{(1)} \quad \text{and} \quad \frac{\partial C}{\partial b_i^{(2)}} = \frac{\partial C}{\partial z_i^{(2)}} \cdot \frac{\partial z_i^{(2)}}{\partial b_i^{(2)}} = \delta_i^{(2)} \cdot 1,$$

which concludes the proof. □

If we want to perform an iteration of gradient descent for our (2×2)-network from page 244, we initially know the values of the parameters $w_{ij}^{(k)}$, $b_i^{(k)}$ and a_{ij} from the previous iteration. Through forward propagation, we obtain the corresponding values of $z_j^{(k)}$, the $n_i^{(k)}$ and the N_i. From these, we can calculate $\delta_i^{(2)}$, and with Proposition 16.29 determine all three derivatives:

$$\frac{\partial C}{\partial w_{ij}^{(2)}}(w_{ij}^{(k)}, b_i^{(k)}), \quad \frac{\partial C}{\partial b_i^{(2)}}(w_{ij}^{(k)}, b_i^{(k)}) \quad \text{and} \quad \frac{\partial C}{\partial a_{ij}} = (N_i - y_i) \cdot n_j^{(2)}.$$

Thus, we are done with the derivatives of the 2nd layer and move on to those of the 1st layer.

Proposition 16.30. *Let $\sigma\colon \mathbb{R} \to \mathbb{R}$ be differentiable, $N \in \mathfrak{D}^{\sigma,2,2}(\mathbb{R}^2, \mathbb{R}^2)$ be a neural network, $D = \{(x, y)\} \subseteq \mathbb{R}^2 \times \mathbb{R}^2$ be a dataset with a single data point and $C\colon \mathbb{R}^{16} \to \mathbb{R}$, $C(w, b, a) = \|N(x) - y\|^2$. Then we have*

$$\frac{\partial C}{\partial w_{ij}^{(1)}} = \delta_i^{(1)} \cdot x_j \quad \text{and} \quad \frac{\partial C}{\partial b_i^{(1)}} = \delta_i^{(1)}$$

with $\delta_i^{(1)} := (\delta_1^{(2)} w_{1i}^{(2)} + \delta_2^{(2)} w_{2i}^{(2)}) \cdot \sigma'(z_i^{(1)})$.

Proof. We start as in the proof of Proposition 16.29, but the dependencies become a bit more complicated. For example, the weights $w_{11}^{(1)}$ and $w_{12}^{(1)}$ do only go into the first neuron of the first layer, but from there they then go into *both* neurons of the second layer:

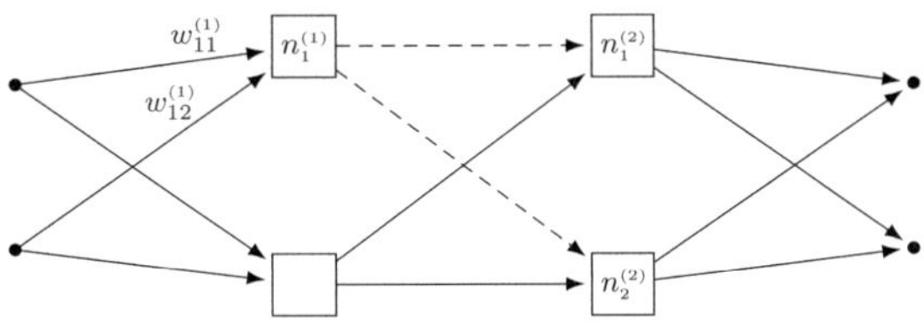

For $j = 1, 2$ we therefore read $C = C(w_{1j}^{(1)})$ as a composition $\mathbb{R} \to \mathbb{R}^2 \to \mathbb{R}$, $C = C\big(z_1^{(2)}(w_{1j}^{(1)}), z_2^{(2)}(w_{1j}^{(1)})\big)$. Then, with the corresponding multidimensional chain rule and the abbreviations and calculations from Proposition 16.29, we get:

$$\begin{aligned}
\frac{\partial C}{\partial w_{1j}^{(1)}} &= \frac{\partial C}{\partial z_1^{(2)}} \frac{\partial z_1^{(2)}}{\partial w_{1j}^{(1)}} + \frac{\partial C}{\partial z_2^{(2)}} \frac{\partial z_2^{(2)}}{\partial w_{1j}^{(1)}} \\
&= \delta_1^{(2)} \frac{\partial}{\partial w_{1j}^{(1)}} \left(\left\langle \begin{bmatrix} w_{11}^{(2)} \\ w_{12}^{(2)} \end{bmatrix}, \begin{bmatrix} n_1^{(1)} \\ n_2^{(1)} \end{bmatrix} \right\rangle + b_1^{(2)} \right) + \delta_2^{(2)} \frac{\partial}{\partial w_{1j}^{(1)}} \left(\left\langle \begin{bmatrix} w_{21}^{(2)} \\ w_{22}^{(2)} \end{bmatrix}, \begin{bmatrix} n_1^{(1)} \\ n_2^{(1)} \end{bmatrix} \right\rangle + b_2^{(2)} \right) \\
&\underset{\substack{\uparrow \\ w_{1j}^{(1)} \text{ only goes} \\ \text{into } n_1^{(1)}}}{=} \delta_1^{(2)} w_{11}^{(2)} \frac{\partial n_1^{(1)}}{\partial w_{1j}^{(1)}} + \delta_2^{(2)} w_{21}^{(2)} \frac{\partial n_1^{(1)}}{\partial w_{1j}^{(1)}} \\
&= \left(\delta_1^{(2)} w_{11}^{(2)} + \delta_2^{(2)} w_{21}^{(2)} \right) \cdot \frac{\partial}{\partial w_{1j}^{(1)}} \sigma \Bigg(\underbrace{\left\langle \begin{bmatrix} w_{11}^{(1)} \\ w_{12}^{(1)} \end{bmatrix}, \begin{bmatrix} x_1 \\ x_2 \end{bmatrix} \right\rangle + b_1^{(1)}}_{= z_1^{(1)}} \Bigg) \\
&= \left(\delta_1^{(2)} w_{11}^{(2)} + \delta_2^{(2)} w_{21}^{(2)} \right) \cdot \sigma'(z_1^{(1)}) \cdot x_j .
\end{aligned}$$

The equation for the derivative with respect to $w_{2j}^{(1)}$ can be obtained analogously. For the derivative with respect to $b_i^{(1)}$, one proceeds as above, but observes in the penultimate line that

$$\frac{\partial C}{\partial b_i^{(1)}} = \left(\delta_1^{(2)} w_{1i}^{(2)} + \delta_2^{(2)} w_{2i}^{(2)} \right) \cdot \sigma'(z_1^{(1)}) \cdot \frac{\partial z_i^{(1)}}{\partial b_i^{(1)}} = \left(\delta_1^{(2)} w_{1i}^{(2)} + \delta_2^{(2)} w_{2i}^{(2)} \right) \cdot \sigma'(z_1^{(1)}) \cdot 1$$

holds. □

Since we have already calculated the $\delta_i^{(2)}$'s when processing the 2nd layer, with Proposition 16.30 we get the derivatives

$$\frac{\partial C}{\partial w_{ij}^{(1)}}(w_{ij}^{(k)}, b_i^{(k)}) \text{ and } \frac{\partial C}{\partial b_i^{(1)}}(w_{ij}^{(k)}, b_i^{(k)}),$$

where $w_{ij}^{(k)}$, $b_i^{(k)}$ and a_{ij} represent still the values of the parameters of the previous iteration. Since we now know all 16 derivatives at $(w_{ij}^{(k)}, b_i^{(k)}, a_{ij})$, we can perform the update according to the gradient method.

In Theorem 16.32 we generalize the above for networks with more than two layers and more than two neurons per layer. To be able to elegantly formulate the corresponding recursion formula for the $\delta_i^{(2)}$'s, we need the following notation.

Definition 16.31. Let $x, y \in \mathbb{R}^d$ be given. We call

$$x \odot y = \begin{bmatrix} x_1 \\ \vdots \\ x_d \end{bmatrix} \odot \begin{bmatrix} y_1 \\ \vdots \\ y_d \end{bmatrix} := \begin{bmatrix} x_1 y_1 \\ \vdots \\ x_d y_d \end{bmatrix}$$

the *Hadamard product* of x and y.

In the following, we consider a deep, fully connected, forward-propagating neural network $N \in \mathfrak{D}^{\sigma,\ell,t}(\mathbb{R}^d, \mathbb{R}^m)$ with linear output, as represented by the picture in Definition 16.23. We combine the weights of each layer into a matrix and the biases of each layer into a vector. Since in general $d \neq \ell$, the weight matrix of the first layer may have a different format than the rest. The network is thus given by the activation function and the matrices

$$W^{(1)} = \begin{bmatrix} w_{11}^{(1)} & \cdots & w_{1d}^{(1)} \\ \vdots & & \vdots \\ w_{\ell 1}^{(1)} & \cdots & w_{\ell d}^{(1)} \end{bmatrix}, \; W^{(k)} = \begin{bmatrix} w_{11}^{(k)} & \cdots & w_{1\ell}^{(k)} \\ \vdots & & \vdots \\ w_{\ell 1}^{(k)} & \cdots & w_{\ell\ell}^{(k)} \end{bmatrix}, \; b^{(k)} = \begin{bmatrix} b_1^{(k)} \\ \vdots \\ b_\ell^{(k)} \end{bmatrix}, \; A = \begin{bmatrix} a_{11} & \cdots & a_{1\ell} \\ \vdots & & \vdots \\ a_{m1} & \cdots & a_{m\ell} \end{bmatrix}$$

for $k = 2, \ldots, t$ and $k = 1, \ldots, t$, respectively. In total, the above matrices contain $\nu := (\ell \cdot d + (t-1) \cdot \ell^2 + t \cdot \ell + m \cdot \ell)$-many scalar parameters. The derivatives necessary for the gradient method can be determined by backpropagation using the following theorem, in which we combine the $\delta_i^{(k)}$'s into vectors $\delta^{(k)} = (\delta_1^{(k)}, \ldots, \delta_\ell^{(k)})$, read $\sigma'(\cdot)$ coordinatewise, and use the auxiliary functions

$$z^{(k)} := \begin{bmatrix} z_1^{(k)} \\ \vdots \\ z_\ell^{(k)} \end{bmatrix} \text{ with } z_i^{(k)} = \left\langle \begin{bmatrix} w_{i1}^{(k)} \\ \vdots \\ w_{i\ell}^{(k)} \end{bmatrix}, \begin{bmatrix} n_1^{(k-1)} \\ \vdots \\ n_\ell^{(k-1)} \end{bmatrix} \right\rangle + b_i^{(k)}$$

from Remark 16.28(iii), but now in higher dimensions.

Theorem 16.32. *Let $\sigma\colon \mathbb{R} \to \mathbb{R}$ be differentiable, $N \in \mathfrak{D}^{\sigma,\ell,t}(\mathbb{R}^d, \mathbb{R}^m)$ be a neural network, $D = \{(x, y)\} \subseteq \mathbb{R}^d \times \mathbb{R}^m$ be a dataset with one single data* *point and* $C\colon \mathbb{R}^\nu \to \mathbb{R}$, $C(w,b,a) = \|N(x) - y\|^2$. *With* $n_j^{(0)} := x_j$ *we have*

$$\frac{\partial C}{\partial a_{ij}} = (N_i - y_i)\cdot n_j^{(t)}, \text{ as well as } \frac{\partial C}{\partial w_{ij}^{(k)}} = \delta_i^{(k)} n_j^{(k-1)} \text{ and } \frac{\partial C}{\partial b_i^{(k)}} = \delta_i^{(k)}$$

for all layers $k = 1, \dots, t$. *The* $\delta_i^{(k)}$ *can be calculated recursively by*

$$\delta^{(t)} = A^\intercal(N - y) \odot \sigma'(z^{(t)}) \text{ and } \delta^{(k)} = (W^{(k+1)})^\intercal \delta^{(k+1)} \odot \sigma'(z^{(k)}),$$

where $k = t-1, \dots, 1$, *i.e., we start with the last layer and work our way back to the first layer. The evaluations* $z^{(k)} = W^{(k)} n^{(k-1)} + b^{(k)}$ *for* $k = 1, \dots, t$ *are all known from the forward propagation.*

Proof. From Propositions 16.29 and 16.30 and their proofs it can be seen that we get the three formulas for the derivatives if we define

$$\delta_i^{(t)} := \sum_{\mu=1}^{m} a_{\mu i}(N_\mu - y_\mu)\cdot \sigma'(z_i^{(t)})$$

and

$$\delta_i^{(k)} := (\delta_1^{(k)} w_{1i}^{(k)} + \dots + \delta_\ell^{(k)} w_{\ell i}^{(2)}) \cdot \sigma'(z_i^{(k-1)})$$

for $k = t-1, \dots, 1$. These are however precisely the i-th entries of the Hadamard products given in the theorem. □

We remind the reader at this point, that for a dataset $D = \{(x^{(p)}, y^{(p)}) \mid p = 1, \dots, n\}$ with $n \geqslant 2$ the cost function becomes

$$C\colon \mathbb{R}^\nu \to \mathbb{R},\ \ C(w_{ij}^{(k)}, b_i^{(k)}, a_{ij}) := \frac{1}{2n}\sum_{p=1}^{n} \|N(x^{(p)}) - y^{(p)}\|^2.$$

On a first glance this looks more complicated, but the derivatives can, by linearity, be easily calculated with Theorem 16.32, cf. Remark 16.28(ii).

As announced at the beginning of this chapter, we will now return to classifiers, especially those with possibly more than two classes. A prototype example of this is handwriting recognition. As illustrated in the following pictures

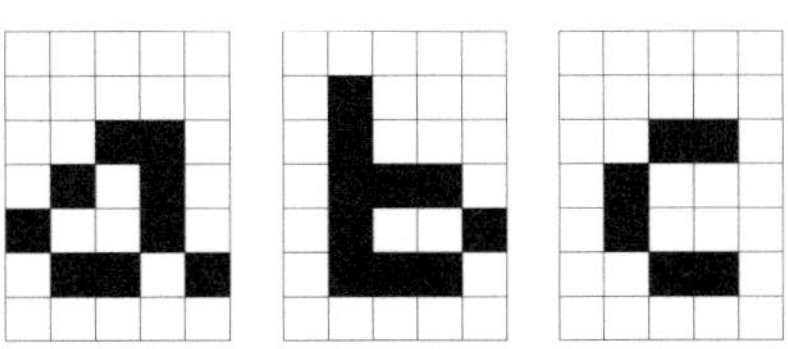

this application deals with datasets whose feature vectors are handwritten letters. Above, these are given as an element of $\mathbb{R}^{7\times 5} \cong \mathbb{R}^{35}$, and thus we deal with $d = 35$. As labels, we use unit vectors, i.e., we assign to each letter a unit vector, for example, the letter "a" to the vector e_1, the letter "b" to the vector e_2, "c" to the vector e_3, etc. The integer m is then the number of letters in our alphabet. This process is called *one-hot encoding*. For the training data, we have knowledge of which letter was written by hand, leading to a dataset

$$D = \{(x^{(p)}, y^{(p)}) \mid p = 1, \dots, n\} \subseteq \mathbb{R}^d \times \{e_1, \dots, e_m\}. \tag{16.2}$$

The obvious goal is to determine a classifier $K\colon \mathbb{R}^d \to \{e_1, \dots, e_m\}$. However, as already discussed in Chapter 2 about logistic regression, it is more sensible to determine a regressor $R\colon \mathbb{R}^d \to (0,1)^m$: Firstly, one can always obtain K by rounding if one has determined R. Secondly, the values of R can be interpreted as probabilities; if, for example, $R(x) = (0.88, 0.01, 0.10, \dots)$ for $x \in \mathbb{R}^{35}\backslash D$, this can be read as x being an "a" with a probability of 0.88, a "b" with a probability of 0.01, a "c" with a probability of 0.10, etc. Thirdly, it will turn out that the approach with R leads to an optimization problem over a class $\mathcal{R}$ of differentiable functions, and therefore an application of backpropagation as in the first part of this subchapter is possible. To formally define the class $\mathcal{R}$, we need the so-called *softmax function*

$$\operatorname{softmax}\colon \mathbb{R}^m \to \mathbb{R}^m, \ \operatorname{softmax}(z_1, \dots, z_m) = \frac{(e^{z_1}, \dots, e^{z_m})}{e^{z_1} + \dots + e^{z_m}}.$$

The elements of $\mathcal{R}$ we define to be compositions of a deep neural network with the softmax function.

Definition 16.33. Let $\sigma\colon \mathbb{R} \to \mathbb{R}$ be an activation function. We put

$$\mathcal{R}^{\sigma,\ell,t}(\mathbb{R}^d, \mathbb{R}^m) := \big\{\operatorname{softmax} \circ N \mid N \in \mathcal{D}^{\sigma,\ell,t}(\mathbb{R}^d, \mathbb{R}^m)\big\}.$$

and refer to the elements of $\mathcal{R}^{\sigma,\ell,t}(\mathbb{R}^d, \mathbb{R}^m)$ as *deep neural networks with softmax output*.

If we denote the coordinate functions of the softmax function with s_i, then $R \in \mathcal{R}^{\sigma,\ell,t}(\mathbb{R}^d, \mathbb{R}^m)$ can be illustrated by the following picture. We point out however that since the $s_i\colon \mathbb{R}^m \to \mathbb{R}$ are multivariable functions, the boxes in the softmax layer are not neurons in the sense of Definition 16.1.

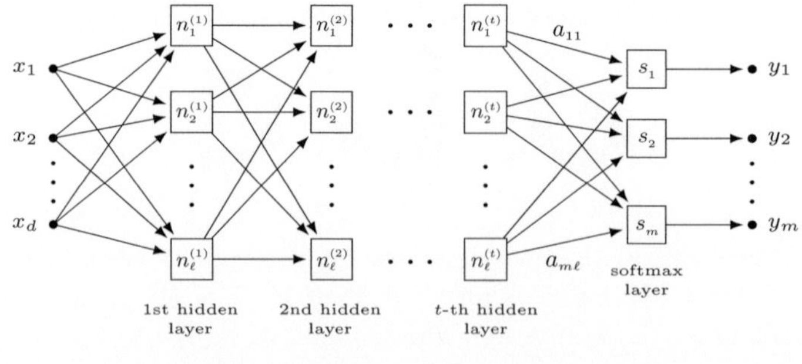

Before we get to backpropagation, we want to note the following: If a dataset with categorical labels is given as in (16.2), one could also try the approximation with a network $N \in \mathcal{D}^{\sigma,\ell,t}(\mathbb{R}^d, \mathbb{R}^m)$. One then has to rely on the optimization process to yield a function N that sends "meaningful" feature vectors — in the example above, for instance, those $x \in \mathbb{R}^{35}$ that actually correspond to a handwritten letter — to labels that are "close to" a unit vector. Although this could of course also fail for $R \in \mathcal{R}^{\sigma,\ell,t}(\mathbb{R}^d, \mathbb{R}^m)$, the approach promoted above does enforce some desirable properties: If we denote by R_i the i-th coordinate function of R, we obtain by design that

$$\forall\, x \in \mathbb{R}^d\colon\ R_i \in (0,1) \ \text{ and } \ \sum_{i=1}^{m} R_i(x) = 1$$

always hold. In particular, for every $R \in \mathcal{R}^{\sigma,\ell,t}(\mathbb{R}^d, \mathbb{R}^m)$ and every $x \in \mathbb{R}^d$ we can consider the $R_i(x)$ as probabilities and apply a maximum likelihood method analogous to the case of logistic regression. Indeed, we will from now on consider the *likelihood function*

$$L\colon \mathcal{R}^{\sigma,\ell,t}(\mathbb{R}^d, \mathbb{R}^m) \to \mathbb{R},\ \ L(R) := \prod_{p=1}^{n} \prod_{i=1}^{m} R_i(x^{(p)})^{y_i^{(p)}}$$

as a cost function. Indeed, L is very suitable for the current problem: By construction, each $y^{(p)}$ is equal to a unit vector. Consequently, for a fixed p in the inner product, the factor with $y^{(p)} = \mathrm{e}_i$ is equal to $R_i(x^{(p)})$ and all other factors are equal to one. It follows that L is large when $R_i(x^{(p)}) \approx 1$ holds for precisely those pairs (i,p) with $y^{(p)} = \mathrm{e}_i$. If this is the case, then $R_i(x^{(p)}) \approx 0$ must hold for pairs (i,p) with $y^{(p)} \neq \mathrm{e}_i$. In summary, L is large if and only if $R(x^{(p)}) \approx y^{(p)}$ holds componentwise.

To maximize L, we parameterize $R \in \mathcal{R}^{\sigma,\ell,t}(\mathbb{R}^d, \mathbb{R}^m)$ as usual by $\nu = (\ell \cdot d + (t-1)\cdot \ell^2 + t\cdot \ell + m \cdot \ell)$-many real parameters and switch to minimizing the *negative log-likelihood function*, which we denote here by

$$C\colon \mathbb{R}^\nu \to \mathbb{R},\ C(w_{ij}^{(k)}, b_i^{(k)}, a_{ij}) := -\log \prod_{p=1}^{n} \prod_{i=1}^{m} R_i(x^{(p)})^{y_i^{(p)}}.$$

By calculating the logarithm, we obtain

$$C(w_{ij}^{(k)}, b_i^{(k)}, a_{ij}) = -\sum_{p=1}^{n} \sum_{i=1}^{m} y_i^{(p)} \log R_i(x^{(p)}) \tag{16.3}$$

and we want to use this formula to determine a minimizer via gradient descent, see Algorithm 16.27. In the next theorem, we therefore establish, analogous to Theorem 16.32, how the derivatives needed can be calculated by backpropagation. According to Remark 16.28(i), we formulate the theorem for the case of a one-hot-encoded dataset with only one single data point.

Theorem 16.34. *Let $\sigma\colon \mathbb{R}\to\mathbb{R}$ be differentiable, $R\in\mathcal{R}^{\sigma,\ell,t}(\mathbb{R}^d,\mathbb{R}^m)$ be a neural network with softmax output, $D=\{(x,y)\}\subseteq\mathbb{R}^d\times\{\mathrm{e}_1,\dots,\mathrm{e}_m\}$ a dataset with only one data point and $C\colon\mathbb{R}^\nu\to\mathbb{R}$, $C(w,b,a)=-(y_1\log R_1(x)+\cdots+y_m\log R_m(x))$. With $n_j^{(0)}:=x_j$ we have*

$$\frac{\partial C}{\partial a_{ij}}=(R_i-y_i)\cdot n_j^{(t)},\quad \textit{as well as}\quad \frac{\partial C}{\partial w_{ij}^{(k)}}=\delta_i^{(k)}n_j^{(k-1)}\quad\textit{and}\quad\frac{\partial C}{\partial b_i^{(k)}}=\delta_i^{(k)},$$

for all layers $k=1,\dots,t$. The $\delta_i^{(k)}$ can be calculated recursively by

$$\delta^{(t)}=A^\intercal(R-y)\odot\sigma'(z^{(t)})\quad\textit{and}\quad\delta^{(k)}=(W^{(k+1)})^\intercal\delta^{(k+1)}\odot\sigma'(z^{(k)})$$

for $k=t-1,\dots,1$.

Proof. We first calculate the partial derivatives of the softmax function, which we abbreviate in this proof with s. Moreover, we write s_i for the softmax's i-th coordinate function. Since $\operatorname{ran}s_i\subseteq(0,1)$ holds, we get

$$\log s_i=\log\frac{\mathrm{e}^{z_i}}{\mathrm{e}^{z_1}+\cdots+\mathrm{e}^{z_m}}=z_i-\log(\mathrm{e}^{z_1}+\cdots+\mathrm{e}^{z_m}).$$

Differentiating gives on the one hand

$$\frac{\partial}{\partial z_j}\log s_i=\tfrac{1}{s_i}\cdot\frac{\partial s_i}{\partial z_j}$$

and, with the Kronecker symbol δ_{ij}, on the other hand

$$\begin{aligned}\frac{\partial}{\partial z_j}\log s_i&=\delta_{ij}-\frac{1}{\mathrm{e}^{z_1}+\cdots+\mathrm{e}^{z_m}}\cdot\frac{\partial}{\partial z_j}(\mathrm{e}^{z_1}+\cdots+\mathrm{e}^{z_m})\\&=\delta_{ij}-\frac{\mathrm{e}^{z_j}}{\mathrm{e}^{z_1}+\cdots+\mathrm{e}^{z_m}}=\delta_{ij}-s_j.\end{aligned}$$

Rearranging and subbing in leads to

$$\frac{\partial s_i}{\partial z_j}=s_i\cdot\frac{\partial}{\partial z_j}\log s_i=s_i(\delta_{ij}-s_j).$$

By definition, $R=s\circ N$ with $N\in\mathcal{D}^{\sigma,\ell,t}(\mathbb{R}^m,\mathbb{R}^n)$. We read C as a function $C=C(N_1,\dots,N_m)$, calculate the corresponding derivatives

$$\begin{aligned}\frac{\partial C}{\partial N_j}&=\frac{\partial}{\partial N_j}\Big(-\sum_{i=1}^m y_i\log s_i\Big)=-\sum_{i=1}^m y_i\frac{\partial}{\partial N_j}\log s_i=-\sum_{i=1}^m\tfrac{y_i}{s_i}\cdot\frac{\partial s_i}{\partial N_j}\\&=-\sum_{i=1}^m\tfrac{y_i}{s_i}\cdot s_i(\delta_{ij}-s_j)=-\sum_{i=1}^m(y_i\delta_{ij}-y_is_j)=s_j\big(\sum_{i=1}^m y_i\big)-y_j\end{aligned}$$

$$= \underset{\substack{\uparrow \\ y \text{ is a unit} \\ \text{vector}}}{} s_j - y_j$$

and obtain for their evaluations

$$\frac{\partial C}{\partial N_j}(N_1, \ldots, N_m) = s_j(N_1, \ldots, N_m) - y_j = R_j - y_j.$$

Furthermore, it holds

$$\frac{\partial N_\mu}{\partial a_{ij}} = \frac{\partial}{\partial a_{ij}}\left[a_{\mu 1} n_1^{(t)} + \cdots + a_{\mu \ell} n_\ell^{(t)}\right] = \begin{cases} n_j^{(t)} & \text{if } \mu = i, \\ 0 & \text{otherwise,} \end{cases}$$

and thus by the multidimensional chain rule

$$\frac{\partial C}{\partial a_{ij}} = \sum_{\mu=1}^{m} \frac{\partial C}{\partial N_\mu} \cdot \frac{\partial N_\mu}{\partial a_{ij}} = (R_i - y_i) \cdot n_j^{(t)}.$$

We now proceed similarly to the proofs of Propositions 16.29 and 16.30, using the auxiliary functions

$$z^{(k)} := \begin{bmatrix} z_1^{(k)} \\ \vdots \\ z_\ell^{(k)} \end{bmatrix} \quad \text{with} \quad z_i^{(k)} = w_{i1}^{(k)} n_1^{(k-1)} + \cdots + w_{i\ell}^{(k)} n_\ell^{(k-1)} + b_i^{(k)},$$

and start by calculating the derivatives with respect to the weights and biases of the last layer. First, we note that

$$\frac{\partial N_\mu}{\partial z_i^{(t)}} = \frac{\partial}{\partial z_i^{(t)}}\left[a_{\mu 1} n_1^{(t)} + \cdots + a_{\mu \ell} n_\ell^{(t)}\right] = a_{\mu i} \cdot \frac{\partial}{\partial z_i^{(t)}} n_i^{(t)} = a_{\mu i} \cdot \sigma'(z_i^{(t)})$$

holds, because $z_i^{(t)}$ only appears in $n_i^{(t)}$. Since $w_{ij}^{(t)}$ and $b_i^{(t)}$ only appear in $z_i^{(t)}$, we further obtain

$$\frac{\partial C}{\partial w_{ij}^{(t)}} = \sum_{\mu=1}^{m} \frac{\partial C}{\partial N_\mu} \cdot \frac{\partial N_\mu}{\partial z_i^{(t)}} \cdot \frac{\partial z_i^{(t)}}{\partial w_{ij}^{(t)}} = \sum_{\mu=1}^{m} (R_\mu - y_\mu) \cdot a_{\mu i} \cdot \sigma'(z_i^{(t)}) \cdot n_j^{(t-1)}$$

and

$$\frac{\partial C}{\partial b_i^{(t)}} = \sum_{\mu=1}^{m} (R_\mu - y_\mu) \cdot a_{\mu i} \cdot \sigma'(z_i^{(t)}) \cdot 1,$$

from which the equations for $\delta^{(t)}$ and for the derivatives in the case $k = t$ follow. We now proceed backwards, and must additionally consider for $1 \leqslant k < t$ that $w_{ij}^{(k)}$ and $b_i^{(k)}$ show up in all neurons of the $(k+1)$-th layer and thus also in all $z_v^{(k+1)}$ for $v = 1, \ldots, \ell$. We therefore obtain by the chain rule for functions

$\mathbb{R} \to \mathbb{R}^\ell \to \mathbb{R}$

$$
\begin{aligned}
\frac{\partial C}{\partial w_{ij}^{(k)}} &= \sum_{v=1}^{\ell} \frac{\partial C}{\partial z_v^{(k+1)}} \cdot \frac{\partial z_v^{(k+1)}}{\partial w_{ij}^{(k)}} \\
&= \sum_{v=1}^{\ell} \frac{\partial C}{\partial z_v^{(k+1)}} \cdot \frac{\partial}{\partial w_{ij}^{(k)}} \left(w_{v1}^{(k+1)} n_1^{(k)} + \cdots + w_{v\ell}^{(k+1)} n_\ell^{(k)} + b_v^{(k+1)}\right) \\
&\underset{\substack{\uparrow \\ w_{ij}^{(k)} \text{ only} \\ \text{appears in} \\ n_i^{(k)}}}{=} \sum_{v=1}^{\ell} \frac{\partial C}{\partial z_v^{(k+1)}} \cdot w_{vi}^{(k+1)} \cdot \frac{\partial n_i^{(k)}}{\partial w_{ij}^{(k)}} \\
&= \sum_{v=1}^{\ell} \frac{\partial C}{\partial z_v^{(k+1)}} \cdot w_{vi}^{(k+1)} \cdot \frac{\partial}{\partial w_{ij}^{(k)}} \sigma\big(\underbrace{w_{i1}^{(k)} n_1^{(k-1)} + \cdots + w_{i\ell}^{(k)} n_\ell^{(k-1)} + b_i^{(k)}}_{=z_i^{(k)}}\big) \\
&\underset{\substack{\uparrow \\ \text{chain} \\ \text{rule}}}{=} \sum_{v=1}^{\ell} \frac{\partial C}{\partial z_v^{(k+1)}} \cdot w_{vi}^{(k+1)} \cdot \sigma'(z_i^{(k)}) \cdot n_j^{(k-1)} \\
&= \Big(\sum_{v=1}^{\ell} w_{vi}^{(k+1)} \frac{\partial C}{\partial z_v^{(k+1)}} \cdot \sigma'(z_i^{(k)})\Big) \cdot n_j^{(k-1)}.
\end{aligned}
$$

For $k = t-1$ we have

$$
\frac{\partial C}{\partial z_v^{(k+1)}} = \frac{\partial C}{\partial z_v^{(t)}} = \sum_{\mu=1}^{m} \frac{\partial C}{\partial N_\mu} \cdot \frac{\partial N_\mu}{\partial z_v^{(t)}} = \sum_{\mu=1}^{m} (R_\mu - y_\mu) \cdot a_{\mu v} \cdot \sigma'(z_v^{(t)}) = \delta_v^{(t)}
$$

and we obtain for $k = t-1, t-2, \ldots, 1$ inductively the formula

$$
\frac{\partial C}{\partial w_{ij}^{(k)}} = \Big(\sum_{v=1}^{\ell} w_{vi}^{(k+1)} \cdot \delta_v^{(k+1)} \cdot \sigma'(z_i^{(k)})\Big) \cdot n_j^{(k-1)},
$$

in which the large bracket coincides with the i-th entry of $(W^{(k+1)})^\intercal \delta^{(k+1)} \odot \sigma'(z^{(k)})$. For the derivative with respect to $b_i^{(k)}$, one proceeds analogously, but after the fifth line in the above long calculation, one obtains

$$
\frac{\partial C}{\partial b_i^{(k)}} = \sum_{v=1}^{\ell} \frac{\partial C}{\partial z_v^{(k+1)}} \cdot w_{vi}^{(k+1)} \cdot \sigma'(z_i^{(k)}) \cdot 1
$$

which finishes the proof. □

Let us finally consider the case of binary labels. Here, of course, one can specialize above $m = 2$, but it may be more natural to use the numbers 0 and 1 instead of the unit vectors e_1 and e_2, compare the logistic regression in Chapter 2. Accordingly, one might then want to use the 1-dimensional sigmoid function instead of the 2-dimensional softmax as output. The class of approximants in this case consists of functions of the form

$$
S := \mathrm{sig} \circ N \colon \mathbb{R}^\nu \to (0,1),
$$

with $N \in \mathcal{D}^{\sigma,\ell,t}(\mathbb{R}^d, \mathbb{R}^1)$. If one now replaces in the cost function (16.3) R_1 with S and R_2 with $1 - S$, one obtains

$$C\colon \mathbb{R}^\nu \to \mathbb{R},\ C(a, w, b) = -\sum_{p=1}^{n} y^{(p)} \log S(x^{(p)}) + (1 - y^{(p)}) \log(1 - S(x^{(p)}))$$

and thus precisely the analogue of the negative log-likelihood function, which we minimized in Chapter 2.4 on logistic regression. In Exercise 16.5 we discuss a practical application and in Exercise 16.9 we show that, when transitioning from $\mathcal{R}^{\sigma,\ell,t}(\mathbb{R}^d, \mathbb{R}^2)$ to the class of functions S as above, expressivity is not affected.

References

We note that the first version of the Stone-Weierstrass Theorem 16.10 can be found, for example, in [Rud91, p. 122]. The second version, Theorem 16.15, follows from the first, by 1-point compactification of $\mathbb{R}$. The Riesz representation theorem in the more general version 16.16 — which runs also under the name of Riesz-Markov theorem — can be found in [Rud87, Theorem 6.19]. The second version is the specialization to a compact domain. For Theorem 16.18 we refer to [Rud87, Theorem 5.19]. Furthermore, we refer to [Rud87, Chapter 6] for definitions and results that go beyond the classic content of a course on measure theory, such as the spaces $\mathrm{M}(\Omega)$ and $\mathrm{M}(\mathbb{R})$ of complex measures; integrals with respect to a complex measure; and Fubini's theorem for complex measures.

The results at the beginning of the chapter on the representation of logical functions are all standard. Our version of Theorem 16.4 is a reformulation of [SSBD14, Claim 20.1]. The history of continuous approximation results begins with the work [Cyb89] from 1989, in which Lemma 16.17 and Theorem 16.20 for $\Omega = [0, 1]^d$ were proven in a direct fashion with a multivariate variant of the proofs we provided. Note the corrigendum [Cyb92]. Several works [HSW89, CL92, Hor91, HSW90, LLPS93, LP93, SC92] followed between 1989 and 1993, in which Cybenko's result was generalized in the sense that (a) the requirements for the activation function were weakened, and (b) approximation within other classes of functions (e.g., L^p, C^m) was considered. To treat the ReLU function, we followed the approach in [Gui18, Lemma 3.15]. The reduction argument we presented as Theorem 16.12 is based on [CL92] and [LP93]. The proof of Theorem 16.22 follows [LLPS93, Remark 3]. Regarding the approximation of continuous functions in terms of uniform convergence on compacts, we mention that in [LLPS93], for $\sigma \in \mathrm{L}^\infty_{\mathrm{loc}}(\mathbb{R})$ such that the closure of the set of discontinuity points of σ is a Lebesgue null set, the equivalence

$$\mathrm{C}(\mathbb{R}^n, \mathbb{R}^m) \subseteq \overline{\mathcal{S}^\sigma(\mathbb{R}^n, \mathbb{R}^m)}^{\mathrm{L}^\infty_{\mathrm{loc}}(\mathbb{R}^n, \mathbb{R}^m)} \iff \sigma \notin \mathbb{R}[X]$$

is shown, compare with Example 16.7(i). Theorem 16.24, allowing the tradeoff between shallow and deep networks, is based on [Han19], but was slightly modified here to handle networks with linear output. In our text, we have only dealt with very simple architectures of neural networks and refer to [Cal20] for a comprehensive treatment.

The technique of backpropagation has its origins in the 1960s; our presentation is based on [Nie15], but was slightly modified and expanded to include also networks with softmax output.

Exercises

Exercise 16.1. In this problem, we consider neurons with Heaviside activation.

(i) Show that the exclusive or function XOR: $\{0,1\}^2 \to \{0,1\}$ cannot be represented by a single neuron with two inputs.

(ii) Provide a representation of XOR by a neural network with (at most) three neurons and Heaviside activation.

Exercise 16.2. Use the conjunctive normal form in order to represent the function $f\colon \{0,1\}^3 \to \{0,1\}$, defined by the following table through a Heaviside-activated neural network.

x_1	x_2	x_3	$f(x_1, x_2, x_3)$
0	0	0	1
0	0	1	1
0	1	0	1
0	1	1	1
1	0	0	0
1	0	1	1
1	1	0	0
1	1	1	0

Exercise 16.3. Show, by applying the Stone-Weierstrass theorem, that the space of shallow neural networks with exponential activation $\mathcal{S}^{\exp}(\mathbb{R}) \subseteq C(\mathbb{R})$ is uniformly dense in the continuous functions on compact sets.

Exercise 16.4. We consider the following neural network,

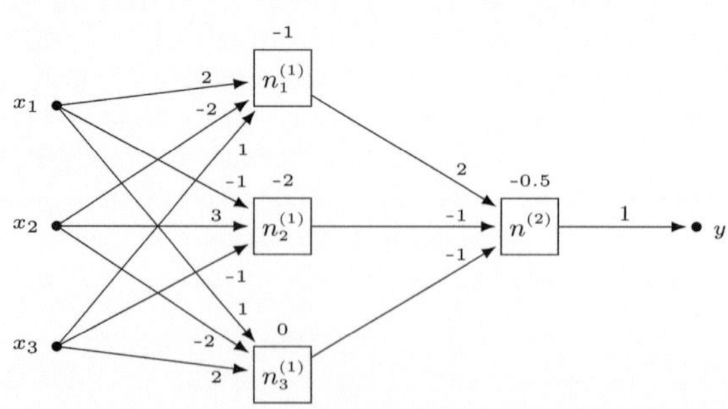

where the numbers above the arrows indicate the weights and the numbers above the boxes indicate the biases of the neurons. Let the neurons $n_1^{(1)}, n_2^{(1)}, n_3^{(1)}$ be ReLU- and let $n^{(2)}$ be sigmoid-activated. For each input given in the following table, calculate successively the outputs of the neurons $n_1^{(1)}, n_2^{(1)}, n_3^{(1)}$, and finally that of the neuron $n^{(2)}$, which is equal to the output of the whole network.

x_1	x_2	x_3	$n_1^{(1)}$	$n_2^{(1)}$	$n_3^{(1)}$	$n^{(2)}$	y
0	0	0					
0	0	1					
0	1	0					
0	1	1					
1	0	0					
1	0	1					
1	1	0					
1	1	1					

Exercise 16.5. "Train" a neural network with one hidden layer using the function `MLPClassifier` from the Python package `sklearn` such that the function

$$f\colon \{0,1\}^3 \to \{0,1\},\ f(x_1,x_2,x_3) := \mathrm{XOR}(x_1,x_2) \vee (x_1 \wedge x_2 \wedge x_3)$$

is approximated as well as possible. Which width for the hidden layer is needed for the above to work out?

Hint: In `MLPClassifier` you only need to specify the width of the hidden layers and their activation function. With `.predict()` you can then print the $\{0,1\}$-valued output for all training data at once. In the background, however, `MLPClassifier` uses an output layer as in Exercise 16.4 and calculates $p \in [0,1]$, so that the entries of the tuple $(p, 1-p)$ can be interpreted as probabilities for the classification with 0 or 1. These tuples can be printed using the command `.predict_proba()`.

Exercise 16.6. We consider the following network,

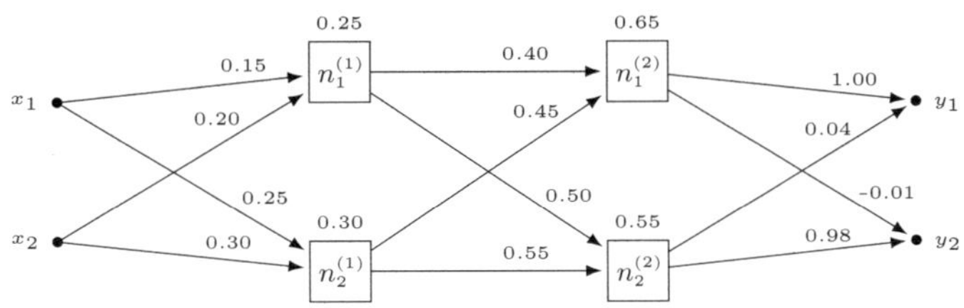

in which the numbers above the arrows indicate the weights and the numbers above the boxes indicate the biases of the neurons. Assume that all neurons are sigmoid-activated. A single training data point (x,y) is given with $x = (0.05, 0.10)$ and $y = (0.05, 1.01)$. Carry out one forward and then one backpropagation *by hand* and determine the first update of the weights, biases and matrix elements, using a learning rate of 0.5.

Hint: Take your time — this problem requires lengthy calculations.

Exercise 16.7. Postcompose the network from Exercise 16.6 with the softmax function and carry out one forward and one backpropagation again.

Exercise 16.8. Write a Python program that classifies handwritten numbers using a deep neural network employing the `MNIST` dataset and Theorem 16.34. Use a part of the dataset as training data and then test the classifier on randomly selected data points that were not part of the training data.

Hint: This is a rather complex task if you code everything "from scratch", but also an opportunity to at least once create and train a neural network all by yourself — alternatively, you may of course use ready-made packages that do the work for you.

Exercise 16.9. Show that the following holds:

$$\mathcal{R}^{\sigma,\ell,t}(\mathbb{R}^d,\mathbb{R}^2) = \left\{ \left[\begin{smallmatrix} S \\ 1-S \end{smallmatrix}\right] \mid S = \mathrm{sig} \circ N \mid N \in \mathcal{D}^{\sigma,\ell,t}(\mathbb{R}^d,\mathbb{R}^1) \right\}.$$

17

Gradient Descent for Convex Functions

In this chapter, we give an introduction to the gradient descent method, which is a classic method to approximate minimizers of real-valued functions numerically. Gradient descent is very versatile and thus extensively treated in the numerical analysis and optimization literature. Since the method is used in many data science and machine learning problems — as we have seen several times in previous chapters — we will now study it for a class of good-natured functions.

The underlying idea of gradient descent can be described as follows. Suppose we are looking for a minimizer of a function $f\colon \mathbb{R}^n \to \mathbb{R}$ as in the following picture.

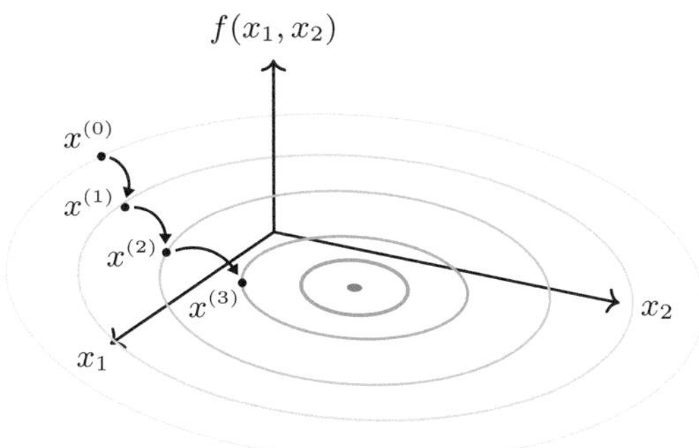

Then one can try to approximate the minimizer by starting at some point, e.g., a randomly chosen $x^{(0)}$, and then move step by step in the direction of steepest descent. Formally, we define a sequence $(x^{(k)})_{k\in\mathbb{N}}$ via

$$x^{(k+1)} := x^{(k)} - \gamma_k \nabla f(x^{(k)}), \tag{17.1}$$

where $\gamma_k > 0$ is called the *step size* or, in the context of machine learning,

S. A. Wegner, *Mathematical Introduction to Data Science*,
https://doi.org/10.1007/978-3-662-69426-8_17

the *learning rate*. The latter may be constant, i.e., $\gamma_k \equiv \gamma$; it may be varied by a given sequence, e.g., $\gamma_k = 1/k$; or it may be chosen anew in each step, e.g., such that $f(x^{(k)} - \gamma_k \nabla f(x^{(k)}))$ becomes as small as possible. A priori, of course, there is no guarantee that in this way $x^{(k)}$ will get close to a minimizer, or that $f(x^{(k)})$ will get close to the minimum value of f. In this chapter, we will show that this is the case under a suitable combination of convexity and differentiability properties on f and suitably small and constant step size. Under these conditions we will further quantify how many iterations are necessary to achieve a certain accuracy of approximation. To ensure that (17.1) is well-defined, we restrict ourselves for the entire chapter to functions that are differentiable and, for simplicity, defined on all of $\mathbb{R}^n$.

Firstly, we clarify our notation.

Definition 17.1. Let $f\colon \mathbb{R}^n \to \mathbb{R}$ be a function. A point $x^* \in \mathbb{R}^n$ is called a *minimum point* or *minimizer* of f, if $f(x^*) \leqslant f(x)$ holds for all $x \in \mathbb{R}^d$. We denote by

$$X^* = \operatorname*{argmin}_{x\in\mathbb{R}^n} f(x)$$

the set of all minimizers of f. If $X^* = \{x^*\}$ is a singleton, we drop the curly brackets and simply write $x^* = \operatorname{argmin}_{x\in\mathbb{R}^n} f(x)$. In particular, the term "minimizer" refers, in the following, always to a global minimum point. The *minimum value* or the *minimum* of f we denote, if it exists, by

$$f^* := \min_{x\in\mathbb{R}^n} f(x).$$

In general, f needs not to have any minimizers. Although the gradient method may still be used sensibly in such cases, see Exercise 17.1, we will below either explicitly assume that minimizers exist, or guarantee this by other conditions on f. Once $X^* \neq \emptyset$ is ensured, another central problem of the gradient descent method is illustrated by the following picture, for a concrete example see Exercise 17.3.

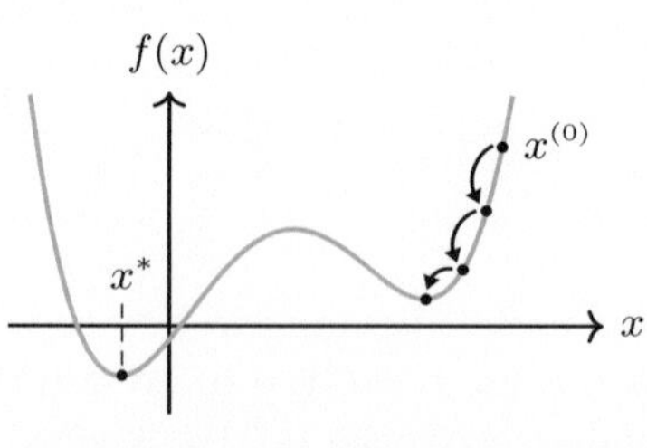

The depicted function has a uniquely determined minimizer and the sequence $(f(x^{(k)}))_{k\in\mathbb{N}}$ is convergent — but its limit is not the minimum f^*. The gradient method can also be successfully applied to such functions; in the picture, an easy fix would be a smaller $x^{(0)}$; for example, $x^{(0)} < 0$. For our theoretical results in this chapter, however, we want to exclude such situations and we will achieve this through the following definition.

Definition 17.2. A function $f\colon \mathbb{R}^n \to \mathbb{R}$ is called *convex* if

$$f(\lambda x + (1-\lambda)y) \leqslant \lambda f(x) + (1-\lambda)f(y)$$

holds for all $\lambda \in [0,1]$ and $x, y \in \mathbb{R}^n$. Note that the inequality trivially holds for $\lambda \in \{0,1\}$ or $x = y$.

For convex functions, the following *local-to-global principle* applies.

Proposition 17.3. *Let $f\colon \mathbb{R}^d \to \mathbb{R}$ be convex. If x is a local minimum point of f, then $x \in X^*$, i.e., x is a (global) minimizer.*

Proof. Let x be a local minimum point. For any $y \in \mathbb{R}^n$, there exists $\lambda_0 \in (0,1)$ such that for all $\lambda \in (0, \lambda_0)$ the estimates

$$f(x) \leqslant f((1-\lambda)x + \lambda y) \leqslant (1-\lambda)f(x) + \lambda f(y) = f(x) - \lambda f(x) + \lambda f(y)$$

hold, where we used the convexity of f for the second estimate. From this we get $f(x) \leqslant f(y)$, because $\lambda > 0$. Since $y \in \mathbb{R}^n$ was arbitrary, this shows that $x \in X^*$. □

Next, we show that for convex differentiable functions the condition $\nabla f(x) = 0$ is not only necessary for x to be a minimizer but even sufficient. Firstly, we characterize convexity using the gradient. Note that in the following lemma, the smaller side of the estimate is precisely the first Taylor polynomial $\mathrm{T}^1_{f,x}(y)$ of f at x, evaluated at y.

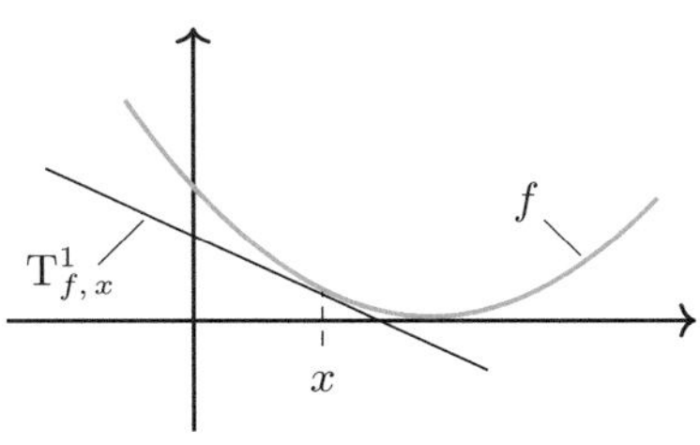

The estimate in the lemma thus says in an intuitive way that the function values of f lie above the tangent to the graph of f at the point x.

Proposition 17.4. *Let $f\colon \mathbb{R}^n \to \mathbb{R}$ be differentiable. Then f is convex if and only if $f(y) \geqslant f(x) + \langle \nabla f(x), y - x\rangle$ holds for all $x, y \in \mathbb{R}^d$.*

Proof. "$\Longrightarrow$" From the definition of convexity, by subtracting $f(x)$, dividing by λ, and taking the limit we get

$$f(y) - f(x) \geqslant \frac{f(x + \lambda(y-x)) - f(x)}{\lambda} \xrightarrow{\lambda \searrow 0} f'(x)(y-x) = \langle \nabla f(x), y - x\rangle.$$

"$\Longleftarrow$" Let $x, y \in \mathbb{R}^n$ and $\lambda \in (0,1)$. We put $z := \lambda x + (1-\lambda)y$. Then by assumption, $f(x) \geqslant f(z) + \langle \nabla f(z), x - z\rangle$ and $f(y) \geqslant f(z) + \langle \nabla f(z), y - z\rangle$ hold. Multiplication with λ, respectively with $(1-\lambda)$, and addition of both equations yields

$$\begin{aligned}
&\lambda f(x) + (1-\lambda) f(y) \\
&\qquad \geqslant \lambda f(z) + \lambda\langle \nabla f(z), x - z\rangle + (1-\lambda) f(z) + (1-\lambda)\langle \nabla f(z), y - z\rangle \\
&\qquad = f(z) + \langle \nabla f(z), \underbrace{\lambda x - \lambda z + (1-\lambda)y - (1-\lambda)z}_{=z-\lambda z-(1-\lambda)z=0}\rangle \\
&\qquad = f(\lambda x + (1-\lambda)y)
\end{aligned}$$

and thus shows the convexity of f. □

As a consequence, we get the result announced earlier.

Corollary 17.5. *Let $f\colon \mathbb{R}^n \to \mathbb{R}$ be convex and differentiable. Then $x \in X^*$ if and only if $\nabla f(x) = 0$.*

Proof. It only needs to be shown that $\nabla f(x) = 0$ is sufficient for a local minimum, because then the desired result follows from Proposition 17.3. Due to the convexity of f, Proposition 17.4 with $\nabla f(x) = 0$ implies that $f(y) \geqslant f(x) + \langle \nabla f(x), y - x\rangle = f(x)$ holds for all $y \in \mathbb{R}^n$ and we are done. □

Now let a function $f\colon \mathbb{R}^n \to \mathbb{R}$ be given, which is convex and differentiable, and additionally let $X^* \neq \emptyset$ hold for this function. For a given initial value $x^{(0)}$ we consider the sequence $(x^{(k)})_{k\in\mathbb{N}}$, defined by (17.1), i.e.,

$$x^{(k+1)} = x^{(k)} - \gamma_k \nabla f(x^{(k)}).$$

Primarily, we are interested in whether, and if so, how quickly, the image sequence $(f(x^{(k)}))_{k\in\mathbb{N}}$ converges to f^*, compare the applications explained in Chapters 2, 4 and 16. The following example illustrates that convexity and differentiability alone are not sufficient to guarantee the convergence of the sequence $(f(x^{(k)}))_{k\in\mathbb{N}}$, but that the choice of step sizes plays a crucial role as well.

Example 17.6. Let $f\colon \mathbb{R} \to \mathbb{R}$, $f(x) = x^2$, i.e., $x^* := \operatorname{argmin}_{x\in\mathbb{R}} f(x) = 0$, and let $x^{(0)} \neq 0$. Then $x^{(k+1)} = (1-2\gamma_k)x^{(k)}$ and we have the following effects, depending on the choice of γ_k.

(i) For $\gamma_k \equiv 1$: $x^{(k)} = \pm x^{(0)}$, and the image sequence $f(x^{(k)}) \equiv f(x^{(0)})$ is constant and not equal to the minimum.

(ii) For $\gamma_k \equiv \gamma > 1$: $(x^{(k)})_{k\in\mathbb{N}}$ is divergent and $f(x^{(k)}) \to \infty$.

(iii) For $\gamma_k \equiv \gamma < 1$: we have $x^{(k)} \to 0 = x^*$ as well as $f(x^{(k)}) \to 0 = f^*$ and thus the desired result.

Now we consider strictly positive null sequences $\gamma_k \searrow 0$. To exclude the uninteresting case that some γ_{k_0} becomes equal to $1/2$, and thus $x^{(k)} = 0$ for $k \geqslant k_0$, and to keep the calculations as simple as possible, we consider the following two examples.

(iv) For $\gamma_k = \frac{1}{k+3}$, we get $x^{(k)} = x^{(0)} \prod_{i=0}^{k} (1 - \frac{2}{i+3}) = x^{(0)} \frac{2}{(k+2)(k+3)}$ by calculating the (telescopic) product. This implies $x^{(k)} \to 0$ and, as in (iii), we obtain the desired result.

(v) Now let $\gamma_k = \frac{1}{2(k+2)^2}$ and thus $x^{(k)} = x^{(0)} \prod_{i=2}^{k+2} (1 - \frac{1}{i^2}) = x^{(0)} \prod_{i=2}^{k+2} (1 - \frac{1}{i^2})$. Taking the logarithm and considering $k \to \infty$ leads to the well-known series (telescopic sum!)

$$\sum_{i=2}^{\infty} \log\left(1 - \frac{1}{i^2}\right) = -\log(2)$$

and therefore shows $x^{(k)} \to x^{(0)} \frac{1}{2} \neq x^*$. In particular, the image sequence converges to a value which, depending on the starting point, can be arbitrarily larger than f^*.

In Example 17.6(i)–(ii) we obtained $f(x^{(k)}) \not\to f^*$ only because we deliberately chose a too large step size. In fact, for any convex and differentiable function f, in the k-th step of the gradient method, a $\gamma_k > 0$ can be chosen such that

$$f(x^{(k+1)}) = f(x^{(k)} - \gamma_k \nabla f(x^{(k)})) < f(x^{(k)})$$

holds as long as $\nabla f(x^{(k)}) \neq 0$: If

$$f(x^{(k+1)}) = f(x^{(k)} - \gamma \nabla f(x^{(k)})) \geqslant f(x^{(k)})$$

would hold for all $\gamma > 0$, then the directional derivative

$$\begin{aligned} -\langle \nabla f(x^{(k)}), \nabla f(x^{(k)}) \rangle &= \partial_{-\nabla f(x^{(k)})} f(x^{(k)}) \\ &= \lim_{h \searrow 0} \frac{f(x^{(k)} - h \nabla f(x^{(k)})) - f(x^{(k)})}{h} \geqslant 0 \end{aligned}$$

would be nonnegative, which is only possible if the gradient at $x^{(k)}$ vanishes. So there is always a sequence of step sizes $(\gamma_k)_{k \in \mathbb{N}}$ such that the image sequence $(f(x^{(k+1)}))_{k \in \mathbb{N}}$ decreases monotonically and therefore converges. As long as we do not land precisely on a minimizer with our $x^{(k)}$'s, the sequence even decreases strictly, which means that our approximation of f^* improves with each iteration. On the other hand, Example 17.6(v) shows that if we let the γ_k decrease too quickly, then $f(x^{(k)}) \not\to f^*$ may occur.

The step sizes therefore on the one hand should neither be a too large constant, but on the other hand, not decrease too quickly towards zero. Our previous arguments and the examples suggest that a constant $\gamma > 0$ should suffice if the derivative of f is bounded. In fact, we now require something stronger, but refer the reader to Exercise 17.10.

Definition 17.7. Let $f\colon \mathbb{R}^n \to \mathbb{R}$ be differentiable and let $L > 0$. Then f is called *L-smooth* if $\|\nabla f(x) - \nabla f(y)\| \leqslant L\|x - y\|$ holds for all $x, y \in \mathbb{R}^n$.

In other words, the above means that ∇f is Lipschitz continuous. We need the following two consequences.

Lemma 17.8. *Let $f\colon \mathbb{R}^n \to \mathbb{R}$ be an L-smooth function with $L > 0$. Then we have $f(x) - f(y) - \langle \nabla f(y), x - y\rangle \leqslant \frac{L}{2}\|x - y\|^2$ for $x, y \in \mathbb{R}^n$.*

Proof. Let $x, y \in \mathbb{R}^n$ be fixed. We define the auxiliary function $h\colon \mathbb{R} \to \mathbb{R}$, $h(t) = f(y + t(x - y))$. Then $h'(t) = \langle \nabla f(y + t(x - y)), x - y\rangle$ by the chain rule. By the fundamental theorem of calculus, we further have

$$f(x) - f(y) = h(1) - h(0) = \int_0^1 h'(t)\,\mathrm{d}t = \int_0^1 \langle \nabla f(y + t(x - y)), x - y\rangle\,\mathrm{d}t.$$

With this, and taking into account that we integrate from 0 to 1, we get

$$\begin{aligned}
f(x) - f(y) - \langle \nabla f(y), x - y\rangle &= \int_0^1 \langle \nabla f(y + t(x - y)), x - y\rangle - \langle \nabla f(y), x - y\rangle\,\mathrm{d}t \\
&\leqslant \int_0^1 |\langle \nabla f(y + t(x - y)) - \nabla f(y), x - y\rangle|\,\mathrm{d}t \\
&\leqslant \int_0^1 \|\nabla f(y + t(x - y)) - \nabla f(y)\|\|x - y\|\,\mathrm{d}t \\
&\leqslant \int_0^1 L\|y + t(x - y) - y\|\|x - y\|\,\mathrm{d}t \\
&= \int_0^1 Lt\|x - y\|^2\,\mathrm{d}t = \tfrac{L}{2}\|x - y\|^2,
\end{aligned}$$

where we first used the Cauchy-Schwarz-Bunyakovsky inequality and then the L-smoothness. □

We now add the assumption that f is convex.

Lemma 17.9. *Let $f\colon \mathbb{R}^n \to \mathbb{R}$ be convex and L-smooth with $L > 0$. Then we have $f(x) - f(y) \leqslant \langle \nabla f(x), x - y\rangle - \frac{1}{2L}\|\nabla f(x) - \nabla f(y)\|^2$ for $x, y \in \mathbb{R}^n$.*

Proof. Let $x, y \in \mathbb{R}^n$ be given. We put $z := y - \frac{1}{L}(\nabla f(y) - \nabla f(x))$ and estimate in the following calculations $f(x) - f(z)$ using Proposition 17.4 and $f(z) - f(y)$ using Lemma 17.8. Then we sub in z and simplify. This gives

$$\begin{aligned}
f(x) - f(y) &= f(x) - f(z) + f(z) - f(y) \\
&\leqslant \langle \nabla f(x), x - z\rangle + \langle \nabla f(y), z - y\rangle + \tfrac{L}{2}\|z - y\|^2
\end{aligned}$$

$$
\begin{aligned}
&= \langle \nabla f(x), x - y + \tfrac{1}{L}(\nabla f(y) - \nabla f(x))\rangle - \langle \nabla f(y), y - y \\
&\quad + \tfrac{1}{L}(\nabla f(y) - \nabla f(x))\rangle + \tfrac{L}{2}\|y - \tfrac{1}{L}(\nabla f(y) - \nabla f(x)) - y\|^2 \\
&= \langle \nabla f(x), x - y\rangle + \langle \nabla f(x) - \nabla f(y), \tfrac{1}{L}(\nabla f(y) - \nabla f(x))\rangle \\
&\quad + \tfrac{1}{2L}\|\nabla f(y) - \nabla f(x)\|^2 \\
&= \langle \nabla f(x), x - y\rangle - \tfrac{1}{2L}\|\nabla f(y) - \nabla f(x)\|^2
\end{aligned}
$$

as claimed. □

Before we prove the first main result on gradient descent, let us note the following equivalence, which is sometimes used in the literature as a definition of convex L-smooth functions.

Corollary 17.10. *Let $f\colon \mathbb{R}^n \to \mathbb{R}$ be differentiable and let $L > 0$. Then the following statements are equivalent.*

(i) f is L-smooth and convex.

(ii) $\forall\, x, y \in \mathbb{R}^n\colon 0 \leqslant f(x) - f(y) - \langle \nabla f(y), x - y\rangle \leqslant \frac{L}{2}\|x - y\|^2$.

Proof. (i) $\Longrightarrow$ (ii): The first inequality follows from convexity with Proposition 17.4 and the second inequality follows from L-smoothness with Lemma 17.8.

(ii) $\Longrightarrow$ (i): The first inequality implies convexity with Proposition 17.4. The proof of Lemma 17.9 shows that convexity and the second inequality, i.e., the inequality in Lemma 17.8, imply that

$$
\begin{aligned}
\tfrac{1}{2L}\|\nabla f(x) - \nabla f(y)\|^2 &\leqslant f(y) - f(x) + \langle \nabla f(x), x - y\rangle \\
&= f(y) - f(x) - \langle \nabla f(x), y - x\rangle \\
&\leqslant \tfrac{L}{2}\|x - y\|^2
\end{aligned}
$$

holds. Here we used the second inequality from (ii) again in the last step, but now with the roles of x and y swapped. Multiplication by $2L$ and taking the square root shows $\|\nabla f(x) - \nabla f(y)\| \leqslant L\|x - y\|$ and thus that f is L-smooth. □

Now we are ready for our first result on the gradient descent method. Below we will show that for an L-smooth and convex function f, and with a suitably small constant step size γ in (17.1), the sequence $(f(x^{(k)}))_{k\in\mathbb{N}}$ converges to the minimum f^* of f.

Theorem 17.11. *Let $f\colon \mathbb{R}^n \to \mathbb{R}$ be convex and L-smooth with $L > 0$. Let $x^* \in X^*$ and $x^{(0)} \in \mathbb{R}^n$ be arbitrary, let $(x^{(k)})_{k\in\mathbb{N}}$ denote the sequence from*

(17.1), *where the step size* $\gamma := 1/L$ *is chosen to be constant. Then we have*

$$0 \leqslant f(x^{(k)}) - f^* \leqslant \frac{2L}{k}\|x^{(0)} - x^*\|^2$$

for $k \geqslant 1$ *and, in particular,* $f(x^{(k)}) \to f^*$ *for* $k \to \infty$.

Proof. ① For $j \geqslant 0$ we plug $x = x^{(j+1)} = x^{(j)} - \frac{1}{L}\nabla f(x^{(j)})$ and $y = x^{(j)}$ into the inequality from Lemma 17.8 to obtain

$$f(x^{(j+1)}) - f(x^{(j)}) - \langle \nabla f(x^{(j)}), -\tfrac{1}{L}\nabla f(x^{(j)})\rangle \leqslant \tfrac{L}{2}\big\| - \tfrac{1}{L}\nabla f(x^{(j)})\big\|^2.$$

Simplifying and rearranging yields

$$\begin{aligned} f(x^{(j+1)}) - f(x^{(j)}) &\leqslant \tfrac{1}{2L}\big\|\nabla f(x^{(j)})\big\|^2 - \tfrac{1}{L}\langle \nabla f(x^{(j)}), \nabla f(x^{(j)})\rangle \\ &= -\tfrac{1}{2L}\big\|\nabla f(x^{(j)})\big\|^2, \end{aligned}$$

from which we see that for this choice of γ the sequence $(f(x^{(j)}))_{j\in\mathbb{N}}$ decreases monotonically and thus converges.

② From Lemma 17.9 and Proposition 17.4 we deduce

$$\begin{aligned} \langle \nabla f(x) - \nabla f(y), x - y\rangle &= \langle f(x), x - y\rangle - \langle \nabla f(y), x - y\rangle \\ &\underset{\substack{\uparrow \\ \text{Lemma} \\ 17.9}}{\geqslant} f(x) - f(y) + \tfrac{1}{2L}\|\nabla f(x) - \nabla f(y)\|^2 \\ &\qquad - \langle \nabla f(y), x - y\rangle \\ &\underset{\substack{\uparrow \\ \text{Prop.} \\ 17.4}}{\geqslant} \tfrac{1}{2L}\|\nabla f(x) - \nabla f(y)\|^2 \end{aligned}$$

for arbitrary $x, y \in \mathbb{R}^n$. We now put $x = x^{(j)}$ and $y = x^*$, so $\nabla f(x^*) = 0$ according to Corollary 17.5, and we obtain

$$\langle \nabla f(x^{(j)}), x^{(j)} - x^*\rangle \geqslant \tfrac{1}{2L}\|\nabla f(x^{(j)})\|^2.$$

After these preparations, we get

$$\begin{aligned} \|x^{(j+1)} - x^*\|^2 &= \|x^{(j)} - \tfrac{1}{L}\nabla f(x^{(j)}) - x^*\|^2 \\ &= \|x^{(j)} - x^*\|^2 + 2\big\langle -\tfrac{1}{L}\nabla f(x^{(j)}), x^{(j)} - x^*\big\rangle + \| - \tfrac{1}{L}\nabla f(x^{(j)})\|^2 \\ &\underset{\substack{\uparrow \\ \text{as} \\ \text{above}}}{\leqslant} \|x^{(j)} - x^*\|^2 - \tfrac{2}{2L^2}\|\nabla f(x^{(j)})\|^2 + \tfrac{1}{L^2}\|\nabla f(x^{(j)})\|^2 \\ &= \|x^{(j)} - x^*\|^2. \end{aligned}$$

③ Now we notice that we only need to show the estimate in the theorem for $k < k_0 := \inf\{j \geqslant 0 \mid x^{(j)} \in X^*\}$. If $k_0 < \infty$ and $k \geqslant k_0$ then the left side of the inequality is zero. We assume that $k_0 \geqslant 1$ (in particular $x^{(0)} \neq x^*$), we fix $0 \leqslant k < k_0$, and we put $\delta_k := f(x^{(k)}) - f(x^*)$. According to what we just noted, $\delta_0, \ldots, \delta_k > 0$. For $0 \leqslant j < k$ we get according to part ①

$$\delta_{j+1} = f(x^{(j+1)}) - f(x^{(j)}) + f(x^{(j)}) - f(x^*) \leqslant \delta_j - \tfrac{1}{2L}\|\nabla f(x^{(j)}\|^2$$

and further

$$\begin{aligned}\delta_j &= f(x^{(j)}) - f(x^*) \underset{\substack{\uparrow \\ \text{Prop.} \\ 17.4}}{\leqslant} -\langle \nabla f(x^{(j)}), x^* - x^{(j)}\rangle \\ &\underset{\substack{\uparrow \\ \text{CSB} \\ \text{inequality}}}{\leqslant} \|\nabla f(x^{(j)})\|\|x^{(j)} - x^*\| \underset{\substack{\uparrow \\ \text{part ②}}}{\leqslant} \|\nabla f(x^{(j)})\|\|x^{(0)} - x^*\|.\end{aligned}$$

It follows $\|\nabla f(x^{(j)})\| \geqslant \delta_j / \|x^{(0)} - x^*\|$ and we obtain

$$\delta_{j+1} \leqslant \delta_j - \tfrac{1}{2L}\|\nabla f(x^{(j)})\|^2 \leqslant \delta_j - \frac{1}{2L\|x^{(0)} - x^*\|^2}\delta_j^2 = \delta_j - \omega\,\delta_j^2,$$

with $\omega := (2L\|x^{(0)} - x^*\|^2)^{-1}$. This means $\omega\delta_j^2 + \delta_{j+1} \leqslant \delta_j$, from which, via division by $\delta_{j+1} \cdot \delta_j$, the inequality $\omega\frac{\delta_j}{\delta_{j+1}} + \frac{1}{\delta_j} \leqslant \frac{1}{\delta_{j+1}}$ follows. By rearranging and estimating we get

$$\frac{1}{\delta_{j+1}} - \frac{1}{\delta_j} \geqslant \frac{\delta_{j+1}}{\delta_j}\Big(\frac{1}{\delta_{j+1}} - \frac{1}{\delta_j}\Big) \geqslant \omega.$$

For $k \geqslant 1$ we now sum up the inequalities for $j = 0, \ldots, k-1$ and obtain the following telescopic sum

$$\frac{1}{\delta_k} \geqslant \frac{1}{\delta_k} - \frac{1}{\delta_0} = \sum_{j=0}^{k-1} \frac{1}{\delta_{j+1}} - \frac{1}{\delta_j} \geqslant k\omega$$

and thus $\delta_k \leqslant \frac{1}{k\omega}$. This implies, together with $f^* - f(x^*) \leqslant f(x^{(k)})$, that

$$0 \leqslant f(x^{(k)}) - f^* = \delta_k \leqslant \frac{1}{k\omega} = \frac{2L\|x^{(0)} - x^*\|^2}{k}$$

holds as desired. □

In the above situation, $2L\|x^{(k)} - x^*\|^2$ may not be explicitly known, but it is for sure constant. The number of iterations of (17.1) that must be performed to achieve an approximation quality of $|f(x^{(k)}) - f^*| < \varepsilon$ is therefore bounded by $K \cdot \frac{1}{\varepsilon}$, where $K \geqslant 0$ is a constant independent of ε. If we choose $\gamma \in (0, 1/L]$, we obtain a similar estimate as in the theorem. If we carry out gradient descent for an L-smooth function, where the concrete value of L is unknown to us, we nevertheless know that for suitably small and constant γ the image sequence will eventually get close to f^*, cf. Exercise 17.6.

Theorem 17.11 does not guarantee the convergence $x^{(k)} \to x^*$ and this cannot generally be expected, consider for instance the trivial example $f \equiv 0$ and $x^* \neq x^{(0)}$, or a situation like in the following picture, in which the $x^{(k)}$ do tend to *a* minimizer, but, e.g., not to the one that is located closest to the starting point.

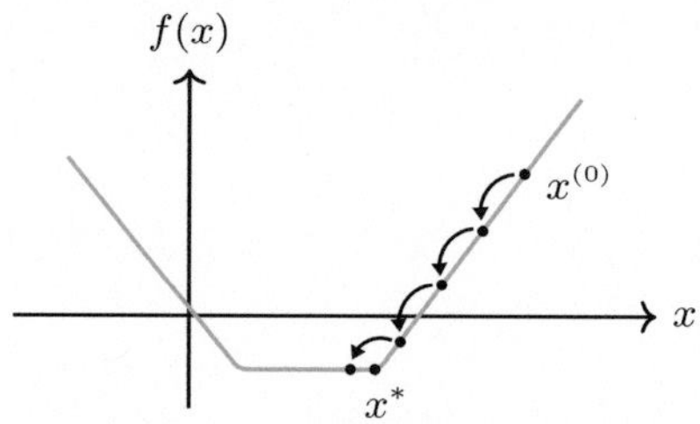

On the other hand, the conditions of Theorem 17.11 certainly ensure the convergence of the sequence $(x^{(k)})_{k\in\mathbb{N}}$ to *some* minimizer — which might, however, be different from x^*.

Corollary 17.12. *Under the assumtions of Theorem 17.11, the gradient descent sequence $(x^{(k)})_{k\in\mathbb{N}}$ converges to a minimizer $z^* \in X^*$ of f.*

Proof. From part ② of the previous proof it follows

$$\forall\, x^* \in X^*,\ k \geqslant 0\colon \|x^{(k+1)} - x^*\| \leqslant \|x^{(k)} - x^*\|.$$

The sequence $(x^{(k)})_{k\in\mathbb{N}}$ is therefore bounded and consequently contains a convergent subsequence $(x^{(k_j)})_{j\in\mathbb{N}}$, whose limit we call z^*. Since f is continuous and since $f(x^{(k)}) \to f^*$ holds according to Theorem 17.11, we get

$$f(z^*) = \lim_{j\to\infty} f(x^{(k_j)}) = f^*,$$

and thus $z^* \in X^*$. According to the above, $(\|x^{(k+1)} - z^*\|)_{k\in\mathbb{N}}$ is monotonically decreasing with a subsequence that converges to zero. But this means that the whole sequence converges to zero, hence $x^{(k)} \to z^*$ must hold. □

The following definition rules out the existence of multiple minimizers.

Definition 17.13. A function $f\colon \mathbb{R}^n \to \mathbb{R}$ is called *strictly convex*, if

$$f(\lambda x + (1-\lambda)y) < \lambda f(x) + (1-\lambda) f(y)$$

holds for all $\lambda \in (0,1)$ and $x \neq y \in \mathbb{R}^n$.

Obviously, every strictly convex function is also convex. Examples of convex and not strictly convex functions are the absolute value $|\cdot|\colon \mathbb{R} \to \mathbb{R}$ or any affine linear function. As announced, strict convexity guarantees the uniqueness of the minimizer, if one exist.

Proposition 17.14. *Let $f\colon \mathbb{R}^n \to \mathbb{R}$ be strictly convex and let $x^* \in X^*$. Then x^* is the only minimizer of f.*

Proof. Let $x \in X^*$ and $x \neq x^*$. Then for arbitrary $\lambda \in (0,1)$ we have

$$\begin{aligned} f(\lambda x + (1-\lambda)x^*) &< \lambda f(x) + (1-\lambda)f(x^*) \\ &= \lambda f(x^*) + (1-\lambda)f(x^*) \\ &= f(x^*), \end{aligned}$$

contradicting that $x^* \in X^*$ is a minimizer. □

We consider the following example, which is central for both the already outlined applications and the subsequent theoretical results on gradient descent.

Example 17.15. The square of the Euclidean norm $\|\cdot\|^2\colon \mathbb{R}^n \to \mathbb{R}$ is strictly convex. In fact, for $\lambda \in (0,1)$ and $x \neq y \in \mathbb{R}^n$ we have

$$\begin{aligned} &\lambda\|x\|^2 + (1-\lambda)\|y\|^2 - \|\lambda x + (1-\lambda)y\|^2 \\ &\quad= \lambda\|x\|^2 + (1-\lambda)\|y\|^2 - \langle \lambda x + (1-\lambda)y, \lambda x + (1-\lambda)y\rangle \\ &\quad= \lambda\|x\|^2 + (1-\lambda)\|y\|^2 - \|\lambda x\|^2 - 2\langle \lambda x, (1-\lambda)y\rangle - \|(1-\lambda)y\|^2 \\ &\quad= (\lambda - \lambda^2)\|x\|^2 + \big[(1-\lambda) - (1-\lambda)^2\big]\|y\|^2 - \lambda(1-\lambda)2\langle x, y\rangle \\ &\quad= \lambda(1-\lambda)\big[\|x\|^2 + \|y\|^2 - 2\langle x, y\rangle\big] \\ &\quad= \lambda(1-\lambda)\|x - y\|^2 > 0. \end{aligned}$$

Rearranging leads to the required estimate.

In combination with the following lemma, Example 17.15 gives rise to an important class of (strictly) convex functions.

Lemma 17.16. *Let $g\colon \mathbb{R}^n \to \mathbb{R}$ be convex and $h\colon \mathbb{R}^m \to \mathbb{R}^n$ be affine linear. Then $g \circ h$ is convex. If g is strictly convex and h is affine linear and injective, then $g \circ h$ is even strictly convex.*

Proof. ① Let h be given by $h(x) = Ax + b$ with $A \in \mathbb{R}^{n\times m}$, $b \in \mathbb{R}^n$, and let $\lambda \in (0,1)$ and $x \neq y \in \mathbb{R}^m$. Then we have

$$\begin{aligned} (g \circ h)(\lambda x + (1-\lambda)y) &= g\big(\lambda Ax + (1-\lambda)Ay + b\big) \\ &= g\big(\lambda Ax + \lambda b + (1-\lambda)Ay + (1-\lambda)b\big) \end{aligned}$$

$$\begin{aligned} &= g\big(\lambda(Ax+b)+(1-\lambda)(Ay+b)\big) \\ &\leqslant \lambda g(h(x))+(1-\lambda)g(h(y)), \end{aligned}$$

where we used $b = \lambda b + (1-\lambda)b$ in the second equation and the convexity of g for the inequality at the end.

② If h is injective, then $Ax + b \neq Ay + b$ holds above and therefore the estimate is strict due to the strict convexity of g. □

We will now reinforce the strict convexity assumption and then show that the existence and uniqueness of a minimizer x^* automatically follows. Moreover, we will see that $x^{(k)} \to x^*$ holds and that we achieve the accuracy $|f(x^{(k)}) - f^*| < \varepsilon$ at the latest after $(K \log \frac{1}{\varepsilon})$-many iterations, where $K \geqslant 0$ is a constant independent of ε.

Definition 17.17. Let $\mu > 0$. The function $f\colon \mathbb{R}^n \to \mathbb{R}$ is called *μ-convex*, if

$$f(\lambda x + (1-\lambda)y) + \lambda(1-\lambda)\tfrac{\mu}{2}\|x-y\|^2 \leqslant \lambda f(x) + (1-\lambda)f(y)$$

holds for all x, $y \in \mathbb{R}^n$ and $\lambda \in [0,1]$. Notice that the inequaility is trivial for $\lambda \in \{0,1\}$ or $x = y$.

Examples of μ-convex functions will be given below. We first note that

$$\mu\text{-convex} \quad \overset{\Longrightarrow}{\not\Longleftarrow} \quad f \text{ strictly convex} \quad \overset{\Longrightarrow}{\not\Longleftarrow} \quad f \text{ convex}.$$

An example for the first non-implication is the exponential function: For $x \to -\infty$, $y := x + 1$ and $\mu > 0$ arbitrary, in Definition 17.17 the right-hand side of the inequality and the first summand on the left tend to zero, while the second summand on the left is strictly positive and independent of x.

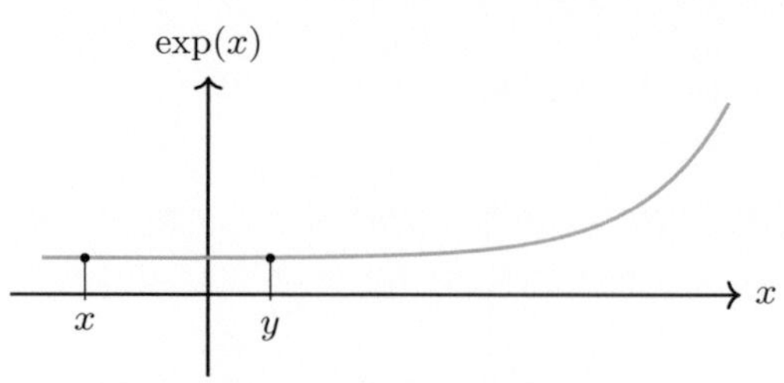

Next, we characterize μ-convex functions as follows.

Lemma 17.18. *A function $f\colon \mathbb{R}^n \to \mathbb{R}$ is μ-convex with $\mu > 0$ if and only if the auxiliary function $\varphi\colon \mathbb{R}^n \to \mathbb{R}$, $\varphi(x) := f(x) - \frac{\mu}{2}\|x\|^2$ is convex. Here, $\|\cdot\|$ denotes the Euclidean norm on $\mathbb{R}^n$.*

Proof. We first note that multiplication of the equation from Example 17.15 with $\mu/2$ leads to the equation

$$\lambda\tfrac{\mu}{2}\|x\|^2 + (1-\lambda)\tfrac{\mu}{2}\|y\|^2 - \tfrac{\mu}{2}\|\lambda x + (1-\lambda)y\|^2 = \tfrac{\mu}{2}\lambda(1-\lambda)\|x-y\|^2,$$

which holds for all $x, y \in \mathbb{R}^n$ and $\lambda \in [0,1]$. Subbing the formula for φ into the definition of convexity shows that φ is convex if and only if, for all x, $y \in \mathbb{R}^n$ and $\lambda \in [0,1]$, the estimate

$$f(\lambda x+(1-\lambda)y) - \tfrac{\mu}{2}\|\lambda x+(1-\lambda)y\|^2 \leqslant \lambda(f(x) - \tfrac{\mu}{2}\|x\|^2) + (1-\lambda)(f(y) - \tfrac{\mu}{2}\|y\|^2)$$

holds. If we now collect all terms containing a norm on the left side, we can apply the first display equation. Doing this leads precisely to the μ-convexity of f. □

As a consequence, we obtain that the square $\|\cdot\|^2 \colon \mathbb{R}^n \to \mathbb{R}$ of the Euclidean norm is not only strictly convex, as stated in Example 17.15, but even μ-convex for $\mu \in (0,2]$. Furthermore, we can now show that μ-convexity implies the existence of a minimizer (which is then unique due to strict convexity).

Proposition 17.19. *Let $f \colon \mathbb{R}^n \to \mathbb{R}$ be μ-convex with $\mu > 0$ and differentiable. Then there exists a uniquely determined minimizer $x^* = \operatorname{argmin}_{x\in\mathbb{R}^n} f(x)$.*

Proof. We show that $f(y) \to \infty$ as $\|y\| \to \infty$. Indeed, according to Lemma 17.18, the auxiliary function $\varphi = f - \frac{\mu}{2}\|\cdot\|^2$ is convex. For $x, y \in \mathbb{R}^n$, Proposition 17.4 therefore implies

$$\begin{aligned} f(y) - \tfrac{\mu}{2}\|y\|^2 = \varphi(y) &\geqslant \varphi(x) + \langle \nabla\varphi(x), y-x\rangle \\ &= f(x) - \tfrac{\mu}{2}\|x\|^2 + \langle \nabla f(x) - \tfrac{\mu}{2}\cdot 2x, y-x\rangle. \end{aligned}$$

We now fix $x \in \mathbb{R}^d$, put $\alpha := \nabla f(x) - \mu x \in \mathbb{R}^n$, add $\frac{\mu}{2}\|y\|^2$ on both sides and obtain

$$\begin{aligned} f(y) &\geqslant \underbrace{f(x) - \tfrac{\mu}{2}\|x\|^2 - \langle\alpha, x\rangle}_{=:K} + \langle\alpha, y\rangle + \tfrac{\mu}{2}\|y\|^2 \\ &\underset{\substack{\uparrow \\ \text{CSB} \\ \text{inequality}}}{\geqslant} K + \underbrace{\big(\tfrac{\mu}{2}\|y\| - \|\alpha\|\big)}_{\substack{>0 \text{ for } \|y\| \\ \text{large enough}}}\|y\| \xrightarrow{\|y\|\to\infty} \infty. \end{aligned}$$

Since f is continuous, the existence of a minimizer follows from well-known theorems covered in real analysis. The uniqueness has already been shown in Proposition 17.14 under even weaker assumptions. □

The next lemma is a characterization of μ-convexity for differentiable functions. The corresponding condition is sometimes used in the literature as a definition for μ-convexity.

Lemma 17.20. *Let $f\colon \mathbb{R}^n \to \mathbb{R}$ be differentiable. Then f is μ-convex with $\mu > 0$ if and only if $f(y) \geqslant f(x) + \langle \nabla f(x), y - x\rangle + \frac{\mu}{2}\|y - x\|^2$ holds for all $x, y \in \mathbb{R}^n$.*

Proof. According to Lemma 17.18, f is μ-convex if and only if $\varphi = f - \frac{\mu}{2}\|\cdot\|^2$ is convex. According to Proposition 17.4, this is equivalent to

$$\forall\, x, y \in \mathbb{R}^n\colon \varphi(y) \geqslant \varphi(x) + \langle \nabla\varphi(x), y - x\rangle.$$

Evaluating φ at x, respectively at y, shows that the inequality is equivalent to

$$f(y) - \tfrac{\mu}{2}\|y\|^2 \geqslant f(x) - \tfrac{\mu}{2}\|x\|^2 + \langle \nabla f(x), y - x\rangle - \mu\langle x, y - x\rangle$$

and thus to

$$\begin{aligned} f(y) &\geqslant f(x) + \langle \nabla f(x), y - x\rangle + \tfrac{\mu}{2}\|y\|^2 - \tfrac{\mu}{2}\|x\|^2 + \mu\|x\|^2 - \mu\langle x, y\rangle \\ &= f(x) + \langle \nabla f(x), y - x\rangle + \tfrac{\mu}{2}\left[\|y\|^2 - 2\langle x, y\rangle + \|x\|^2\right] \\ &= f(x) + \langle \nabla f(x), y - x\rangle + \tfrac{\mu}{2}\|x - y\|^2 \end{aligned}$$

as claimed. □

A comparison with the estimate in Proposition 17.4 shows again that μ-convexity is stronger than (strict) convexity, since a term which is always strictly positive for $x \neq y$ is added to the smaller side of the convexity estimate.

Remark 17.21. Swapping x and y in Lemma 17.20 shows that $f\colon \mathbb{R}^n \to \mathbb{R}$ is μ-convex with $\mu > 0$ if and only if

$$\forall\, x, y \in \mathbb{R}^n\colon f(x) - f(y) - \langle \nabla f(y), x - y\rangle \geqslant \tfrac{\mu}{2}\|x - y\|^2$$

holds. This estimate looks very similar to those in Lemma 17.8 and Corollary 17.10. However, it should be noted that we are estimating *downwards* here—and with the same term $\|x - y\|^2$ as for L-smoothness, but with a different multiplicative constant. If we combine Corollary 17.10 with Lemma 17.20, it is easy to see that a differentiable function $f\colon \mathbb{R}^n \to \mathbb{R}$ is simultaneously μ-convex and L-smooth if and only if

$$\forall\, x, y \in \mathbb{R}^n\colon \tfrac{\mu}{2}\|x - y\|^2 \leqslant f(x) - f(y) - \langle \nabla f(y), x - y\rangle \leqslant \tfrac{L}{2}\|x - y\|^2$$

holds. From this it immediately follows that, if a function is simultaneously μ-convex and L-smooth, necessarily $L \geqslant \mu$ must hold.

The following lemma is our last preparation before we can prove our final theorem on gradient descent.

Lemma 17.22. *Let* $f\colon \mathbb{R}^n \to \mathbb{R}$ *be* μ*-convex and* L*-smooth with* $\mu, L > 0$. *Then* $\langle \nabla f(y) - \nabla f(x), y - x\rangle \geqslant \frac{\mu L}{\mu+L}\|x-y\|^2 + \frac{1}{\mu+L}\|\nabla f(x) - \nabla f(y)\|^2$ *holds for all* $x, y \in \mathbb{R}^n$.

Proof. Let $x, y \in \mathbb{R}^n$ be arbitrary and let $\varphi = f - \frac{\mu}{2}\|\cdot\|^2$ be the auxiliary function as used in previous proofs. We first use the same arguments and calculations as in the proof of the previous lemma, but then we will employ the L-smoothness and estimate further upwards. We get

$$\begin{aligned} 0 &\underset{\substack{\uparrow\\ \text{Lemma 17.18}\\ \text{and Prop. 17.4}}}{\leqslant} \varphi(x) - \varphi(y) - \langle \nabla\varphi(y), x - y\rangle \\ &\underset{\substack{\uparrow\\ \text{as in}\\ \text{Lemma 17.20}}}{=} f(x) - f(y) - \langle \nabla f(y), x - y\rangle - \tfrac{\mu}{2}\|x-y\|^2 \\ &\underset{\substack{\uparrow\\ \text{Lemma}\\ 17.8}}{\leqslant} \tfrac{L}{2}\|x-y\|^2 - \tfrac{\mu}{2}\|x-y\|^2 \\ &\leqslant (L-\mu)\|x-y\|^2. \end{aligned}$$

① We now first deal with the case $L = \mu$. Above, all inequalities then become equalities and it follows $f(x) = f(y) + \langle \nabla f(y), x - y\rangle + \frac{\mu}{2}\|x-y\|^2$ for all $x, y \in \mathbb{R}^n$. If we fix y and differentiate with respect to x, we obtain $\nabla f(x) = \nabla f(y) + \mu(x - y)$ and, by rearranging, we get

$$\forall\, x, y \in \mathbb{R}^n\colon \nabla f(x) - \nabla f(y) = \mu(x-y).$$

Putting $L = \mu$ into the statement of the lemma, we see that the assertion reduces to

$$\forall\, x, y \in \mathbb{R}^n\colon \langle \nabla f(y) - \nabla f(x), y - x\rangle \geqslant \tfrac{\mu}{2}\|x-y\|^2 + \tfrac{1}{2\mu}\|\nabla f(x) - \nabla f(y)\|^2.$$

Substituting $\nabla f(x) - \nabla f(y)$, according to our previous findings, shows that we get $\mu\|x-y\|^2$ on both sides.

② Now let $L > \mu$. Then our initial chain of inequalities, together with Corollary 17.10, applied to the differentiable function φ states that the latter is $(L-\mu)$-smooth. Applying Lemma 17.9 twice (once with x and y swapped) yields the inequalities

$$\begin{aligned} \varphi(x) - \varphi(y) &\leqslant \langle \nabla\varphi(x), x - y\rangle - \frac{1}{2(L-\mu)}\|\nabla\varphi(x) - \varphi(y)\|^2, \\ \varphi(y) - \varphi(x) &\leqslant \langle \nabla\varphi(y), y - x\rangle - \frac{1}{2(L-\mu)}\|\nabla\varphi(y) - \varphi(x)\|^2. \end{aligned}$$

Now we add them together and sub in $\varphi = f - \frac{\mu}{2}\|\cdot\|^2$. It follows

$$\begin{aligned}
0 &\leqslant \langle \nabla\varphi(x) - \nabla\varphi(y), x-y\rangle - \frac{1}{L-\mu}\|\nabla\varphi(x) - \nabla\varphi(y)\|^2 \\
&= \langle \nabla f(x) - \nabla f(y) - \mu(x-y), x-y\rangle \\
&\qquad - \frac{1}{L-\mu}\|\nabla f(x) - \nabla f(y) - \mu(x-y)\|^2 \\
&= \langle \nabla f(x) - \nabla f(y), x-y\rangle - \mu\|x-y\|^2 - \frac{1}{L-\mu}\Big(\|\nabla f(x) - \nabla f(y)\|^2 \\
&\qquad - 2\mu\langle \nabla f(x) - \nabla f(y), x-y\rangle + \mu^2\|x-y\|^2\Big) \\
&= \big(1 + \frac{2\mu}{L-\mu}\big)\langle \nabla f(x) - \nabla f(y), x-y\rangle - \big(\mu + \frac{\mu^2}{L-\mu}\big)\|x-y\|^2 \\
&\qquad - \frac{1}{L-\mu}\|\nabla f(x) - \nabla f(y)\|^2 \\
&= \frac{L+\mu}{L-\mu}\langle \nabla f(x) - \nabla f(y), x-y\rangle - \frac{\mu L}{L-\mu}\|x-y\|^2 \\
&\qquad - \frac{1}{L-\mu}\|\nabla f(x) - \nabla f(y)\|^2.
\end{aligned}$$

Rearranging and multiplying by $\frac{L-\mu}{L+\mu}$ yields the claimed inequality. □

The following theorem is our final result on gradient descent. Having reinforced our assumptions compared to Theorem 17.11, by requiring μ-convexity in addition to L-smoothness, we are now able to prove not only the convergence of the recursively defined sequence (17.1) and quantify its speed, but we also obtain an improved result on the accuracy estimate.

Theorem 17.23. *Let $f\colon \mathbb{R}^n \to \mathbb{R}$ be μ-convex and L-smooth with μ, $L > 0$. Let $x^{(0)} \in \mathbb{R}^n$ be arbitrary and let $(x^{(k)})_{k\geqslant 0}$ be the sequence from (17.1) with step size $\gamma := \frac{2}{\mu+L}$. Then, according to Proposition 17.19, there exists a uniquely determined minimizer $x^* = \operatorname{argmin}_{x\in\mathbb{R}^n} f(x)$, we have $L \geqslant \mu$, and for $k \geqslant 1$ it holds*

(i) $\|x^{(k)} - x^*\| \leqslant \left|\frac{L-\mu}{L+\mu}\right|^k \|x^{(0)} - x^*\|$ *and*

(ii) $0 \leqslant f(x^{(k)}) - f(x^*) \leqslant \frac{L}{2}\left|\frac{L-\mu}{L+\mu}\right|^{2k} \|x^{(0)} - x^*\|^2$.

In particular, it follows that $x^{(k)} \to x^*$ and $f(x^{(k)}) \to f(x^*)$ as $k \to \infty$.

Proof. First, we note that $L \geqslant \mu$ follows from Remark 17.21.

(i) We use the recursion formula (17.1) and then calculate the left-hand side, where we artificially add the term $-\nabla f(x^*)$ in order to be able to apply Lemma 17.22:

$$\begin{aligned}
&\|x^{(k+1)} - x^*\|^2 \\
&\quad = \|x^{(k)} - \gamma\nabla f(x^{(k)}) - x^*\|^2
\end{aligned}$$

$$
\begin{aligned}
&= \|x^{(k)} - x^*\|^2 - 2\gamma\langle\nabla f(x^{(k)}) \underbrace{-\nabla f(x^*)}_{\substack{=0 \text{ due to} \\ \text{Corollary 17.5}}}, x^{(k)} - x^*\rangle + \gamma^2\|\nabla f(x^{(k)})\|^2 \\
&\underset{\substack{\uparrow \\ \text{Lemma} \\ 17.22}}{\leqslant} \|x^{(k)} - x^*\|^2 + \frac{2}{\mu+L}\cdot\Big[\frac{\mu L}{\mu+L}\|x^{(k)} - x^*\|^2 + \\
&\qquad \frac{1}{\mu+L}\|\nabla f(x^{(k)}) - \underbrace{\nabla f(x^*)}_{=0}\|^2\Big] + \frac{1}{(\mu+L)^2}\|\nabla f(x^{(k)})\|^2 \\
&= \Big(1 - \frac{4\mu L}{(\mu+L)^2}\Big)\|x^{(k)} - x^*\|^2 + \underbrace{\Big(-\frac{4}{(\mu+L)^2} + \frac{1}{(\mu+L)^2}\Big)\|\nabla f(x^{(k)})\|^2}_{\leqslant 0} \\
&\leqslant \frac{\mu^2 + 2\mu L + L^2 - 4\mu L}{(\mu+L)^2}\|x^{(k)} - x^*\|^2 = \Big(\frac{L-\mu}{L+\mu}\Big)^2\|x^{(k)} - x^*\|^2.
\end{aligned}
$$

Now we take the square root on both sides and then apply the inequality successively to get the desired estimate.

(ii) After the preliminary work above, we now have

$$
\begin{aligned}
f(x^{(k)}) - f(x^*) &\underset{\substack{\uparrow \\ \text{Lemma} \\ 17.8}}{\leqslant} \langle\nabla f(x^*), x^{(k)} - x^*\rangle + \tfrac{L}{2}\|x^{(k)} - x^*\|^2 \\
&\underset{\substack{\uparrow \\ \nabla f(x^*)=0}}{=} \tfrac{L}{2}\|x^{(k)} - x^*\|^2 \\
&\underset{\substack{\uparrow \\ \text{(i)}}}{\leqslant} \tfrac{L}{2}\Big(\frac{L-\mu}{L+\mu}\Big)^{2k}\|x^{(0)} - x^*\|^2.
\end{aligned}
$$

The convergence of the recursively defined sequence itself, as well as of the image sequence, follows now immediately as $\frac{L-\mu}{L+\mu} \in [0, 1)$. □

If we put $\kappa := L/\mu$, then $\kappa \geqslant 1$ holds, and it follows

$$
\Big(\frac{L-\mu}{L+\mu}\Big)^2 = \left(\frac{\frac{L}{\mu} - 1}{\frac{L}{\mu} + 1}\right)^2 = \Big(\frac{\kappa-1}{\kappa+1}\Big)^2 = \Big(1 - \frac{2}{\kappa+1}\Big)^2 \leqslant \mathrm{e}^{-\frac{4}{\kappa+1}}.
$$

Under the assumptions of Theorem 17.23 we thus obtain

$$
0 \leqslant f(x^{(k)}) - f(x^*) \leqslant \tfrac{L}{2}\|x^{(0)} - x^*\|^2\,\mathrm{e}^{-\frac{4}{\kappa+1}k} =: C\mathrm{e}^{-ck}
$$

and as $C\mathrm{e}^{-ck} < \varepsilon$ holds if and only if $k > \frac{1}{c}(\log\frac{1}{C} + \log\frac{1}{\varepsilon})$, we see that the number of iterations of (17.1), that must be performed to achieve the accuracy $|f(x^{(k)}) - f(x^*)| < \varepsilon$, is asymptotically bounded by $K \cdot \log\frac{1}{\varepsilon}$ for $\varepsilon \searrow 0$, where K is a constant independent of ε. We advise the reader to compare this with the much slower convergence in Theorem 17.11.

Remark 17.24. In many applications, one minimizes a function of the form

$$f\colon \mathbb{R}^n \to \mathbb{R},\ f(x) = \sum_{j=1}^{N} f_j(x)$$

with a large number of functions f_j, compare Chapter 16.2. Due to the special form of f, the gradient ∇f is the sum of the ∇f_j's. To accelerate (17.1), one can switch to *stochastic gradient descent* and put

$$x^{(k+1)} := x^{(k)} - \gamma \sum_{i=1}^{m} \nabla f_{j_i}(x^{(k)}),$$

where the $j_1, \ldots, j_m \in \{1, \ldots, N\}$ are chosen randomly in each step.

References

Our presentation is mainly based on [Bub15, Lot22, Bec17, Nes18, BV04, NW06]. Exercise 17.10 is a variation of the so-called *projected gradient method.* The statement of the built-in remark can be found, e.g., in [Roc70, Corollary 25.5.1].

Alternatively to the approach in this chapter, many authors initially assume that the function f to be minimized is twice differentiable. L-smoothness and μ-convexity can then be described by upper and lower estimates of $\langle \nabla^2 f(\cdot)h, h\rangle$. For the exact relationship to the notions used above, we refer to [Nes18].

This chapter only provides a very small excerpt of what can be said about gradient descent. For further algorithms, e.g., for accelerating (17.1), strategies for choosing step sizes, classic (counter-)examples, generalizations to non-differentiable functions using subgradients, the treatment of function $f\colon A \subset \mathbb{R}^n \to \mathbb{R}$, and also the question of how gradients can be calculated numerically, we refer to the literature mentioned above.

There exist also theoretical results for the stochastic gradient method, which estimate the expectation of $\|x^{(k)} - x^*\|$. We refer to [Wen17] and the original literature cited therein.

Exercises

Exercise 17.1. Let $f\colon \mathbb{R} \to \mathbb{R}$, $f(x) = \mathrm{e}^x$. By which choice of γ_k can you ensure that $(f(x^{(k)}))_{k\in\mathbb{N}}$ with $x^{(k+1)} = x^{(k)} - \gamma_k f'(x^{(k)})$ converges to $\inf_{x\in\mathbb{R}} f(x)$?

Exercise 17.2. Let $f\colon \mathbb{R} \to \mathbb{R}$, $f(x) = x^2$. Examine the sequence $(x^{(k)})_{k\in\mathbb{N}}$ with $x^{(k+1)} = x^{(k)} - 2^{-k} f'(x^{(k)})$ for convergence. Are you surprised by the result?

Exercise 17.3. Let $f\colon \mathbb{R} \to \mathbb{R}$, $f(x) = 3x^4 + 4x^3 - 36x^2 + 42$, be given. Execute gradient descent with constant step size and show that the corresponding recursively defined sequence may converge to a local minimum point, which is not a (global) minimizer, if the initial value is chosen unfavorably.

Exercise 17.4. Show that the derivative of the following function decreases exponentially fast for $x \searrow 0$. Thus, the gradient method, once it is near zero, only progresses very slowly.

$$f\colon \mathbb{R} \to \mathbb{R},\ f(x) = \begin{cases} \mathrm{e}^{-1/x}, & \text{if } x \in (0, \frac{1}{4}], \\ (x + 8e^{-4} - \frac{1}{4})^2 - 64\,\mathrm{e}^{-8} + \mathrm{e}^{-4}, & \text{if } x \in (\frac{1}{4}, \infty), \\ 0, & \text{otherwise.} \end{cases}$$

Do you believe that one can still get a sequence (17.1) with constant step size that converges to a minimizer regardless of the starting value?

Exercise 17.5. Show that $f\colon \mathbb{R}^n \to \mathbb{R}$, $f(x) = \frac{1}{2}\langle x, Ax\rangle + \langle b, x\rangle + c$ with $A \in \mathbb{R}^{n\times n}$ positive definite, as well as $b \in \mathbb{R}^n$ and $c \in \mathbb{R}$ arbitrary, is μ-convex and L-smooth and determine for which μ and L this is the case.

Exercise 17.6. Let $f\colon \mathbb{R}^n \to \mathbb{R}$ be convex and L-smooth with $L > 0$. Let $x^* \in X^*$ and $x^{(0)}$ be arbitrary, let $(x^{(k)})_{k\in\mathbb{N}}$ be the sequence from (17.1) with constant step size $\gamma \in (0, 1/L)$. Show that then $f(x^{(k)}) - f(x^*) \leqslant \frac{2}{k\gamma}\|x^{(0)} - x^*\|^2$ holds for $k \geqslant 0$.

Exercise 17.7. In Lemma 17.22, the auxiliary function φ is in fact even $\frac{L-\mu}{2}$-smooth, as we have "given away" a factor of 1/2 in the very first estimate. Go through the proof again and find out how the estimate in Theorem 17.23 improves if you take the 1/2 into account.

Exercise 17.8. When implementing the gradient method, one can incorporate a termination criterion so that the algorithm runs only until the distance between $x^{(k+1)}$ and $x^{(k)}$ does not change much anymore. On the other hand, it may be useful to limit the number of overall iterations. This is particularly true when dealing with functions for which no runtime bound is known.

function GRADIENT DESCENT $(f, (\gamma_k)_{k\in\mathbb{N}}, K, \varepsilon)$
    $x^{(0)} \leftarrow$ random point in $\mathbb{R}^d$
    for $k \leftarrow 0$ to K **do**
        $x^{(k+1)} \leftarrow x^{(k)} - \gamma_k \nabla f(x^{(k)})$
        if $\|x^{(k+1)} - x^{(k)}\| < \varepsilon$ **then**
            break
    return $x^{(k+1)}$

Implement the gradient method according to the given pseudocode for the function $f\colon \mathbb{R}^2 \to \mathbb{R}$, $f(x, y) = x^2 + xy + y^2$, where you calculate the gradient analytically by hand. Test it for various constant step sizes $\gamma_k \equiv \gamma > 0$ and null sequences $\gamma_k \to 0$. Calculate manually the uniquely determined minimum and compare.

Exercise 17.9. Show that the function $f\colon \mathbb{R}^n \to \mathbb{R}$, $f(x) = \|x\|^{1+\alpha}$, $\alpha \in (0, 1)$, is convex and continuously differentiable, but fails to be L-smooth for any $L > 0$. Then, for $n = 1$ and $\alpha = 1/2$, carry out gradient descent with $\gamma_k \equiv 1$, using different initial values and different numbers of iterations. Come up with a conjecture about the convergence behavior of the sequence $(x^{(k)})_{k\in\mathbb{N}}$ and prove it.

Exercise 17.10. Let $f\colon \mathbb{R}^n \to \mathbb{R}$ be continuously differentiable and convex, let x^* be a minimizer and let $x^{(0)} \in \mathbb{R}^n$ be arbitrary. In this problem, we show that there always exists a sequence of step sizes $(\gamma_k)_{k\in\mathbb{N}} \subseteq [0,\infty)$ such that the sequence $(f(x^{(k)}))_{k\in\mathbb{N}}$ defined by (17.1) converges to f^*. To this end, let $\eta_0 := 1$ and $\eta_k := 1/\sqrt{k}$ for $k \geqslant 1$, and let $B := \overline{\mathrm{B}}_R(x^*)$ with $R > \|x^* - x^{(0)}\|$. Now we proceed as follows: Given $x^{(k)}$, we put

$$y^{(k+1)} := x^{(k)} - \eta_k \nabla f(x^{(k)}).$$

If then $y^{(k+1)} \in B$, we define $\gamma_k := \eta_k$, i.e., $x^{(k+1)} = y^{(k+1)}$. If not, we choose $\gamma_k < \eta_k$ such that $x^{(k+1)} = x^{(k)} - \gamma_k \nabla f(x^{(k)}) \in \partial B$.

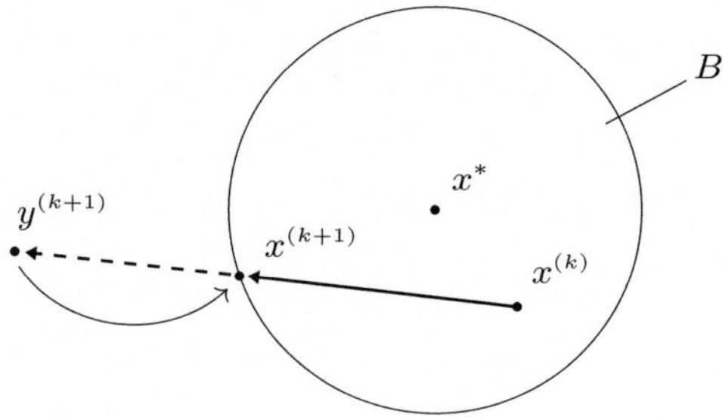

(i) Show, by means of Proposition 17.4 and via the identity $2\langle a,b\rangle = \|a\|^2 + \|b\|^2 - \|a-b\|^2$, that for $k \geqslant 0$ the following estimate holds:

$$f(x^{(k)}) - f(x^*) \leqslant \frac{1}{2\eta_k}\big(\|x^{(k)} - x^*\|^2 + \|y^{(k)} - y^{(k+1)}\|^2 - \|x^{(k+1)} - x^*\|^2\big).$$

(ii) Show that the series $\sum_{k=1}^{\infty} \frac{1}{k}(f(x^{(k)}) - f(x^*))$ converges.

(iii) Conclude with a contradiction argument that $f(x^{(k)}) \to f^*$ holds.

Remark: It can be shown that a differentiable and convex function on $\mathbb{R}^n$ is automatically continuously differentiable. Hence, if we merely assume, as at the very beginning of the chapter, that $f\colon \mathbb{R}^n \to \mathbb{R}$ is convex and differentiable and has at least one minimizer, then this exercise shows that step sizes can always be chosen in such a way that $f(x^{(k)}) \to f^*$ holds.

A

Selected Results of Probability Theory

In this appendix, we first present a summary of those results from probability theory that we used in the main text. In particular, we establish our notation for random variables, expected values, variances, etc. In the second part, we prove the characterization of normally distributed random vectors used in Chapters 8–12, as well as formulas for linear combinations of random vectors, which we noted in Fact 10.1 without proof.

A.1 Probability

For unexplained notation from measure theory, we refer to the literature recommended at the end of the chapter. The following definitions are all standard, although the expected value is often defined differently in basic probability courses.

Definition A.1. Let $\Omega \neq \emptyset$ be a non-empty set. A triple $(\Omega, \Sigma, \mathrm{P})$ is called a *probability space* if the following conditions are met.

(P1) Σ is a *σ-algebra* over Ω, i.e., $\Sigma \subseteq \mathcal{P}(\Omega)$, such that

(i) $\Omega \in \Sigma$,

(ii) $A \in \Sigma \implies \Omega \backslash A \in \Sigma$,

(iii) $(A_n)_{n\in\mathbb{N}} \subseteq \Sigma \implies \bigcup\limits_{n=1}^{\infty} A_n \in \Sigma$.

(P2) $\mathrm{P}\colon \Sigma \to [0,1]$ is a *probability measure*, that is

(iv) $\mathrm{P}(\Omega) = 1$,

(v) P is *σ-additive*, i.e., for $(A_n)_{n\in\mathbb{N}} \subseteq \Sigma$ with pairwise disjoint A_n it holds

$$\mathrm{P}\big(\bigcup_{n=1}^{\infty} A_n\big) = \sum_{n=1}^{\infty} \mathrm{P}(A_n).$$

S. A. Wegner, *Mathematical Introduction to Data Science*,
https://doi.org/10.1007/978-3-662-69426-8_18

If $(\Omega, \Sigma, \mathrm{P})$ is a probability space, then Ω is called the *sample space* and the elements of Σ are called *events*.

Definition A.2. Let $(\Omega, \Sigma, \mathrm{P})$ be a probability space.

(i) Two events $A, B \in \Sigma$ are called *independent*, if $\mathrm{P}(A \cap B) = \mathrm{P}(A) \cdot \mathrm{P}(B)$ holds.

(ii) For $A, B \in \Sigma$, $\mathrm{P}(A|B) := \dfrac{\mathrm{P}(A \cap B)}{\mathrm{P}(B)}$ is called the *conditional probability of A given B*.

Proposition A.3. (Law of total probability) *Let* $(\Omega, \Sigma, \mathrm{P})$ *be a probability space and let* $A, B \in \Sigma$. *Then it holds*

$$\mathrm{P}(A) = \mathrm{P}(A|B) \cdot \mathrm{P}(B) + \mathrm{P}(A|B^{\mathrm{c}}) \cdot \mathrm{P}(B^{\mathrm{c}}).$$

Definition A.4. Let $(\Omega, \Sigma, \mathrm{P})$ be a probability space. A measurable map $X\colon \Omega \to \mathbb{R}^d$ is called a *random vector* or *random variable* if $d = 1$. Here, $\mathbb{R}^d$ is equipped with the σ-algebra $\mathcal{B}^d$ of Borel sets. We use the following notation

$$\begin{aligned} \mathrm{P}\big[X \in B\big] &:= \mathrm{P}(X^{-1}(B)) = P(\{\omega \in \Omega \mid X(\omega) \in B\}) \text{ for } B \in \mathcal{B}^d, \\ P\big[X \geqslant a\big] &:= \mathrm{P}(\{\omega \in \Omega \mid X(\omega) \geqslant a\}) \text{ for } a \in \mathbb{R}, \\ P\big[a \leqslant \|X\| \leqslant b\big] &:= \mathrm{P}(\{\omega \in \Omega \mid a \leqslant \|X(\omega)\| \leqslant b\}) \text{ for } a, b \in \mathbb{R}, \\ P\big[X_i \geqslant c_i \text{ for all } i\big] &:= \mathrm{P}(\{\omega \in \Omega \mid X_i(\omega) \geqslant c_i \text{ for all } i\}) \text{ for } c_i \in \mathbb{R} \end{aligned}$$

for random variables and random vectors, respectively. If X is a random vector and $\omega \in \Omega$, we call $X(\omega)$ a *realization* of X.

Definition A.5. Let $(\Omega, \Sigma, \mathrm{P})$ be a probability space, let $X, Y\colon \Omega \to \mathbb{R}$ be random variables and let $k \geqslant 2$.

(i) If the integral

$$\mathrm{E}(X) := \int_\Omega X \,\mathrm{dP}$$

exists, its value is called the *expected value* or *expectation* of X. Similarly, $\mathrm{E}(X^k)$ is called the *k-th moment* of X, if the corresponding integral exists.

(ii) If X is integrable, and thus $\mathrm{E}(X) \in \mathbb{R}$, then

$$\mathrm{V}(X) := \mathrm{E}((X - \mathrm{E}(X))^2) \quad \text{and} \quad \sigma(X) := \sqrt{\mathrm{V}(X)}$$

are the *variance* and the *standard deviation* of X; for the latter we assume $\mathrm{V}(X) < \infty$.

(iii) If X and Y are integrable and the (well-defined!) random variables $X -$

$\mathrm{E}(X)$ and $Y - \mathrm{E}(Y)$ are integrable again, we call

$$\operatorname{Cov}(X, Y) := \mathrm{E}(X - \mathrm{E}(X))\,\mathrm{E}(Y - \mathrm{E}(Y))$$

the *covariance* of X and Y.

Proposition A.6. *Let $(\Omega, \Sigma, \mathrm{P})$ be a probability space, $X, Y\colon \Omega \to \mathbb{R}$ be integrable random variables with $\mathrm{V}(X), \mathrm{V}(Y), \operatorname{Cov}(X, Y) < \infty$ and $a, b \in \mathbb{R}$. Then we have*

(i) $\mathrm{E}(aX + bY) = a\,\mathrm{E}(X) + b\,\mathrm{E}(Y)$,

(ii) $X \leqslant Y$ *pointwise* $\Longrightarrow \mathrm{E}(X) \leqslant \mathrm{E}(Y)$,

(iii) $\mathrm{V}(X) = \mathrm{E}(X^2) - \mathrm{E}(X)^2$,

(iv) $\mathrm{V}(aX + b) = a^2\,\mathrm{V}(X)$,

(v) $\mathrm{V}(aX + bY) = a^2\,\mathrm{V}(X) + b^2\,\mathrm{V}(Y) + 2ab\operatorname{Cov}(X, Y)$,

(vi) $|\operatorname{Cov}(X, Y)| \leqslant \sqrt{\mathrm{V}(X)}\sqrt{\mathrm{V}(Y)}$.

Definition A.7. Let $(\Omega, \Sigma, \mathrm{P})$ be a probability space, $X\colon \Omega \to \mathbb{R}^d$ a random vector and $\rho\colon \mathbb{R}^d \to [0, \infty)$ a measurable function. If

$$\forall\, A \in \mathcal{B}^d\colon\ \mathrm{P}\big[X \in A\big] = \int_A \rho(x)\,\mathrm{d}\lambda^d(x),$$

holds, then we say that the random vector X is *ρ-distributed*. Important examples are Gaussian distributed random vectors and uniformly distributed random vectors with

$$\rho(x) = \tfrac{1}{(2\pi\sigma^2)^{d/2}}\,\mathrm{e}^{-\frac{\|x-\mu\|^2}{2\sigma^2}} \quad \text{respectively} \quad \rho(x) = \frac{1}{\lambda(B)} \cdot \mathbb{1}_B(x),$$

where $\mu \in \mathbb{R}^d$, $\sigma > 0$ and $B \in \mathcal{B}^d$ with $\lambda^d(B) \in (0, \infty)$ are fixed.

Proposition A.8. *Let $(\Omega, \Sigma, \mathrm{P})$ be a probability space, $X\colon \Omega \to \mathbb{R}^d$ a ρ-distributed random variable and $k \geqslant 1$. Then for moments and expected value the following holds:*

$$\mathrm{E}(X^k) = \int_{\mathbb{R}} x^k \rho(x)\,\mathrm{d}\lambda(x).$$

In fact, the generalization $\mathrm{E}(f(X)) = \int_{\mathbb{R}} f(x)\rho(x)\,\mathrm{d}\lambda(x)$ holds for a large class of functions f and is sometimes referred to as the "law of the unconscious statistician".

Theorem A.9. (Markov's inequality) *Let $(\Omega, \Sigma, \mathrm{P})$ be a probability space and let $X\colon \Omega \to [0, \infty)$ be a random variable. Then for $a > 0$ we have*

$$\mathrm{P}\big[X \geqslant a\big] \leqslant \frac{\mathrm{E}(X)}{a}.$$

Theorem A.10. (Chebyshev's inequality) *Let* $(\Omega, \Sigma, \mathrm{P})$ *be a probability space and* $X\colon \Omega \to \mathbb{R}$ *a random variable, such that* $\mathrm{E}(X)$ *and* $\mathrm{V}(X)$ *are finite and* $\mathrm{V}(X)$ *is not zero. Then for* $a > 0$ *we have*

$$\mathrm{P}\big[|X - \mathrm{E}(X)| \geqslant a\big] \leqslant \frac{\mathrm{V}(X)}{a^2}.$$

Definition A.11. Let $(\Omega, \Sigma, \mathrm{P})$ be a probability space and let random variables $X_1, \ldots, X_n\colon \Omega \to \mathbb{R}$ be given. The $X_1, \ldots, X_n$ are called *independent* if the equation

$$P\Big[(X_1, \ldots, X_n) \in \prod_{i=1}^{n} A_i\Big] = \prod_{i=1}^{n} P\big[X_i \in A_i\big]$$

holds for arbitrary Borel sets $A_i \in \mathcal{B}$. A sequence $(X_i)_{i\in\mathbb{N}}$ of random variables is called *independent* if every finite subsequence is independent.

Proposition A.12. *Let* $(\Omega, \Sigma, \mathrm{P})$ *be a probability space and* $X, Y\colon \Omega \to \mathbb{R}$ *be independent random variables with finite expected value and finite variance. Then the following holds true:*

(i) $\mathrm{E}(X \cdot Y) = \mathrm{E}(X) \cdot \mathrm{E}(Y)$,

(ii) $\mathrm{V}(X + Y) = \mathrm{V}(X) + \mathrm{V}(Y)$.

Theorem A.13. (Cloning theorem) *For* $i \in \mathbb{N}$, *let probability spaces* $(\Omega_i, \Sigma_i, \mathrm{P}_i)$ *and random variables* $Y_i\colon \Omega_i \to \mathbb{R}$ *be given. Then there exists a probability space* $(\Omega, \Sigma, \mathrm{P})$ *and independent random variables* $X_1, X_2, \ldots\colon \Omega \to \mathbb{R}$ *such that each* X_i *has the same distribution as* Y_i, *i.e.,* $\mathrm{P}[X_i \in A] = \mathrm{P}_i[Y_i \in A]$ *for every Borel set* $A \in \mathcal{B}$ *and every* $i \in \mathbb{N}$.

Theorem A.14. (Weak law of large numbers) *Let* $(\Omega, \Sigma, \mathrm{P})$ *be a probability space and let* $X\colon \Omega \to \mathbb{R}$ *be a random variable, such that* $\mathrm{E}(X)$ *and* $\mathrm{V}(X)$ *are finite. Let* $(X_i)_{i\in\mathbb{N}}$ *be a sequence of independent copies of* X. *Then for* $\varepsilon > 0$ *and* $n \in \mathbb{N}$ *we have*

$$\mathrm{P}\Big[\Big|\frac{X_1 + \cdots + X_n}{n} - \mathrm{E}(X)\Big| \geqslant \varepsilon\Big] \leqslant \frac{\mathrm{V}(X)}{n\varepsilon^2}.$$

For $n \to \infty$, $\mathrm{P}[\,\cdots]$ *thus tends to zero and we say that*

$$\frac{X_1 + \cdots + X_n}{n} \xrightarrow{n\to\infty} \mathrm{E}(X)$$

converges in probability, *where* $\mathrm{E}(X)$ *on the right-hand side is to be understood as a constant random variable.*

Remark A.15. The elements X_i of a sequence as in Theorem A.14 are called *samples* of the random variable X. Note that for *theoretical* statements about multiple executions of a random experiment described by X, samples in this sense are considered and then, for example, their mean value (as a new random variable!) can be studied, cf. Theorem A.14. For *simulations*, realizations $X(\omega_1), \ldots, X(\omega_n)$, cf. Definition A.4, of X are used and then, for example, their mean value (as a real number!) can be considered.

Definition A.16. Let $\rho, \tau \in \mathrm{L}^1(\mathbb{R})$. Then

$$\rho * \tau \colon \mathbb{R} \to \mathbb{R},\ (\rho * \tau)(s) = \int_{\mathbb{R}} \rho(s-t)\tau(t)\,\mathrm{d}\,\lambda(t)$$

exists for almost all $s \in \mathbb{R}$. We have $\rho * \tau \in \mathrm{L}^1(\mathbb{R})$ and $\rho * \tau$ is called the *convolution* of ρ and τ.

Theorem A.17. *Let $(\Omega, \Sigma, \mathrm{P})$ be a probability space and $X, Y \colon \Omega \to \mathbb{R}$ independent ρ- respectively τ-distributed random variables with $\rho, \tau \in \mathrm{L}^1(\mathbb{R})$. Then $X + Y \colon \Omega \to \mathbb{R}$ is a $\rho * \tau$-distributed random variable.*

A.2 Gaussian distribution

In the following, we compile indispensable facts about Gaussian distributed random variables and vectors for the main text. Firstly, we show that the terms *mean* and *variance* are justified for the parameters of a Gaussian distributed random variable.

Theorem A.18. *Let $(\Omega, \Sigma, \mathrm{P})$ be a probability space and let $X \colon \Omega \to \mathbb{R}$ be a Gaussian distributed random variable, that is $X \sim \mathcal{N}(\mu, \sigma^2)$ holds with $\mu \in \mathbb{R}$ and $\sigma > 0$. Then we have*

$$\mathrm{E}(X) = \mu \quad \textit{and} \quad \mathrm{V}(X) = \sigma^2.$$

Proof. Indeed, for the expected value we may calculate

$$\begin{aligned}
\mathrm{E}(X) &= \int_{\mathbb{R}} x \frac{1}{\sqrt{2\pi}\sigma}\, \mathrm{e}^{-\frac{(x-\mu)^2}{2\sigma^2}}\, \mathrm{d}\lambda(x) \underset{u:=\frac{x-\mu}{\sqrt{2}\sigma}}{\overset{\uparrow}{=}} \frac{1}{\sqrt{2\pi}\sigma} \int_{\mathbb{R}} (\sqrt{2}\sigma u + \mu)\, \mathrm{e}^{-u^2} \sqrt{2}\sigma \,\mathrm{d}\lambda(u) \\
&= \frac{1}{\sqrt{\pi}} \Big[\sqrt{2}\sigma \underbrace{\int_{\mathbb{R}} u\, \mathrm{e}^{-u^2}\, \mathrm{d}\lambda(u)}_{=0} + \mu \underbrace{\int_{\mathbb{R}} \mathrm{e}^{-u^2}\, \mathrm{d}\lambda(u) \Big]}_{=\sqrt{\pi}} = \mu,
\end{aligned}$$

where the first integral is zero due to symmetry and the second one can be computed, for example, by squaring, then applying Fubini, and finally as an improper Riemann integral in polar coordinates. The variance is given by

$$\mathrm{V}(X) = \mathrm{E}(X^2) - \mathrm{E}(X)^2 = \int_{\mathbb{R}} x^2 \rho(x)\, \mathrm{d}\lambda(x) - \mu^2$$

$$
\begin{aligned}
&\underset{u:=\frac{x-\mu}{\sqrt{2}\sigma}}{\overset{\uparrow}{=}} \frac{1}{\sqrt{2\pi}\sigma}\int_{\mathbb{R}}(\sqrt{2}\sigma u+\mu)^2\,\mathrm{e}^{-u^2}\,\sqrt{2}\sigma\,\mathrm{d}\lambda(u)-\mu^2\\
&= \frac{1}{\sqrt{\pi}}\Big[2\sigma^2\underbrace{\int_{\mathbb{R}}u^2\,\mathrm{e}^{-u^2}\,\mathrm{d}u}_{=\sqrt{\pi}/2}+2\sqrt{2}\sigma\mu\underbrace{\int_{\mathbb{R}}u\,\mathrm{e}^{-u^2}\,\mathrm{d}u}_{=0}+\mu^2\underbrace{\int_{\mathbb{R}}\mathrm{e}^{-u^2}\,\mathrm{d}u}_{=\sqrt{\pi}}\Big]-\mu^2\\
&= \sigma^2,
\end{aligned}
$$

where the first integral is calculated via integration by parts and the other two can be obtained as in the first part of the proof. Notice that we have omitted the "λ" for space reasons. □

We note the following very simple result.

Proposition A.19. *Let $(\Omega,\Sigma,\mathrm{P})$ be a probability space and let $X\colon\Omega\to\mathbb{R}$ be a random variable with $X\sim\mathcal{N}(\mu,\sigma^2)$. Let $a\in\mathbb{R}$. Then $a+X\sim\mathcal{N}(a+\mu,\sigma^2)$.*

Proof. For $A\in\mathcal{B}$ we have

$$
\mathrm{P}\big[a+X\in A\big]=\frac{1}{\sqrt{2\pi}\sigma}\int_{A-a}\mathrm{e}^{\frac{(x-\mu)^2}{2\sigma^2}}\,\mathrm{d}\lambda(x)=\frac{1}{\sqrt{2\pi}\sigma}\int_{A}\mathrm{e}^{\frac{(y-(a+\mu))^2}{2\sigma^2}}\,\mathrm{d}\lambda(y)
$$

by substituting $y:=x+a$. □

Next, we characterize (spherical) normally distributed random vectors via their coordinate functions — this equivalence was used many times in the main text.

Theorem A.20. *Let $(\Omega,\Sigma,\mathrm{P})$ be a probability space and $X\colon\Omega\to\mathbb{R}^d$ be a random vector with coordinate functions $X_1,\dots,X_d$. Then $X\sim\mathcal{N}(\mu,\sigma^2,\mathbb{R}^d)$ if and only if $X_i\sim\mathcal{N}(\mu,\sigma^2)$ holds for $i=1,\dots,d$ and with independent $X_1,\dots,X_d$.*

Proof. For $A=A_1\times\cdots\times A_d\subseteq\mathbb{R}^d$ with $A_i\in\mathcal{B}^d$ we calculate

$$
\begin{aligned}
\mathrm{P}\big[X\in A\big]&\overset{X\sim\mathcal{N}(\mu,\sigma^2,\mathbb{R}^d)}{\overset{\downarrow}{=}}\frac{1}{(2\pi\sigma^2)^{d/2}}\int_A\mathrm{e}^{-\frac{\|x-\mu\|^2}{2\sigma^2}}\,\mathrm{d}\lambda^d(x)\\
&=\frac{1}{(2\pi\sigma^2)^{d/2}}\int_A\mathrm{e}^{\frac{(x_1-\mu_1)^2}{2\sigma^2}}\cdots\mathrm{e}^{-\frac{(x_d-\mu_d)^2}{2\sigma^2}}\,\mathrm{d}\lambda(x_1,\dots,x_d)\\
&=\frac{1}{\sqrt{2\pi}\sigma}\int_{A_1}\mathrm{e}^{\frac{(x_1-\mu_1)^2}{2\sigma^2}}\,\mathrm{d}\lambda(x_1)\cdots\frac{1}{\sqrt{2\pi}\sigma}\int_{A_d}\mathrm{e}^{\frac{(x_d-\mu_d)^2}{2\sigma^2}}\,\mathrm{d}\lambda(x_d)\\
&\underset{X_i\sim\mathcal{N}(\mu,\sigma^2)}{\overset{\uparrow}{=}}\mathrm{P}\big[X_1\in A_1\big]\cdots\mathrm{P}\big[X_d\in A_d\big],
\end{aligned}
$$

where the first and last equalities hold under the conditions indicated there. The claimed equivalence now follows from the above calculation:

"$\Longrightarrow$" For $B \in \mathcal{B}$ and fixed $1 \leqslant i \leqslant d$, put above $A := \mathbb{R} \times \cdots \times B \times \cdots \times \mathbb{R}$, where B is the i-th factor. Then it follows $\mathrm{P}[X_i \in B] = \mathrm{P}[X \in \mathbb{R}^{i-1} \times B \times \mathbb{R}^{d-i}]$, and we have shown the equation needed for independence.

"$\Longleftarrow$" Let $A = A_1 \times \cdots \times A_d \subseteq \mathbb{R}^d$ be a cuboid. Since the X_i's are independent, it follows $\mathrm{P}[X \in A] = \mathrm{P}[X_1 \in A_1] \cdots \mathrm{P}[X_d \in A_d]$. Reading our initial calculation backwards, we obtain

$$\mathrm{P}[X \in A] = \frac{1}{(2\pi\sigma^2)^{d/2}} \int_A \mathrm{e}^{-\frac{\|x-\mu\|^2}{2\sigma^2}} \, \mathrm{d}\lambda(x)$$

for cuboids A, and thus also for arbitrary Borel sets in $\mathbb{R}^d$. □

Lastly, we turn to linear combinations of independent normally distributed random variables.

Theorem A.21. *Let $(\Omega, \Sigma, \mathrm{P})$ be a probability space, let $X_1, \dots, X_n \to \mathbb{R}$ be independent random variables with $X_i \sim \mathcal{N}(0,1)$ and let $\lambda_1, \dots, \lambda_n \in \mathbb{R}\backslash\{0\}$. Then the random variable*

$$X := \sum_{i=1}^{n} \lambda_i X_i \sim \mathcal{N}(0, \sigma^2)$$

is Gaussian distributed with mean zero and variance $\sigma^2 := \lambda_1^2 + \cdots + \lambda_d^2 > 0$.

Proof. We first note that, for $\lambda \neq 0$, $X \sim \mathcal{N}(0,1)$ and $A \in \mathcal{B}$,

$$\mathrm{P}\big[\lambda X \in A\big] = \mathrm{P}\big[X \in \tfrac{1}{\lambda} A\big] = \int_{\frac{1}{\lambda}A} \rho(x) \, \mathrm{d}x = \int_A \rho(\tfrac{x}{\lambda}) \tfrac{1}{\lambda} \, \mathrm{d}x$$

holds, where $\rho(x) = \frac{1}{\sqrt{2\pi}} \mathrm{e}^{-x^2/2}$. Thus, λX is $\rho(\frac{\cdot}{\lambda})\frac{1}{\lambda}$-distributed. Subbing in ρ shows

$$\rho(\tfrac{t}{\lambda})\tfrac{1}{\lambda} = \frac{1}{\sqrt{2\pi}} \mathrm{e}^{-(\frac{t}{\lambda})^2/2} \, \frac{1}{\lambda} = \frac{1}{\sqrt{2\pi}\lambda} \mathrm{e}^{-\frac{t^2}{2\lambda^2}}$$

for $t \in \mathbb{R}$ and thus $\lambda X \sim \mathcal{N}(0, \lambda^2)$. Applying this to our initial situation, we obtain $\lambda_i X_i \sim \mathcal{N}(0, \lambda_i^2)$ for $i = 1, \dots, d$. It is easy to see that the convolution of L^1-functions is associative, and that Theorem A.17 also holds for more than two random variables. To complete our proof, it thus suffices to show

$$(\rho_1 * \cdots * \rho_d)(x) = \frac{1}{\sqrt{2\pi}\sigma} \mathrm{e}^{-\frac{t^2}{2\sigma^2}}, \quad \text{with} \quad \rho_i(x) = \frac{1}{\sqrt{2\pi}\lambda_i} \mathrm{e}^{-\frac{t^2}{2\lambda_i^2}}.$$

We will prove this only for $d = 2$, and use abbreviations $a := \lambda_1^2$, $b := \lambda_2^2$ and

$c := a + b$. This yields

$$\begin{aligned}
(\rho_1 * \rho_2)(s) &= \frac{1}{2\pi\sqrt{ab}} \int_{\mathbb{R}} \exp\big(-\frac{(s-t)^2}{2a}\big) \exp\big(-\frac{t^2}{2b}\big)\, \mathrm{d}t \\
&= \frac{1}{2\pi\sqrt{ab}} \int_{\mathbb{R}} \exp\big(-\frac{b(s^2 - 2st + t^2) + at^2}{2ab}\big)\, \mathrm{d}t \\
&= \frac{1}{2\pi\sqrt{ab}} \int_{\mathbb{R}} \exp\big(-\frac{t^2(b+a) - 2stb + bs^2}{2ab}\big)\, \mathrm{d}t \\
&= \frac{1}{2\pi c^2\sqrt{ab/c}} \int_{\mathbb{R}} \exp\big(-\frac{t^2(b+a)/c - 2stb/c + bs^2/c}{2ab/c}\big)\, \mathrm{d}t \\
&= \frac{1}{2\pi c^2\sqrt{ab/c}} \int_{\mathbb{R}} \exp\big(-\frac{(t-(bs)/c)^2 - (sb/c)^2 + s^2(b/c)}{2ab/c}\big)\, \mathrm{d}t \\
&= \frac{1}{\sqrt{2\pi c}} \exp\big(-\frac{(sb/c)^2 - s^2(b/c)}{2ab/c}\big) \frac{1}{\sqrt{2\pi(ab/c)}} \int_{\mathbb{R}} \exp\big(-\frac{(t-(bs)/c)^2}{2ab/c}\big)\, \mathrm{d}t \\
&= \frac{1}{\sqrt{2\pi c}} \exp\big(-\frac{(sb/c)^2c^2 - s^2(b/c)c^2}{2abc}\big) \\
&= \frac{1}{\sqrt{2\pi c}} \exp\big(-\frac{s^2(b^2 - bc)}{2abc}\big) \\
&= \frac{1}{\sqrt{2\pi c}} \exp\big(-\frac{s^2}{2c}\big),
\end{aligned}$$

where we have again omitted the "λ" for reasons of space. The case of d-many factors in the convolution is similar. □

References

We refer to [GW14] for an introduction, and to [Shi16] for deeper results. The version of the cloning theorem (Theorem A.13) stated above, is taken from the German source [Beh13, Chapter 4.5].

Bibliography

[AA84] A. Albert and J. A. Anderson. On the existence of maximum likelihood estimates in logistic regression models. *Biometrika*, 71:1–10, 1984.

[Agg16] C. C. Aggarwal. *Recommender Systems.* Springer Cham, 2016.

[AM12] Y. Abu-Mostafa. Lecture 14 — Support Vector Machines. Youtube-Video, Caltech, https://www.youtube.com/watch?v=eHsErlPJWUU, 2012.

[BC99] C. Burges and D. Crisp. Uniqueness of the SVM solution. In *NIPS Vol. 99*, 1999.

[Bec17] A. Beck. *First-order Methods in Optimization*, volume 25 of *MOS/SIAM Ser. Optim.*, 2017.

[Beh13] E. Behrends. *Elementare Stochastik. Ein Lernbuch — von Studierenden mitentwickelt.* Springer Spektrum, Heidelberg, 2013.

[Ber17] B. Bernstein. On the uniqueness of the SVM solution. https://davidrosenberg.github.io/mlcourse/Labs/UniquenessOfSVM.pdf, 2017.

[BHK20] A. Blum, J. Hopcroft, and R. Kannan. *Foundations of Data Science.* Cambridge University Press, Cambridge, 2020.

[Bis06] C. M. Bishop. *Pattern Recognition and Machine Learning.* Inf. Sci. Stat. Springer, New York, 2006.

[Bub15] S. Bubeck. Convex optimization: algorithms and complexity. *Found. Trends Mach. Learn.*, 8(3-4):231–357, 2015.

[Bul06] Y. Bulatov. Curse of dimensionality and intuition. http://yaroslavvb.blogspot.com/2006/05/curse-of-dimensionality-and-intuition.html, 2006.

[BV04] S. Boyd and L. Vandenberghe. *Convex Optimization.* Cambridge University Press, 2004.

[Cal20] O. Calin. *Deep Learning Architectures. A Mathematical Approach.* Springer Ser. Data Sci. Springer, Cham, 2020.

[Chu07] F. Chung. Random walks and local cuts in graphs. *Linear Algebra and its Applications*, 423(1):22–32, 2007.

[CL92] C. K. Chui and X. Li. Approximation by ridge functions and neural networks with one hidden layer. *J. Approx. Theory*, 70(2):131–141, 1992.

[Con90] J. B. Conway. *A Course in Functional Analysis.* 2nd ed. Graduate Texts in Mathematics 96. Springer-Verlag, New York, 1990.

[CLRS22] T. H. Cormen, C. E. Leiserson, R. L. Rivest, and C. Stein. *Introduction to Algorithms.* 4th ed., MIT Press, Cambridge, 2022.

S. A. Wegner, *Mathematical Introduction to Data Science*,
https://doi.org/10.1007/978-3-662-69426-8

[Col12] M. Collins. Convergence proof for the perceptron algorithm. https://www.cs.columbia.edu/~mcollins/courses/6998 – 2012/notes/perc.converge.pdf, 2012.

[Cyb89] G. Cybenko. Approximation by superpositions of a sigmoidal function. *Math. Control Signals Syst.*, 2(4):303–314, 1989.

[Cyb92] G. Cybenko. Correction to: Approximation by superpositions of a sigmoidal function. *Math. Control Signals Syst.*, 5(4):455, 1992.

[DH08] P. Deuflhard and A. Hohmann. *Numerical Analysis. A First Course in Scientific Computing.* de Gruyter, Berlin, 1995.

[Don04] D. L. Donoho. High-dimensional data analysis: The curses and blessings of dimensionality. https://dl.icdst.org/pdfs/files/236e636d7629c1a53e6ed4cce1019b6e.pdf, 2004.

[DS07] S. Dasgupta and L. Schulman. A probabilistic analysis of EM for mixtures of separated, spherical gaussians. *The Journal of Machine Learning Research*, 8:203–226, 2007.

[Dub21] Y. Dubois. Curse of dimensionality. https://yanndubs.github.io/machine-learning-glossary/concepts/curse, 2021.

[For20] M. Fornasier. Foundations of data analysis. Exercises, TU München (not anymore available online), 2020.

[Fun06] S. Funk. Netflix update: Try this at home. https://sifter.org/~simon/journal/20061211.html, 2006.

[Giu03] E. Giusti. *Direct methods in the calculus of variations.* World Scientific, Singapore, 2003.

[Gow14] S. Gower. Netflix Prize and SVD. http://buzzard.ups.edu/courses/2014spring/420projects/math420-UPS-spring-2014-gower-netflix-SVD.pdf, 2014.

[GW14] G. Grimmett and D. Welsh. *Probability—An Introduction.* 2nd ed. Oxford University Press, Oxford, 2014.

[Gui18] L. F. Guilhoto. An overview of artificial neural networks for mathematicians. https://math.uchicago.edu/~may/REU2018/REUPapers/Guilhoto.pdf, 2018.

[Ham] M. Hampton. SVD computation example. Lecture Notes, University of Minnesota Duluth, https://www.d.umn.edu/~mhampton/m4326svd_example.pdf.

[Han19] B. Hanin. Universal function approximation by deep neural nets with bounded width and ReLU activations. *Mathematics*, 7(10), 2019.

[Hor91] K. Hornik. Approximation capabilities of multilayer feedforward networks. *Neural Networks*, 4(2):251–257, 1991.

[HSW89] K. Hornik, M. Stinchcombe, and H. White. Multilayer feedforward networks are universal approximators. *Neural Networks*, 2(5):359–366, 1989.

[HSW90] K. Hornik, M. Stinchcombe, and H. White. Universal approximation of an unknown mapping and its derivatives using multilayer feedforward networks. *Neural Networks*, 3(5):551–560, 1990.

[JL84] W. B. Johnson and J. Lindenstrauss. Extensions of Lipschitz mappings into a Hilbert space. Contemp. Math. 26, 189-206, 1984.

[Kal96] D. Kalman. A singularly valuable decomposition: The SVD of a matrix. *The College Mathematics Journal*, 27(1):2–23, 1996.

[Köp13] M. Köppen. The curse of dimensionality. *5th Online World Conference on Soft Computing in Industrial Applications*, 2013.

[Lef19] M. Lefkowitz. Professor's perceptron paved the way for AI — 60 years too soon. *Cornell Cronicle* https://news.cornell.edu/stories/2019/09/professors-perceptron-paved-way-ai-60-years-too-soon, 2019.

[LL16] I. E. Leonard and J. E. Lewis. *Geometry of Convex Sets.* John Wiley & Sons, Hoboken, NJ, 2016.

[LLPS93] M. Leshno, V. Ya. Lin, A. Pinkus, and S. Schocken. Multilayer feedforward networks with a nonpolynomial activation function can approximate any function. *Neural Networks*, 6(6):861–867, 1993.

[Lot22] M. Lotz. Mathematics of Machine Learning. https://homepages.warwick.ac.uk/staff/Martin.Lotz/files/learning/lectnotes-all.pdf, 2022.

[LP93] V. Ya. Lin and A. Pinkus. Fundamentality of ridge functions. *J. Approx. Theory*, 75(3):295–311, 1993.

[LRU12] J. Leskovec, A. Rajaraman, and J. D. Ullman. *Mining of Massive Datasets.* Cambridge University Press, Cambridge, 2012.

[MC] MIT-CSAIL. Quadratic programming with Python and CVXOPT. https://courses.csail.mit.edu/6.867/wiki/images/a/a7/Qp-cvxopt.pdf.

[Nes18] Yu. Nesterov. *Lectures on Convex Optimization*, Volume 137 in *Springer Optim. Appl.,* 2nd ed., Springer, Cham, 2018.

[Nie15] M. A. Nielsen. *Neural Networks and Deep Learning.* Determination Press, 2015.

[Nos16] J. Noss. 6.034 Recitation 7: Support Vector Machines (SVMs). https://www.youtube.com/watch?v=ik7E7r2a1h8&t=2422s, 2016.

[NW06] J. Nocedal and S. J. Wright. *Numerical Optimization.* Springer Ser. Oper. Res. Financ. Eng., 2nd ed., Springer, New York, 2006.

[Ola96] M. Olazara. A sociological study of the official history of the perceptrons controversy. *Social Studies of Science*, 26(3):611–659, 1996.

[Pin20] I. Pinelis. Asymptotically tight concentration of norms of subgaussian random vectors with independent coordinates, as the dimension $n \to \infty$? https://mathoverflow.net/questions/357484/asymptotically-tight-concentration-of-norms-of-subgaussian-random-vectors-with-i, 2020.

[Ras14] The Raspberry Pi Foundation. Getting started with Mathematica. https://projects.raspberrypi.org/en/projects/getting-started-with-mathematica/8, 2014.

[Rin08] C. M. Ringel. Funktionen (GHR). Lecture Notes, University of Bielefeld, https://www.math.uni-bielefeld.de/~sek/funktion, 2008.

[Roc70] R. T. Rockafellar. *Convex Analysis*, volume 28 of *Princeton Math. Ser.* Princeton University Press, Princeton, NJ, 1970.

[Ros59] F. Rosenblatt. The design of an intelligent automaton. *Research Trends, Cornell Aeronautical Laboratory, inc.*, VI(2), 1959.

[Rud87] W. Rudin. *Real and Complex Analysis.* 3rd ed., McGraw-Hill, New York, 1987.

[Rud91] W. Rudin. *Functional Analysis.* 2nd ed., McGraw-Hill, New York, 1991.

[SKKR01] B. Sarwar, G. Karypis, J. Konstan, and J. Riedl. Item-based collaborative filtering recommendation algorithms. *WWW10, May 1–5*, Hong Kong, 2001.

[Sha08] A. Shashua. Introduction to Machine Learning. arXiv: 0904.3664v1, 2008.

[Sha15] C. Shalizi. Lecture notes on modern regression. Lecture Notes, Carnegie Mellon University, https://www.stat.cmu.edu/~cshalizi/mreg/15, 2015.

[SSBD14] S. Shalev-Shwartz and S. Ben-David. *Understanding Machine Learning. From Theory to Algorithms.* Cambridge University Press, Cambridge, 2014.

[Shi16] A. N. Shiryaev. *Probability. 1.* Graduate Texts in Mathematics 95. Translated from the fourth (2007) Russian edition by R. P. Boas and D. M. Chibisov Springer, New York, 2016.

[SC08] I. Steinwart and A. Christmann. *Support Vector Machines.* Inf. Sci. Stat. Springer, New York, 2008.

[SC92] X. Sun and E. W. Cheney. The fundamentality of sets of ridge functions. *Aequationes Math.*, 44(2-3):226–235, 1992.

[Ver18] R. Vershynin. *High-Dimensional Probability*, volume 47 of *Cambridge Series in Statistical and Probabilistic Mathematics.* Cambridge University Press, Cambridge, 2018.

[Weg21] S.-A. Wegner. Lecture notes on high-dimensional data. https://arxiv.org/abs/2101.05841, 2021.

[Wen17] C. Wendler. Das stochastische Gradientenverfahren. Bachelor Thesis, University of Innsbruck, https://acl.inf.ethz.ch/people/chrisw/bsc_thesis.pdf, 2017.

Index

S. A. Wegner, *Mathematical Introduction to Data Science*,
https://doi.org/10.1007/978-3-662-69426-8

MIX
Papier aus verantwortungsvollen Quellen
Paper from responsible sources
FSC® C105338

If you have any concerns about our products,
you can contact us on
ProductSafety@springernature.com

In case Publisher is established outside the EU,
the EU authorized representative is:
Springer Nature Customer Service Center GmbH
Europaplatz 3, 69115 Heidelberg, Germany

Printed by Libri Plureos GmbH
in Hamburg, Germany